AF754049

DAIRY SCIENCES

DAIRY SCIENCES

Editor

Wetzel Eilerts

scitus
academics

Dairy Sciences

Edited by **Wetzel Eilerts**

This Book is Printed on Acid-Free Paper. ♾

This book contains information obtained from highly regarded resources. Copyright for individual articles remains with the authors as indicated. All chapters are distributed under the terms of the Creative Commons Attribution License, which permits unrestricted use, distribution, and reproduction in any medium, provided the original author and source are credited.

ISBN: 9781681179247

Library of Congress Control Number: 2017948953

© 2018 by

SCITUS Academics LLC,

616, Corporate Way, Suite 2, 4766,

Valley Cottage, NY 10989

www.scitusacademics.com

Notice

Reasonable efforts have been made to publish reliable data and views articulated in the chapters are those of the individual contributors, and not necessarily those of the editors or publishers. Editors or publishers are not responsible for the accuracy of the information in the published chapters or consequences of their use. The publisher believes no responsibility for any damage or grievance to the persons or property arising out of the use of any materials, instructions, methods or thoughts in the book. The editors and the publisher have attempted to trace the copyright holders of all material reproduced in this publication and apologize to copyright holders if permission has not been obtained. If any copyright holder has not been acknowledged, please write to us so we may rectify.

Printed in United States of America.

Preface

Dairy husbandry is one of the strategic sectors of agriculture that has a major impact on public health, the environment and economy. Dairy science and technology provides knowledge and expertise on the production of highly nutritious and safe milk and dairy products in a sustainable manner. It focuses on the biological, chemical, physical, and microbiological aspects of milk itself, and on the technological (processing) aspects of the transformation of milk into its various consumer products, including beverages, fermented products, concentrated and dried products, butter and ice cream. Milk is as ancient as mankind itself, as it is the substance created to feed the mammalian infant. Technological advances have only come about very recently in the history of milk consumption, and our generations will be the ones credited for having turned milk processing from an art to a science. The availability and distribution of milk and milk products today in the modern world is a blend of the centuries old knowledge of traditional milk products with the application of modern science and technology.

'Dairy Sciences' presents cutting edge research work on all aspects of dairy sciences and technology with interests in biochemistry, breeding, economics, engineering, environment, food science, genetics, microbiology, nutrition, pathology, physiology, processing, public health, quality assurance, and sanitation. It delivers information on all aspects of dairy science and technology including animal husbandry, welfare and behavior, animal health, physiology, biochemistry and endocrinology of lactation, milk composition and processing. It also present the coverage of a wide range of dairy products, including fluid and condensed milks, cheese, butter, fermented milk products and ice cream, as well as dry or condensed milk and whey products. It aims to explore the mechanisms and issues involved in the transformation of milk using a multidisciplinary and multiscale approach, from the native molecules to their final functionality, including nutritional and biological properties. This book is designed to support the entire dairy supply chain including farmers, producer co-operatives, manufacturers, processors and distributors of dairy products.

Table of Contents

CHAPTER 1

Analytical Methods used to Quantify Isoflavones in Cow's Milk: A Review

Frédéric Daems[1] & Jean-Michel Romnee[1] & Stéphanie Heuskin[2] & Éric Froidmont [3] & Georges Lognay[2]

[1]*Valorisation of Agricultural Products Department, Walloon Agricultural Research Center, Chaussée de Namur 24, B-5030 Gembloux, Belgium*

[2] *Agro-Bio Chem Department, Gembloux Agro-Bio Tech, University of Liège, Passage des Déportés 2, B-5030 Gembloux, Belgium*

[3] *Production and Sectors Department, Walloon Agricultural Research Center, Rue de Liroux 8, B-5030 Gembloux, Belgium*

ABSTRACT

This paper provides an update and comprehensive review of the analytical methods used for quantifying isoflavones and their metabolites in cow's milk. Isoflavones are secondary plant metabolites that are similar to 17 β-estradiol in chemical structure. They form one of the most common categories of phytoestrogens. Numerous health benefits have been attributed to isoflavones, but many of these compounds are also considered to be endocrine disruptors, with adverse effects on health. These contradictory trends offer an attractive prospect for future research, and therefore, sensitive and reliable analytical methods are required to clarify various issues about isoflavones. For this review, a structured methodology was used to select 26 relevant articles published between 2005 and 2015 from the Scopus and CAB Abstract databases. The review discusses individual steps of the analytical procedures described in these articles, including sample preparation, instrumental analysis and validation. The most commonly used analytical procedure is sample preparation involving liquid-liquid extraction and an enzymatic hydrolysis step followed by liquid chromatography with mass spectrometry analysis. Currently, however, there is no standardized procedure for the sample preparation and analysis of isoflavones in milk.

Keywords: Isoflavones Phytoestrogens Milk Quantification Analysis

1 INTRODUCTION

Isoflavones are classified as phytoestrogens, which are widely distributed in the plant kingdom. These plant secondary metabolites are structurally similar to 17 β-estradiol and bind selectively, but weakly, to mammalian estrogen receptors (ER), with a preference for ERβ (Vitale et al. 2013; Baber 2013; Mostrom and Evans 2012). These diphenolic compounds appear to have antioxidant activity and could offer alternative therapies for a range of hormone-dependent conditions, including cancer, menopausal symptoms, cardiovascular disease, and osteoporosis (Ko 2014; Vitale et al. 2013; Baber 2013; Mostrom and Evans 2012; Mortensen et al. 2009). They also can be considered as endocrine disruptors, however, with the potential to have adverse health effects (Wielogorska et al. 2015; Sirotkin and Harrath 2014; Maggioni et al. 2013; Wocławek-Potocka et al. 2013; Patisaul and Jefferson 2010; Afssa 2005). The highest concentrations are found in plants of the Fabaceae family, such as soybean (*Glycine max* L.) and clover (*Trifolium pratense* L.) (Ko 2014; Mostrom and Evans 2012). In plants, isoflavones are usually found in conjugated forms as glucosides, acetylglucosides, and malonylglucosides (Kalač 2013; Mostrom and Evans 2012). When ingested by cows, isoflavones appear to be metabolized mainly in the rumen, and the main route of excretion is through feces and urine, with only a small proportion being excreted in milk (Njåstad et al. 2014). A diagram of isoflavone metabolization in cows is shown in Fig. 1. Biochanin A (BA) is demethylated mainly into genistein (GE) and via a ring cleavage into *para*-ethyl phenol (a compound that apparently has no estrogenic activity) and organic acids (Njåstad et al. 2014; Kalač 2013; Mostrom and Evans 2012). Formononetin (FO) is demethylated into daidzein (DA) and then reduced via hydrogenation and ring scission to equol (EQ) (a microbial metabolite of isoflavone with high estrogenic activity). FO can also be metabolized into other metabolites, such as *O*-desmethylangolensin (*O*-DMA) (compounds with low estrogenic activity) (Njåstad et al. 2014; Kalač 2013; Mostrom and Evans 2012; Setchell and Clerici 2010). The aglycones seem to be the most biologically active forms and are quickly absorbed by the rumen and gut mucosa (Vitale et al. 2013; Kalač 2013; Mostrom and Evans 2012; Lundh 1990). Most of them are conjugated by glucuronic acid (a limited amount is conjugated with sulfate) during this absorption process. A small proportion of aglycones reaches the blood circulatory system, but the aglycones are rapidly conjugated in the liver and other tissues (Mostrom and Evans 2012; King et al. 1998; Lundh 1990). These isoflavone metabolites then circulate throughout the body and are excreted in feces and urine or transferred to the milk.

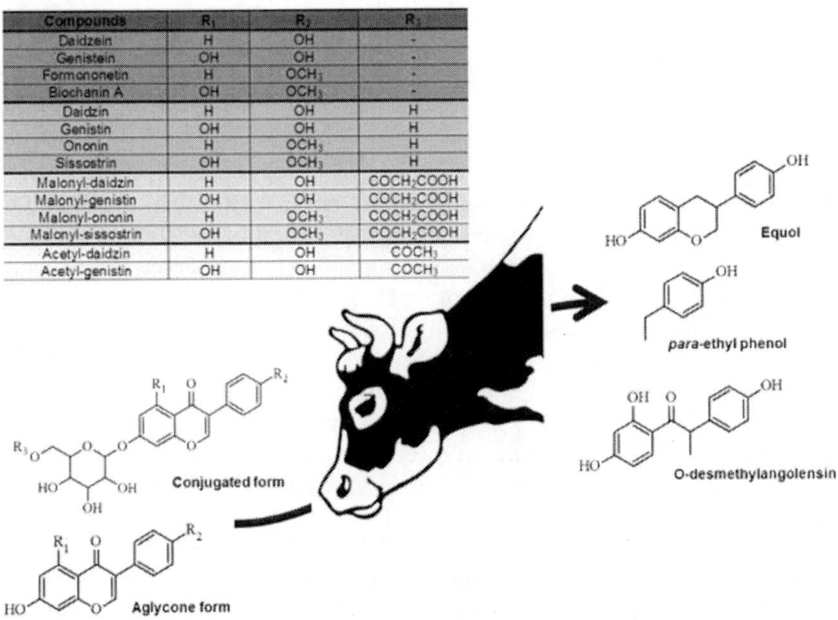

Compounds	R_1	R_2	R_3
Daidzein	H	OH	-
Genistein	OH	OH	-
Formononetin	H	OCH_3	-
Biochanin A	OH	OCH_3	-
Daidzin	H	OH	H
Genistin	OH	OH	H
Ononin	H	OCH_3	H
Sissostrin	OH	OCH_3	H
Malonyl-daidzin	H	OH	$COCH_2COOH$
Malonyl-genistin	OH	OH	$COCH_2COOH$
Malonyl-ononin	H	OCH_3	$COCH_2COOH$
Malonyl-sissostrin	OH	OCH_3	$COCH_2COOH$
Acetyl-daidzin	H	OH	$COCH_3$
Acetyl-genistin	OH	OH	$COCH_3$

Fig. 1. Chemical structures of some of the main isoflavones occurring in forages and their microbial metabolites found in milk (Kalač 2013; Vitale et al. 2013; Mostrom and Evans 2012; Saviranta et al. 2010)

In recent decades, due to the growing interest in isoflavones, a diverse range of analytical methods has been developed for isoflavone determination and quantification in plant material, soil, water, food, and food supplements, as well as in biological matrices, generating a huge amount of scattered information (Fig. 2). The overall analytical method used for isoflavones is highly dependent upon the matrix characteristics, the availability of the techniques, the desired selectivity, and the need to obtain information about the chemical structure of these compounds or simply to unambiguously identify previously targeted compounds. Initially, samples are usually freeze-dried or simply frozen. Depending on the matrix analyzed, isoflavones are extracted by classical maceration or liquid-liquid extraction (LLE), but a promising approach that is increasingly being used is to combine extraction/clean-up techniques such as ultrasound-assisted extraction (UAE), solid-phase extraction (SPE), microwave-assisted extraction (MAE) or matrix solid-phase dispersion (MSPD) (Rostagno et al. 2009). Given that isoflavones are often in conjugated forms, a hydrolysis step is sometimes used when the aim is to find aglycones. The number of analytical separation and detection techniques proposed is also numerous. Depending on the matrix and the information required, many of these techniques have been developed using variants of immunoassays (enzyme-linked immunosorbent assay, ELISA; radioimmunoassay, RIA; and time-resolved fluorescence immunoassay, TR-FIA), gas chromatography (GC), liquid chromatography (LC), and capillary electrophoresis (CE) coupled with various detection modes (mass spectrometry, MS; diode array detection, DAD; fluorescence detection, FLD; and electrochemical detection, ED) (Ko 2014; Rostagno et al. 2009; Valls et al. 2009; Vacek et al. 2008; Stalikas 2007; De Rijke et al. 2006; Grynkiewicz et al. 2005; Wu et al. 2004). Immunoassays are non-chromatographic methods with high sensitivity and specificity for single-component detection, high throughput screening and relatively low cost per sample. For this approach, however, preparing the antibodies for corresponding analytes is a challenging task and it suffers from the likelihood of cross-reactivity between the similar compounds, leading to a possible overestimation of the targeted analyte (Ko 2014; Wu et al. 2004). CE is also a promising technique with a high separation resolution and a low amount of sample required, but it is restricted to laboratories that have the necessary technical capacity (Wu et al. 2004; Wang et al. 2002). The most common analytical method used for quantifying isoflavones remains LC coupled with MS or ultra-violet detection (LC-MS, LC-UV). Thanks to its high sensitivity, MS detection is used preferentially when low amounts of isoflavones are thought to be in the matrix. UV detection is used when isoflavones are thought to be present in larger amounts (Ko 2014).

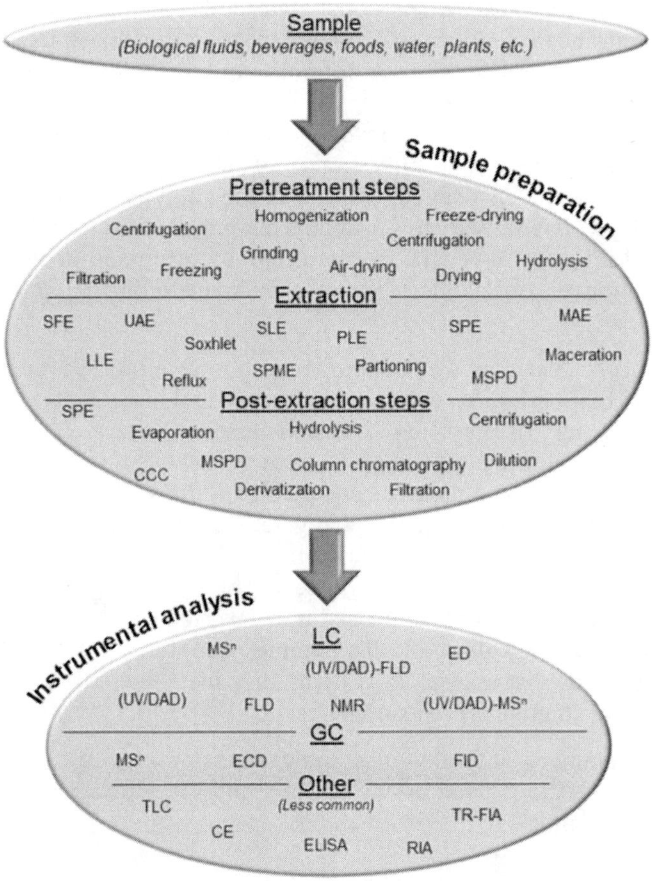

Fig. 2. Main sample preparation and instrumental analysis methods for the determination of isoflavones and their related compounds. Abbreviations: *UAE* ultrasound-assisted extraction, *SFE* supercritical fluid extraction, *LLE* liquid-liquid extraction, *SLE* solid-liquid extraction, *SPME* solid-phase micro-extraction, *SPE* solid-phase extraction (incorporating the Quechers approach), *MAE* microwave-assisted extraction, *MSPD* matrix solid-phase dispersion, *CCC* counter-current chromatography, *LC* liquid chromatography (*HPLC* high-performance liquid chromatography, *UPLC* ultra-performance liquid chromatography, and *UHPLC* ultra high-performance liquid chromatography), *GC* gas chromatography, *MS* mass spectrometry, *UV* ultra-violet, *DAD* diode array detector (also called photodiode array detector, *PDA*), *FLD* fluorescence detector, *NMR* nuclear magnetic resonance, *ED* electrochemical detection, *ECD* electron capture detection, *FID* flame ionization detection, *TLC* thin layer chromatography, *CE* capillary electrophoresis, *ELISA* enzyme-linked immunosorbent assay, *RIA* radioimmunoassay, and *TR-FIA* time-resolved fluorescence immunoassay

Milk is one of the most widely consumed foods in the world and several studies have been conducted in recent years to estimate its isoflavone content and to assess whether or not milk can be considered as a useful or dangerous source of isoflavones for humans. An initial answer was presented in two articles, especially with regard to the impact of cattle feed on the EQ content of milk (Kalač 2011, 2013). As far as we know, there has not yet been a specific review of the analytical methods used for quantifying isoflavones in cow's milk. This paper, therefore, reviews the current situation with regard to sample preparation and analysis for the quantification of isoflavones in cow's milk, including the most recent advances.

The literature overview was conducted online using the Scopus and CAB Abstracts bibliographic databases. Four keywords were identified: "phytoestrogens", "isoflavones", "milk", and "quantitative analysis". These keywords and their descriptors were used to create a Boolean equation in the two databases. The most recent studies (2005–2015) were selected. Key data of the selected articles are given in Tables 1 and 2. Where the authors of a selected article refer to a method published before 2005, the analytical protocol from the reference is summarized and presented in these tables. All the analytical procedures described in these articles were analyzed chronologically: initially, sample preparation procedures were compared and discussed, and then the instrumental analysis and validation parameters were examined.

Table 1. Summary of isoflavones analyzed, sample storage, and sample preparation methods used between 2005 and 2015 for quantification of isoflavones in milk

Isoflavone *(milk origin)*	Storage	Sample preparation			Reference
		Pre-extraction steps	Extraction	Post-extraction steps	
BA, DA, EQ, FO, GE, and PR *(feeding experiments and health impact experiments)*	Frozen (−20 °C)		- Deproteinized and defatted (2.5 mL milk with 0.25 mL acetate buffer [pH 5–5.2 and 2 mol L^{-1}], 1 mL heptane and 2 mL acetone) - Centrifugation - Evaporation to dryness (30 °C) - Reconstitution with 0.8 mL water	- Enz. hydrolysis (8 µL of β-glucuronidase/sulfatase from *H pomatia* type H2, 40 °C, and 3–4 h) - 0.2 mL of MeOH was added - Centrifugation - Supernatant was analyzed	Adler et al. (2015), Adler et al. (2014), Njåstad et al. (2014), Höjer et al. (2012), Nielsen et al. (2012), Andersen et al. (2009a), and Steinshamn et al. (2008).
EQ *(commercial milks)*	Frozen (−18 °C)	- Thawed overnight at 4–6 °C, equilibrated to room T°, and homogenized (2 mL) - Enz. hydrolysis (0.1 mL β-glucuronidase/sulfatase from *H. pomatia* type H2 [≥45,000/3 750 units.mL^{-1}], 37 °C, and 2 h) - Cooled - Centrifugation	- Double L/L extraction (n-hexane and ethyl acetate, 3×3 mL) - Evaporation to dryness - Reconstitution MeOH:H_2O (80:20, *v/v*)	- Filtration (0.2 µm)	Daems et al. (2015)
DA and GE *(commercial milks)*	Frozen (−20 °C) or directly analyzed		- Deproteinized (10 g of milk mixed with 12 mL of 1% (*v/v*) acetic acid in acetonitrile, 1 g NaCl, and 2.5 g $MgSO_4$) - Centrifugation - Evaporation to dryness - Reconstitution (5 mL water)	- SPE (OASIS-HLB) - Evaporation - Reconstitution in 200 µL of H_2O/MeOH (3:1, *v/v*) - Filtration (0.2 µm)	Wielogórska et al. (2015)
EQ *(technological processing)*	N.C.		- Deproteinized (10 mL of milk mixed with 3 mL of acetonitrile and centrifuged at 5 °C)	- Supernatant stored at −80 °C - Filtration (0.2 µm)	Tsen et al. (2014)
PU *(technological processing)*	Frozen (≤−10 °C)	- Freeze-drying - Defatted three times with hexane (1:5, *w/v*)	- Extraction with MeOH (1:3, *w/v*) - Evaporation to dryness	- Filtration (0.45 µm)	Rastogi et al. (2013)

Isoflavone (*milk origin*)	Storage	Sample preparation: Pre-extraction steps	Sample preparation: Extraction	Sample preparation: Post-extraction steps	Reference
			- Reconstitution in MeOH (30 g.mL^{-1})		
GE and GE-in (*commercial milks*)	N.C.		- Double UAE (10 and 5 mL of acetonitrile, 15 min) - Centrifugation	- Filtration (0.45 μm) - Concentration and dilution - SPE (OASIS-HLB) - Concentration - Dilution	Maggioni et al. (2013)
DA, EQ, FO, and GE (*feeding experiments*)	Frozen (−18 °C)		- L/L extraction (20 mL milk/20 mL acetone/acetate buffer pH 5.2 [4:1, v/v]) - Centrifugation - Evaporation (halved)	- Enz. hydrolysis (0.2 mL sulfatase/glucuronidase from *H. pomatia*, 52 °C, and 2 h) - Centrifugation - SPE (C18) - Evaporation to dryness - Reconstitution in MeOH (80%, v/v) - Filtration (0.45 μm)	Flachowsky et al. (2011)
BA, DA, EQ, FO, and GL (*feeding experiments and health impact experiments*)	Frozen (−20 °C)	- Equilibration at room temperature - Enz. hydrolysis (β-glucuronidase, 100 μL.4 mL^{-1} of milk, 37 °C, and 1 h) - Centrifugation (cream and precipitate were discarded)	- Double L/L extraction (hexane and ethyl acetate, 3×3 mL) - Evaporation to dryness - Re-suspension (0.5 mL acetonitrile)	- Filtration (0.2 μm)	Nielsen et al. (2011), Skaanild and Nielsen (2010), Nielsen et al. (2009) and Andersen et al. (2009b)
DA, EQ, GE, and GL (*technological processing and feeding experiments*)	Frozen (−20 °C)	- Enz. hydrolysis (100 μL β-glucuronidase [11,400 U.mL^{-1}]/sulfatase [3200 U.mL^{-1}] from *H. pomatia* type H2 were added at 2 mL of milk and 3 mL of acetate buffer [0.1 M and pH 5], 37 °C, and 12 h)	- L/L extraction (3×2 mL ethylacetate) - Centrifugation - Evaporation to dryness - Reconstitution (0.5 mL methanol:water [1:1, v/v])		Křížová et al. (2011a), Křížová et al. (2011b), Krajčová et al. (2010), and Třináctý et al. (2009)
BA, DA, EQ, FO, GE, and GL (*commercial milks*)	N.C.		- L/L extraction (10 mL milk/10 mL acetone/acetate buffer [2 M, pH 5.2, 9:1, v/v])	- Enz. hydrolysis (0.3 mL *H. pomatia* type H5 [500000 units, 20 mL^{-1} water], 52 °C, and 16 h)	Antignac et al. (2009)
				- Reverse-phase SPE (C18) - Normal phase SPE (SiOH)	
BA, DA, EQ, FO, GE, and O-DMA (*commercial milks and feeding experiments*)	Frozen (−80 °C)	- Enz. hydrolysis (100 μL β-glucuronidase [500 units] and 80 μL sulfatase [40 units] were added at 5 mL milk, 37 °C, and 2 h)	- L/L extraction (1.5 mL ammonium acetate buffer 1.5 M+7.5 mL diethyl ether, chilled at −80 °C, centrifugation, and repeated 3 times) - Evaporation to dryness (40 °C) - Reconstitution in 2 mL of sodium acetate buffer (0.2 M and pH 5) containing MeOH 20%	- Centrifugation - SPE (OASIS-HLB 3 cc) - Evaporation to dryness (60 °C) - Reconstitution 200 μL MeOH/sodium acetate buffer (pH 5, 0.2 M, 3:1, v/v) - Centrifugation - Filtration (0.45 μm)	Mustonen et al. (2009) and Hoikkala et al. (2007)
BA, DA, EQ, FO, GE, and GL (*commercial milks*)	Frozen (−20 °C)	- Freeze-dried	- SLE (100 mg/2 mL of 10% MeOH in sodium acetate (0.1%, pH 5), three time)	- Enz. hydrolysis (purified β-glucuronidase from *H. pomatia*, cellulase from *Trichoderma reesi*, and β-glucosidase from almonds, 37 °C and 16 h) - SPE (Strata C-18E, 50 mg.mL^{-1}) - Dried - Reconstitution (200 μL MeOH 40%)	Kuhnle et al. (2008)
DA, DA-in, GE, GE-in, GL, and GL-in (*technological processing*)	N.C.		- Deproteinized (2 mL of milk mixed with 50 μL IS and 3 mL acetonitrile and stored overnight at 10 °C)	- Equilibrated at room T° - Evaporation to dryness - Reconstitution (2.4 mL 50% acetonitrile) Filtration (0.22 μm)	Uzzan et al. (2007)

The underlined reference corresponds to the referenced method used and cited in the other articles

Abbreviations: *BA* biochanin A, *DA* daidzein, *DA-in* daidzin (glycoside form of DA), *EQ* equol, *FO* formononetin, *GE* genistein, *GE-in* genistin (glycoside form of GE), *GL* glycitein, *GL-in* glycitin (glycoside form of GL), *PR* prunetin, *PU* puerarin, and *O-DMA* *O*-desmethylangolensin

The underlined reference corresponds to the referenced method used and cited in the other articles

Abbreviations: *BA* biochanin A, *DA* daidzein, *DA-in* daidzin (glycoside form of DA), *EQ* equol, *FO* formononetin, *GE* genistein, *GE-in* genistin (glycoside form of GE), *GL* glycitein, *GL-in* glycitin (glycoside form of GL), *PR* prunetin, *PU* puerarin, and *O-DMA* *O*-desmethylangolensin

Table 2. Summary of separation and detection methods used between 2005 and 2015 for quantification of isoflavones in milk

Instrumental analysis		Validation parameters									Reference
Separation	Detection	Quantification mode	LOQ	LOD	Sel.	Lin.	Prec.	Acc.	Stab.	Rob.	
HPLC Column: Zorbax XDB (150× 2.1 mm i.d., 3.5 µm) Mobile phases: (A) 0.5% acetic acid (B) MeOH Elution: gradient Column T°: N.C. Flow: 0.4 mL.min^{-1} Run time: 14 min Inj. vol.: 50 µL	MS/MS ES$^+$, MRM mode	Standard addition	N.C.	0.05–0.17 ng.mL^{-1}	N.C.	N.C.	N.C.	N.C.	N.C.	N.C.	Adler et al. (2015), Adler et al. (2014), Njåstad et al. (2014), Höjer et al. (2012), Nielsen et al. (2012), Andersen et al. (2009a), Steinshamn et al. (2008)
UPLC Column: Acquity UPLC® HSS T3 (100×2.1 mm, 1.8 µm) Mobile phases: (A) water/acetonitrile (95/5, v/v) with 0.01% formic acid (B) acetonitrile/water (95/5, v/v) with 0.01% formic acid Elution: gradient Column T°: 40 °C Flow: 0.6 mL.min^{-1} Run time: 5 min Inj. vol.: 10 µL	MS/MS ESI$^+$, MRM mode	External calibration	5 ng.mL^{-1}	0.3 ng.mL^{-1}	V	V	V	V	V	V	Daems et al. (2015)
UHPLC Column: Acquity HSS T3 (100× 2.1 mm, 1.8 µm) Mobile phases: (A) 5 mM ammonium acetate in H_2O/acetonitrile (9:1, v/v) (B) MeOH:acetonitrile (1:1, v/v) Elution: gradient Column T°: 45 °C Flow: 0.4 mL.min^{-1} Run time: 10.5 min Inj. vol.: 10 µL	MS/MS ESI$^+$, MRM mode	External calibration	0.35–0.36 µg.g^{-1}	0.20–0.21 µg.g^{-1}	V	V	V	V	V	N.C.	Wielogórska et al. (2015)
LC	IT-TOF-MS$^-$	External calibration	N.C.	N.C.	N.C.	N.C.	N.C.	N.C.	N.C.	N.C.	Tsen et al. (2014)
HPLC Column: Supelcosil LC-8-DB (250× 4.6 mm i.d., 5 µm)+guard column (40×4.6 mm i.d.) Mobile phases: (A) water (B) MeOH Elution: gradient Column T°: 25 °C Flow: 0.8 mL.min^{-1} Run time: 45 min Inj. vol.: 10 µL	PDA UV: 254 nm	External calibration	500 ng	300 ng	V	V	V	V	N.C.	N.C.	Rastogi et al. (2013)
HPLC Column: LUNA C8 (50×2.0 mm i.d., 5 µm) Mobile phases: (A) 0.05% acetic acid in Milliq water (B) acetonitrile Elution: gradient Column T°: N.C. Flow: 200 µL.min^{-1} Run time: 18 min Inj. vol.: 10 µL	MS/MS ES$^-$, MRM mode	External calibration	N.C.	0.6–0.9 ng.g^{-1}	N.C.	V	V	V	N.C.	N.C.	Maggioni_2013
HPLC Column: Purospher STAR RP-18® (250× 4.0 mm i.d., 5 µm particle size) Mobile phases: (A) aqueous 1% formic acid (B) acetonitrile 1% formic acid Elution: gradient Column T°: 35 °C Flow: 0.5 mL.min^{-1} Run time: 41 min Inj. vol.: 20 µL	DAD-MS/MS UV: 200–600 nm MS: ESI$^+$, SIM mode	External calibration	0.006–0.7 ng.mL^{-1}	N.C.	N.C.	V	N.C.	V	N.C.	N.C.	Nielsen et al. (2011) and Skaanild and Nielsen (2010), Nielsen et al. (2009) and Andersen et al. (2009b)
HPLC	MS/MS	N.C.	N.C.	N.C.	N.C.	N.C.	N.C.	N.C.	N.C.	N.C.	Flachowsky et al. (2011)

Instrumental analysis		Validation parameters									Reference
Separation	Detection	Quantification mode	LOQ	LOD	Sel.	Lin.	Prec.	Acc.	Stab.	Rob.	
Column: YMC-Pack ODS-AM 12S05 column (150×4.6 mm i.d., 5 μm) Mobile phases: (A) acetic acid 0.1% (B) acetonitrile/acetic acid [99.9:0.1, v/v] Elution: gradient Column T°: N.C. Flow: 1.0 mL.min^{-1} Run time: 60 min Inj. vol.: N.C.	ESI$^-$, N.C. mode										
HPLC Column: discovery C18 (150×3 mm, 5 μm)+discovery C18 guard column (20×4 mm, 5 μm) Mobile phases: (A) MeOH (B) 0.1% acetic acid water (v/v) Elution: gradient Column T°: 45 °C Flow: 0.7 mL.min^{-1} Run time: 16 min Inj. vol.: 10 μL	MS/MS APCI$^+$, SRM mode	External calibration	2–15 ng.mL^{-1}	0.5–5 ng.mL^{-1}	N.C.	N.C.	V	V	N.C.	N.C.	Křížová et al. (2011a), Křížová et al. (2011b), Krajčová et al. (2010) and Třináctý et al. (2009)
HPLC Column: RP octadecyl grafted silica stationary phase Nucleosil® C_{18}AB (50×2.1 mm, 5 μm+guard column 10×2.1 mm) Mobile phases: (A) methanol (B) water 0.5% acetic acid Elution: gradient Column T°: N.C. Flow: 0.3 mL.min^{-1} Run time: 30 min Inj. vol.: 10 μL	MS/MS ESI$^+$, MRM mode	External calibration	0.05–0.50 ng.g^{-1}	N.C.	V	V	V	N.C.	N.C.	V	Antignac et al. (2009)
HPLC Column: Zorbax Eclipse XDB-C18 (250×4.6 mm+precolumn Zorbax R-P 4×4 mm) Mobile phases: (A) MeOH/10 mM $Na_2HPO_4.2H_2O$ pH 6.5 (B) MeOH 100% Elution: gradient Column T°: 40 °C Flow: 1 mL.min^{-1} Run time: 31 min Inj. vol.: 10 μL	DAD-FLD UV: 260 nm FL: 254 nm (ex) 465 and 310 nm (em)	External calibration	3.3–117.2 ng.mL^{-1}	1.3–58.6 ng.mL^{-1}	V	V	N.C.	V	N.C.	N.C.	Mustonen et al. (2009) and Hoikkala et al. (2007)
HPLC Column: diphenyl column (Varian Pursuit, 150×2.0 mm i.d., 3 μm) Mobile phases: (A) 40% ammonium acetate (0.1%, pH 4.8) in MeOH (B) 100% MeOH Elution: gradient Column T°: 50 °C Flow: 250 μL.min^{-1} Run time: N.C. Inj. vol.: N.C.	MS/MS ESI$^-$, N.C. mode	External calibration	N.C.	15 ng.g^{-1}ww	V	V	V	V	N.C.	N.C.	Kuhnle et al. (2008)
HPLC Column: C18 YMC-ODS-AM (250×3 mm S5)+C_{18} guard column (AJ0-7597) Mobile phases: (A) acetic acid:acetonitrile:water (0.1:9.95:89.95) (B) acetic acid:acetonitrile:water (0.1:89.95:9.95) Elution: gradient Column T°: 40 °C Flow: 0.8 mL.min^{-1} Run time: N.C. Inj. vol.: 10 μL	UV	External calibration	N.C.	N.C.	N.C.	V	N.C.	N.C.	N.C.	N.C.	Uzzan et al. (2007)

The underlined reference corresponds to the referenced method used and cited in the other articles

Abbreviations: *LOQ* limit of quantification, *LOD* limit of detection, *Sel.* selectivity, *Lin.* linearity, *Prec.* precision, *Acc.* accuracy, *Stab.* stability in extract or sample, *Rob.* robustness, *inj. vol.* injection volume, *N.C.* not communicated, and *V* communicated

2 SAMPLE PREPARATION

Broadly, an analytical method can be split into two distinct parts: sample preparation and sample analysis. Sample preparation is undertaken before instrumental analysis and should ensure that: (1) the target compounds can be extracted from the matrix; (2) interference that could disrupt the detection of the target analyte or damage the chromatographic column can be eliminated or reduced; (3) to allow analysis of samples with low concentrations of the target compound; and (4) the analyte of interest can be transformed into a more suitable form that could be easily separated, detected and quantified. In Table 1, all the information about sample preparation found in the 26 selected articles is detailed and split into four chronological steps. The table also presents the origin of the milk and the isoflavones quantified. Studies using the same sample preparation procedure were grouped together. Articles where the sample preparation procedure was based on a previous study, but with some modifications, were not grouped (Daems et al. 2015; Flachowsky et al. 2011; Antignac et al. 2009). Thirteen sample preparation procedures were identified and are discussed in the next section.

2.1 Sample conditioning

As shown in Table 1, most milk samples in the experiments reviewed were simply frozen between the sampling and analysis days. In other articles, sample condition was not mentioned. Often, information about the handling of milk samples after collection and before actual analysis was limited. An exception was Wielogórska et al. (2015), who reported that isoflavones (DA and GE) were stable in the matrix for only 2 weeks, indicating that all samples were analyzed immediately after collection or stored at −20 °C for no longer than 2 weeks before being thawed for analytical purposes. Poor storage can lead to errors that cannot be corrected later and that will affect the outcome of the final analysis. How should samples be stored, and for how long, without affecting their original isoflavone profile? For matrices such as soybean, some isoflavones (mainly the conjugated forms) have a fairly unstable character (Rostagno et al. 2009). In ruminants, glucuronide-conjugated compounds are the major form found in biological fluids (Mostrom and Evans 2012; Nielsen et al. 2012). The degradation problem could, therefore, also arise during milk storage and the information about the conjugated form content could be lost. More attention should be given to the storage and handling of milk samples between sampling and analysis in future studies.

If freeze-drying is an easy method for milk-sample storage, it would be interesting to check isoflavone stability under a range of temperature and storage-time conditions. The freezing method could be also compared with other techniques, such as freeze-drying. Isoflavones can be found in powdered milks (Maggioni et al. 2013; Antignac et al. 2009). Therefore, freeze-drying, which is often used for plant matrices because it does not seem to affect these compounds (Hoerger et al. 2011), could be an interesting alternative to the traditional freezing process. These findings need to be taken into account in future studies.

2.2 Hydrolysis

Among the 13 sample preparation procedures listed in Table 1, only four did not include a hydrolysis step. Wielogórska et al. (2015) proposed a method without a hydrolysis step for the quantification of 19 endocrine disruptors in milk, including DA and GE. Rostagi et al. (2013) did not use a hydrolysis step because they sought to quantify puerarin (PU, the 8-C-glucoside of daidzein) from pasteurized toned milk enriched with extracts from *Pueraria tuberosa* Linn (commonly called "kudzu"). According to these authors, this glycoside form of isoflavone that is abundant in kudzu might be beneficial in the treatment and prevention of some diseases and cancers. The two other procedures were used by Maggioni et al. (2013) and Rastogi et al. (2013) to quantify aglycones in infant powdered milk formulas and glycosides in soy-fortified milk. Integrating a hydrolysis step into the analytical protocol depends on the objectives sought. Hydrolysis is often used when the exact nature and composition of isoflavone glycosides in the matrix analyzed are unknown or when target aglycone compounds are sought. It allows the time required for instrumental analysis to be considerably reduced and facilitates the separation of target compounds. It also makes it unnecessary to use standard references that are difficult to obtain, not available, or often expensive. Introducing an additional step, however, can increase analytical variability.

As shown in Table 1, most of the studies reviewed used a hydrolysis step. Unlike other matrices where three procedures are reported (enzymatic, acidic and, least common, basic hydrolysis) (Rostagno et al. 2009; Schwartz et al. 2009), for milk only enzymatic hydrolysis is carried out to hydrolyze isoflavones. The authors focused on aglycones because these forms appear to be the most biologically active and can be absorbed by the intestinal tract of humans and animals (Ko 2014; Mostrom and Evans 2012). In addition, some isoflavones, such as EQ, seem to be sensitive to acidic conditions (Setchel and Clerici 2010). Enzymatic hydrolysis involves incubating the sample with enzymes at a fixed temperature (within a range of 37 to 52 °C, depending on the study) and over different incubation times (between 1 and ≥16 h). A solution buffer (pH of about 5) is sometimes used to optimize hydrolysis conditions.

Unlike plant matrices (Rostagno et al. 2009), only β-glucuronidase/sulfatase from *Helix pomatia* was involved in the hydrolysis of isoflavones in milk. This enzymatic juice contains glucuronidase and sulfatase activities. The former catalyzes the breakdown of complex carbohydrates (glucuronides) and the latter catalyzes the hydrolysis of sulfate esters, the two main forms of isoflavones found in biological fluids (Ko 2014; Nielsen et al. 2012; Rostagno et al. 2009). Only Kuhnle et al. (2008) reported using a mixed solution of three enzymes: β-glucuronidase from *H. pomatia*; cellulase from *Trichoderma reesi*; and β-glucosidase from almonds.

The hydrolysis temperature varied between 37 and 40 °C, but it did reach 52 °C in some studies (Flachowsky et al. 2011; Antignac et al. 2009). The temperature generally used by suppliers to evaluate the glucuronidase and sulfatase activities from these enzymes was 37 °C. Where a buffer solution was used, the pH of the solution varied between 4.6 and 5.2. The optimal pH values given by suppliers were 5.0 and 6.2 for glucuronidase and sulfatase activities, respectively. Hydrolysis time varied, ranging from 1 h (Andersen et al. 2009b) to overnight (Antignac et al. 2009; Kuhnle et al. 2008). During an evaluation of the robustness of their hydrolysis protocol based on a one-variable-at-a-time technique, Daems et al. (2015) demonstrated that an incubation time of 1 h was sufficient for hydrolyzing conjugated forms of EQ. Another important point, but one on which there is little information in the studies, is the minimum enzyme activity needed to hydrolyze all the conjugated forms of isoflavones present in milk. When enzyme activity values were provided by authors or when estimating them was possible, it was found that they varied greatly across the studies. For example, for their analysis of isoflavones (containing EQ) in commercial milk, Hoikkala et al. (2007) used 100 units of glucuronidase activity per ml of sample, whereas Daems et al. (2015) used 2250 units. In the product information data sheets for other biological fluids, Sigma-Aldrich recommends using between 1000 and 20,000 units per mL of β-glucuronidase type H2 from *H. pomatia* (EC 3.2.1.31), but no specific information is given on the hydrolysis of milk samples. One disadvantage of using *H. pomatia* juice is that it usually contains appreciable levels of some isoflavones (according to Taylor et al. (2005): 13.38 ng GE.mL^{-1} enzyme, 0.25 ng DA.mL^{-1}, and 0.58 ng EQ.mL^{-1}) which can affect the quantification. Daems et al. (2015) and Grace and Teale (2006) did not detect any trace of EQ, and Grace and Teale (2006) found that GE was the major isoflavone in *H. pomatia* juice. These observations show that isoflavone concentrations appear to fluctuate from one enzyme batch to another. The hydrolysis with this enzymatic solution also showed interfering peaks with isoflavone signals in the chromatogram of human milk extracts (Franke and Cluster 1996). There are two ways of addressing these problems. One is a pre-cleanup of the enzymatic preparation by SPE (Kuhnle et al. 2007; Grace and Teale 2006). The second way, probably the easier of the two, involves working with a blank sample where the milk has been replaced by a buffer solution in order to estimate the isoflavone content in the enzymatic mixture (Taylor et al. 2005).

2.3 Isolation of target compounds

Solvent extraction is the main step used for the recovery and isolation of bioactive compounds from a matrix before instrumental analysis and it can be performed before or after the hydrolysis step. Additional steps are sometimes necessary to remove lipids or proteins, which can cause interferences during instrumental analysis. Fat is one of the major causes of the matrix effect in LC-MS (Jiang et al. 2012; Trontelj 2012) and most of the protocols in Table 1therefore indicate defatting and deproteinizing steps. Lipids can be removed by LLE with an apolar solvent (e.g., n-heptane or n-hexane) and centrifugation. Proteins are removed by precipitation in acidic conditions and centrifugation. Deproteinization is often used at the same time that the defatting or extraction of target compounds is done because organic solvents can also precipitate the proteins (Trontelj 2012).

Liquid-liquid extraction is the most widely used technique for extracting isoflavones from milk. The extractions are repeated two to three times and the extracts are pooled. Among the analytical protocols listed in Table 1, only three used SLE to recover isoflavones in milk samples (Maggioni et al. 2013; Rastogi et al. 2013; Kuhnle et al. 2008). Only Rastogi et al. (2013) deliberately used the freeze-drying process, whereas two other studies used this extraction technique because they were analyzing milk samples in powder form. Optimizing extraction is a challenging task given the wide polarity range of isoflavones. As an example, compared with DA, GE has an additional hydroxyl group and therefore DA and GE are expected to have different solubilities (Harjo et al. 2007). In general, conjugated forms are more polar than aglycones and are extracted with polar solvents such as alcohols (MeOH, EtOH) or acetonitrile, often mixed with varying proportions of water. For isoflavone quantification in milk, a hydrolysis step is almost always used in order to remove the polar groups from conjugated forms, leading to the formation of isoflavone aglycones, which are less polar and almost insoluble in water (Vitale et al. 2013; Ishii et al. 2013; Harjo et al. 2007). This is why less polar solvents (diethyl ether or ethyl acetate), with no water added, are used for extraction in milk. The choice of solvent composition should be determined empirically according to the target molecules that the researchers wish to extract.

Looking more closely at the order in which the hydrolysis and isolation steps are done, some authors conducted the hydrolysis step first and then the extraction step, whereas others used the reverse order. This raises the question: is one protocol better than the other? If the less polar solvents are more efficient for extracting aglycone forms, when the hydrolysis is performed after the extraction step the conjugated isoflavones might not be completely extracted, leading to an underestimation of the overall isoflavone content. Some authors (Daems et al. 2015; Křížová et al. 2011a) reported that isoflavones are distributed more widely in the aqueous fraction than in the lipid fraction. In contrast, Tsen et al. (2014) reported that isoflavones are distributed mostly in the lipid fraction and that the defatting process, therefore, reduces the overall amount of isoflavones in a sample. These are all matters that need to be addressed in future studies.

2.4 Purification and concentration extracts

After the extraction of isoflavones in milk, the extract is subjected to a series of post-treatment steps before instrumental analysis. In almost all protocols, centrifugation is performed. As well as eliminating the cream (fat) and the precipitated proteins, it also isolates other small solid impurities. A filtration step is often performed just before the sample extract injection (Wielogórska et al. 2015; Daems et al. 2015; Tsen et al. 2014; Rastogi et al. 2013; Flachowsky et al. 2011; Andersen et al. 2009b; Nielsen et al. 2009; Hoikkala et al. 2007; Uzzan et al. 2007) to prevent damage to the analytical columns.

In milk, isoflavones are usually present in low concentrations (a few ppb or, sometimes, higher for microbial metabolites) (Kalač 2013) and, therefore, it is necessary to concentrate the target analyte before instrumental analysis. SPE and evaporation were the most commonly used enrichment techniques. SPE using C_{18} bonded silica (Flachowsky et al. 2011; Antignac et al. 2009; Kuhnle et al. 2008) or a universal polymeric reversed-phase sorbent called OASIS-HLB from Waters (Wielogórska et al. 2015; Maggioni et al. 2013; Mustonen et al. 2009; Hoikkala et al. 2007) were reported to provide a clean concentrated isoflavone extract. Probably due to its typical chemistry, OASIS-HLB has the advantage of greater retention and good recoveries even if the sorbent runs dry (Rostagno et al. 2009). The final extracts are then evaporated until dry and re-suspended with the polar solvent used in the instrumental analysis.

3 INSTRUMENTAL ANALYSIS

After these preparation steps, the sample extract can be injected into instruments that will allow the target compounds to be separated, detected and quantified. The data in the 26 selected articles are presented in Table 2. When different authors used the same analytical methods, the data presented are derived from all the information in the articles. Thirteen analytical methods were identified and are discussed below.

3.1 Analyte separation

Unlike in other matrices (plants, foods or other biological fluids), there are few separation methods reported in the literature for the quantification of isoflavones in milk. Over the past 10 years, the only analytical separation technique used has been reverse-phase LC with a binary solvent system that contains acidified water and an acidified polar organic solvent, such as methanol or acetonitrile (Table 2).

With the LC technique, a guard column is often placed before the analytical column in order to retain all impurities remaining in the sample extract (Rastogi et al. 2013; Křížová et al. 2011a, b; Antignac et al. 2009; Třináctý et al. 2009; Hoikkala et al. 2007; Uzzan et al. 2007). This also enables the compounds from the extract to start being separated. These small columns are regularly changed and allow the lifetime of the main analytical column to be increased. They often have a stationary phase similar to that of the main column. Columns chosen for isoflavone separation in milk are exclusively packed with reversed-phases, mainly C_{18} with an internal diameter of 2.0–4.6 mm, a length of 50–250 mm, and a particle size of 3–5 µm. Two other stationary phase chemistries are sometimes also mentioned in the articles: C_8 (Rastogi et al. 2013; Maggioni et al. 2013) and diphenyl (Kuhnle et al. 2008). Two recent studies report on the use of ultra high-performance LC (UHPLC) columns with a particle size of 1.8 µm, an internal diameter of 2.1 mm, and a length of 100 mm (Wielogórska et al. 2015; Daems et al. 2015). This technology allows separations that have positive effects on both resolution and accuracy. It also increases sample-throughput due to reduced analysis times (Gumustas et al. 2013; Valls et al. 2009). The reduction of analysis time is illustrated in Table 2. For classical LC analysis, the average analysis time is about 35 min, ranging from a minimum of 14 min (Steinshamn et al. 2008) to a maximum of 60 min (Flachowsky et al. 2011). In the two studies using UHPLC it was between 5 and 10.5 min. These times could be further reduced by manipulating the column temperature, flow rate, mobile phase composition, and solvent gradients.

The mobile phase flow rate depends upon its composition, column type and temperature. In the selected articles, the flow rates varied between 0.2 and 1 $mL.min^{-1}$, the temperature from 25 to 45 °C, and the injection volume from 10 to 50 μL. The mobile phase is usually a mixture of water and an organic solvent (methanol or acetonitrile) with a small proportion of formic or acetic acids (up to a max. of 1%). The differing hydrophobicities of isoflavones indicate that the gradient elution mode should be used (Rostagno et al. 2009). Since isoflavones exhibit a weak acidic nature, the use of acids in the mobile phase can enhance chromatographic resolution and peak shape (Rostagno et al. 2009; Wu et al. 2004). Acidification of the mobile phase also significantly increases the isoflavone limit of detection (LOD) by promoting positive ion formation in the MS source (Wu et al. 2004).

3.2 Analyte detection

As shown in Table 2, the most commonly used detection methods combined with LC are MS and UV detectors. They can be used alone, in combination (Andersen et al. 2009b; Nielsen et al. 2009) or with other types of detectors (Mustonen et al. 2009; Hoikkala et al. 2007). Hoikkala et al. (2007) used UV-DAD at 262 nm for GE, *O*-DMA and BA and they used FLD for DA and FO (λ_{ex} at 254 nm and λ_{em} at 465 nm), as well as for EQ (λ_{ex} at 254 nm and λ_{em} at 310 nm). Although the LC-FLD method is often more sensitive than the UV absorption methods, the number of isoflavones that are naturally fluorescent is fairly limited (Ko 2014). In addition, the acidification of the mobile phase has a quenching effect that can reduce FLD sensitivity (Vacek et al. 2008).

Overall, LC-UV and LC-MS are the two main detection modes used for quantifying isoflavones. In milk, however, MS is preferred to UV detection. UV detection uses the isoflavone characteristic of having at least one aromatic ring that strongly absorbs UV light, with a maximum wavelength ranging from 230 to 280 nm (Ko 2014; Rostagno et al. 2009; Stalikas 2007; Wu et al. 2004). A range of 300 to 550 nm was sometimes also used as a second maximum wavelength and attributed to the substitution pattern and conjugation of the C-ring (Stalikas 2007). This is not the case with all isoflavones. For example, the microbial metabolites EQ and *O*-DMA exhibit poor UV absorption characteristics, making UV detection unsuitable for their measurement (Ko 2014; Setchel and Clerici 2010). For isoflavones that exhibit this UV absorption characteristic, the UV–visible spectra of many of them are very similar (Vacek et al. 2008). Another disadvantage of this method is its low sensitivity, which restricts its use to quantification in matrices likely to have a large amount of isoflavones, such as plants, plant-derived samples and supplemented foods (Ko 2014; Wu et al. 2004). Among the studies presented in Table 2, only two used UV as the sole mode of detection (Rastogi et al. 2013; Uzzan et al. 2007) and both focused on analyzing fortified milk.

In order to overcome the problems of non-specificity and lack of sensitivity, researchers have focused on MS detection. Usually, this method is highly sensitive, with low limits of quantification (LOQ) under or around the ppb level (see Table 2). Electrospray ionization (ESI) and atmospheric pressure chemical ionization (APCI) are the two sources most commonly used for quantifying isoflavones in milk. In the selected articles, these sources were used in positive and/or negative mode, except for APCI, which was used exclusively in positive mode with select reaction monitoring (SRM). In the ESI, multiple reaction monitoring (MRM) was used. ESI and APCI interfaces appeared to be suitable for the analysis of isoflavones, but ESI is better suited for the conjugated forms. ESI is the softest and most suitable method for ionization without the in-source fragmentation of the relatively weak glucuronide ether or ester bonds. Thermally labile conjugates form molecular ions in ESI, but with APCI these conjugates decompose because they are not stable enough (Trontelj 2012; Hoikkala et al. 2003). In their comparison of ESI and APCI sources for the quantitative determination of phytoestrogens in urine, Rybak et al. (2008) showed that, for EQ, ESI was better in terms of measurement precision, sensitivity and LOD.

The benefits of LC-MS include high precision and sensitivity, suitability for determining isoflavones and their metabolites or conjugated forms, less samples manipulation than for GC, and applicability to non-volatile analytes with direct injection of the liquid samples. The higher price of this technology remains its major disadvantage compared with other separation and detection technologies. LC-MS, however, is now a fast-moving field, with many new detection and chromatographic techniques being developed. With their shorter analysis time, lower solvent consumption, better resolution properties, and better sensitivity than classical LC, UPLC and UHPLC with MS appear to be the fastest and easiest methods to implement and to produce reliable results.

3.3 Validation parameters

As far as we could ascertain, method validations were seldom discussed in the previous review articles on isoflavones and related compounds. Where they were mentioned, the validation parameters are shown in Table 2 (when several works used the same analytical procedure, the various data in each study were collected and summarized in this table).

The validation parameters taken into account in this table come from the EMA (2015) guidelines, which describe the suitable criteria for the validation of analytical methods used in veterinary drug residue depletion studies. The guidelines could apply to the quantification of isoflavones in milk because these estrogenic compounds can be active in very low amounts. As shown in Table 2, most of the studies used external calibration as the quantification mode for determining the isoflavone content of milk, whereas Steinshamn et al. (2008) and related studies used a standard addition protocol. Among these studies, Andersen et al. (2009a) reported that the LOD for FO and EQ were 0.047 and 0.170 $ng.mL^{-1}$, respectively. In the other selected articles, the validation parameters most often omitted were sample or extract stability and the robustness of the analytical method proposed. The most comprehensive validation procedures were reported by Wielogórska et al. (2015) (for DA, GE and other endocrine disruptors) and Daems et al. (2015) (for EQ).

4 CONCLUSION

This paper summarizes the current analytical methods used for quantifying isoflavones and their metabolites in cow's milk. Careful examination of methodological aspects showed that most of the studies used LLE to extract isoflavones from milk. Given that aglycones are the most biologically active forms of isoflavones, most of the analytical protocols used a hydrolysis step with β-glucuronidase/sulfatase from *H. pomatia*.

Although non-chromatographic immunoassays and CE with MS or ED detectors are seen as promising techniques for quantifying isoflavones, reverse-phase LC-MS is the only analytical method used in the past 10 years. LC-MS is now a fast-moving field, with many new detection and chromatographic techniques being developed over the past decade that achieve high sensitivity and reduce analysis time. The low isoflavone content (about ppb value) naturally present in milk, except in the case of some isoflavone metabolites, justifies the use of MS as a detection method. The new chromatographic technologies, UPLC and UHPLC, reduce the time needed and achieve good resolution.

Although some analytical protocols have been used frequently by researchers, there is still no standardized procedure available for the sample preparation and determination of these phytoestrogenic compounds in milk. A rigorous validation referring to a guideline would be enough to ensure the reliability of the chromatographic results. For sample preparation, however, a standardized procedure is interesting to compare the results from the different studies.

ACKNOWLEDGMENTS

The authors wish to thank the Public Service of Wallonia (PhytoHealth Project, Moerman funded) for providing financial assistance during the course of this research.

REFERENCES

1. Adler SA, Purup S, Hansen-Møller J, Thuen E, Gustavson AM, Steinshamn H (2014) Phyto-oestrogens and their metabolites in milk produced on two pastures with different botanical compositions. Livest Sci 163:62–68
2. Adler AS, Purup S, Hansen-Møller J, Thuen E, Steinshamn H (2015) Phytoestrogens and their metabolites in bulk-tank milk: effects of farm management and season. PLoS One 10(5):e0127187
3. Afssa (2005) Sécurité et bénéfices des phyto-estrogènes apportés par l'alimentation—Recommandations, France. https://www.anses.fr/sites /default/files /documents/NUT-Ra-Phytoestrogenes.pdf (accessed March 2015)
4. Andersen C, Weisbjerg MR, Hansen-Møller J, Sejrsen K (2009a) Effect of forage on the content of phyto-oestrogens in bovine milk. Animal 3(4):617–622
5. Andersen C, Nielsen TS, Purup S, Kristensen T, Eriksen J, Søegaard K, Sørensen J, Fretté XC (2009b) Phyto-oestrogens in herbage and milk from cows grazing white clover, red clover, lucerne or chicory-rich pastures. Animal 3(8):1189–1195
6. Antignac J-P, Gaudin-Hirret I, Naegeli H, Cariou R, Elliott C, Le Bizec B (2009) Multi-functional sample preparation procedure for measuring phytoestrogens in milk, cereals, and baby-food by liquid-chromatography tandem mass spectrometry with subsequent determination of their estrogenic activity using transcriptomic assay. Anal Chim Acta 637:55–63
7. Baber RJ (2013) Phytoestrogens in health: the role of isoflavones. In: VR Preedy (ed.) Isoflavones: Chemistry, Analysis, Function and Effects (No. 5), UK, Chapter 1, 1-13
8. Daems F, Jasselette C, Romnee J-M, Planchon V, Lognay G, Froidmont É (2015) Validating the use of an ultra-performance liquid chromatography with tandem mass spectrometry method to quantify equol in cow's milk. Dairy Sci Technol 95(3):303–319
9. De Rijke E, Out P, Niessen WMA, Ariese F, Gooijer C, Brinkman UAT (2006) Analytical separation and detection methods for flavonoids. J Chromatogr A 1112:31–63
10. EMA (2015) VICH GL49: studies to evaluate the metabolism and residue kinetics of veterinary drugs in food-producing animals—validation of analytical methods used in residue depletion studies, UK. http://www.ema.europa.eu /docs/en_GB /document_library/Scientific_guideline/2011/04/WC500105053.pdf (accessed February 2015)
11. Flachowsky G, Hünerberg M, Meyer U, Kammerer DR, Carle R, Goerke M, Eklund M (2011) Isoflavone concentration of soybean meal from various origins and transfer of isoflavones into milk of dairy cows. J Verbr Lebensm 6:449–456
12. Franke AA, Custer LJ (1996) Daidzein and genistein concentrations in human milk after soy consumption. Clin Chem 42(6):955–964
13. Grace PB, Teale P (2006) Purification of the crude solution from Helix pomatia for use as β-glucuronidase and aryl sulfatase in phytoestrogen assays. J Chromatogr B 832:158–161

14. Grynkiewicz G, Hsycińska H, Ramza J, Zagrodzka J (2005) Chromatographic quantification of isoflavones (why and how). Acta Chromatogr 15:31–65
15. Gumustas M, Kurbanoglu S, Uslu B, Ozkan SA (2013) UPLC versus HPLC on drug analysis: advantageous, applications and their validation parameters. Chromatographia 76:1365–1427
16. Harjo B, Wibowo C, Zhang EJN, Luo KQ, Ng KM (2007) Development of process alternatives for separation and purification of isoflavones. Ind Eng Chem Res 46:181–189
17. Hoerger CC, Praplan AP, Becker L, Wettstein FE, Hungerbühler K, Bucheli TD (2011) Quantification of five isoflavones and coumestrol in various solid agroenvironmental matrices using 13C3-labeled internal standards. J Agric Food Chem 59:847–856
18. Hoikkala AA, Schiavoni E, Wähälä K (2003) Analysis of phyto-oestrogens in biological matrices. Br J Nutr 89:S5–S18
19. Hoikkala A, Mustonen E, Saastamoinen I, Jokela T, Taponen J, Saloniemi H, Wähälä K (2007) High levels of equol in organic skimmed Finnish cow milk. Mol Nutr Food Res 51:782–786
20. Höjer A, Adler S, Purup S, Hansen-Møller J, Martinsson K, Steinshamn H, Gustavsson AM (2012) Effects of feeding dairy cows different legume-grass silages on milk phytoestrogen concentration. J Dairy Sci 95(5):4526–4540
21. Ishii K, Hosoda K, Furuta T (2013) Isoflavones: high-performance liquid chromatographic analysis of glucuronic acid- and sulfuric acid-conjugated metabolites of daidzein and genistein in human plasma and urine. In: VR Preedy (ed) Isoflavones: chemistry, analysis, function and effects (No. 5), UK, Chapter 13, 196-217
22. Jiang H, Cao H, Zhang Y, Fast DM (2012) Systematic evaluation of supported liquid extraction in reducing matrix effect and improving extraction efficiency in LC–MS/MS based bioanalysis for 10 model pharmaceutical compounds. J Chromatogr B 891–892:71–80
23. Kalač P (2011) The effects of silage feeding on some sensory and health attributes of cow's milk: a review. Food Chem 125:307–317
24. Kalač P (2013) Fresh and ensiled forages as a source of estrogenic equol in bovine milk: a review. Czech J Anim Sci 58(7):296–303
25. King RA, Mano MM, Head RJ (1998) Assessment of isoflavonoid concentrations in Australian bovine milk samples. J Dairy Res 65:479–489
26. Ko K-P (2014) Isoflavones: chemistry, analysis, functions and effects on health and cancer. Asian Pac J Cancer Prev 15(17):7001–7010
27. Krajčová A, Schulzová V, Lojza J, Křížova L, Hajšlová J (2010) Phytoestrogens in bovine plasma and milk–LC-MS/MS analysis. Czech J Food Sci 28(4):264–274
28. Křížová L, Třináctý J, Hajšlová J, Havlíková Š (2011a) The effect of technological processing on the content of isoflavones in bovine milk and dairy products. In: Tzi-Bun Ng (ed.) Soybean – Applications and Technology, Croatia, Chapter 6, 95-110
29. Křížová L, Veselý A, Třináctý J, Schulzová V, Hurajová A, Hajšlová J, Kvasničková E, Havlíková Š (2011b) Changes in isoflavones concentrations in cheese during processing and ripening. Acta Univ Agric et Silvic Mendel Brun 59:153–162
30. Kuhnle GGC, Dell'Aquila C, Low Y-L, Kussmaul M, Bingham SA (2007) Extraction and quantification of phytoestrogens in foods using automated solid-phase extraction and LC/MS/MS. Anal Chem 79:9234–9239
31. Kuhnle GG, Dell'Aquila C, Aspinall SM, Runswick SA, Mulligan AA, Bingham SA (2008) Phytoestrogen content of animal origin: dairy products, eggs, meat, fish, and seafood. J Agric Food Chem 56(21):10099–10104
32. Lundh TJ-O (1990) Conjugation of the plant estrogens formononetin and daidzein and their metabolite equol by gastrointestinal epithelium from cattle and sheep. J Agric Food Chem 38:1012–1016

33. Maggioni S, Bagnati R, Pandelova M, Schramm K-W, Benfenati E (2013) Genistein and dicarboximide fungicides in infant formulae from the EU market. Food Chem 136:116–119
34. Mortensen A, Kulling SE, Schwartz H, Rowland I, Ruefer CE, Rimbach G, Cassidy A, Magee P, Millar J, Hall WL, Birkved FK, Sorensen IK, Sontag G (2009) Analytical and compositional aspects of isoflavones in food and their biological effects. Mol Nutr Food Res 53:S266–S309
35. Mostrom M, Evans TJ (2012) Phytoestrogens. In: Gupta RC (ed) Veterinary toxicology—basic and clinical principles (2nd edn), Netherlands, Chapter 76, 1012-1028
36. Mustonen EA, Tuori M, Saastamoinen I, Taponen J, Wähälä K, Saloniemi H, Vanhatalo A (2009) Equol in milk of dairy cows is derived from forage legumes such as red clover. Brit J Nutr 102:1552–1556
37. Nielsen TS, Nørgaard JV, Purup S, Fretté XC, Bonefeld-Jørgensen EC (2009) Estrogenic activity of bovine milk high or low in equol using immature mouse uterotrophic responses and an estrogen receptor transactivation assay. Cancer Epidemiol 33:61–68
38. Nielsen TS, Purup S, Wärri A, Godschalk RW, Hilakivi-Clarke L (2011) Effects of maternal exposure to cow's milk high or low in isoflavones on carcinogen-induced mammary tumorigenesis among rat offspring. Cancer Prev Res 4(5):694–701
39. Nielsen TS, Höjer A, Gustavsson A-M, Hansen-Møller J, Purup S (2012) Proliferative effect of whey from cow's milk varying in phyto-oestrogens in human breast and prostate cancer cells. J Dairy Res 79:143–149
40. Njåstad KM, Adler SA, Hanse-Møller J, Thuen E, Gustavsson A-M, Steinshamn H (2014) Gastrointestinal metabolism of phytoestrogens in lactating dairy cows fed silages with different botanical composition. J Dairy Sci 97:7735–7750
41. Patisaul HB, Jefferson W (2010) The pros and cons of phytoestrogens. Front Neuroendocrin 31:400–419
42. Rastogi S, Katara A, Pandey MM, Arora S, Singh RRB, Rawat AKS (2013) Physical stability and HPLC analysis of Indian kudzu (Pueraria tuberosa Linn.) fortified milk. Evid Based Complement Alternat Med, vol. 2013, Article ID 368248, 6 pages
43. Rostagno MA, Villares A, Guillamón A, García-Lafuente A, Martínez JA (2009) Sample preparation for the analysis of isoflavones from soybeans and soy foods. J Chromatogr A 1216:2–29
44. Rybak ME, Parker DL, Pfeiffer CM (2008) Determination of urinary phytoestrogens by HPLC–MS/MS: a comparison of atmospheric pressure chemical ionization (APCI) and electrospray ionization (ESI). J Chromatogr B 861:145–150
45. Saviranta NMM, Julkunen-Tiitto R, Oksanen E, Karjalainen RO (2010) Leaf phenolic compounds in red clover (Trifolium pratense L.) induced by exposure to moderately elevated ozone. Environ Pollut 158:440–446
46. Schwartz H, Sontag G, Plumb J (2009) Inventory of phytoestrogen databases. Food Chem 113:736–747
47. Setchell KDR, Clerici C (2010) Equol: history, chemistry and formation. J Nutr 140:1355S–1362S
48. Sirotkin AV, Harrath AH (2014) Phytoestrogens and their effects. Eur J Pharmacol 741:230–236
49. Skaanild MT, Nielsen TS (2010) The influence of bovine milk high or low in isoflavones on hepatic gene expression in mice. J Toxicol, vol. 2010, Article ID 423179, 5 pages
50. Stalikas CD (2007) Extraction, separation, and detection methods for phenolic acids and flavonoids. J Sep Sci 30:3268–3295
51. Steinshamn H, Purup S, Thuen E, Hansen-Møller J (2008) Effects of clover-grass silages and concentrate supplementation on the content of phytoestrogens in dairy cow milk. J Dairy Sci 91:2715–2725

52. Taylor JI, Grace PB, Bingham SA (2005) Optimization of conditions for the enzymatic hydrolysis of phytoestrogen conjugates in urine and plasma. Anal Biochem 341:220–229
53. Třináctý J, Křížová L, Schulzová V, Hajšlová J, Hanuš O (2009) The effect of feeding soybean-derived phytoestogens on their concentration in plasma and milk of lactating dairy cows. Arch Anim Nutr 63(3):219–229
54. Trontelj J (2012) Quantification of glucuronide metabolites in biological matrices by LC-MS/MS. In: Prasain JK (ed) Tandem mass spectrometry–applications and principles USA, Chapter 23, 531-558
55. Tsen SY, Siew J, Lau EKL, Roslee FAB, Chan HM, Loke WM (2014) Cow's milk as a dietary source of equol and phenolic antioxidants: differential distribution in the milk aqueous and lipid fractions. Dairy Sci Technol 94:625–632
56. Uzzan M, Nechrebeki J, Labuza TP (2007) Thermal and storage stability of nutraceuticals in a milk beverage dietary supplement. J Food Sci 72(3):E109–E114
57. Vacek J, Klejdus B, Lojková L, Kubán V (2008) Current trends in isolation, separation, determination and identification of isoflavones: a review. J Sep Sci 31:2054–2067
58. Valls J, Millán S, Martí MP, Borràs E, Arola L (2009) Advanced separation methods of food anthocyanins, isoflavones and flavanols. J Chromatogr A 1246:7143–7172
59. Vitale DC, Piazza C, Melilli B, Drago F, Salomone S (2013) Isoflavones: estrogenic activity, biological effect and bioavailability. Eur J Drug Metab Pharmacokinet 38:15–25
60. Wang C-C, Prasain JK, Barnes S (2002) Review of the methods used in the determination of phytoestrogens. J Chromatogr B 777:3–28
61. Wielogórska E, Elliot CT, Danaher M, Chevallier O, Connolly L (2015) Validation of an ultra high performance liquid chromatography-tandem mass spectrometry method for detection and quantification of 19 endocrine disruptors in milk. Food Control 48:48–55
62. Wocławek-Potocka I, Mannelli C, Boruszewska D, Kowalczyk-Zieba I, Waśniewski T, Skarżyński DJ (2013) Diverse effects of phytoestrogens on the reproductive performance: cow as a model. Int J Endocrinol, vol. 2013, Article ID 650984, 15 pp
63. Wu Q, Wang M, Simon JE (2004) Analytical methods to determine phytoestrogenic compounds. J Chromatogr B 812:325–355

CHAPTER 2

Organization of Lipids in Milks, Infant Milk Formulas and Various Dairy Products: Role of Technological Processes and Potential Impacts

Christelle Lopez[1,2] & Chantal Cauty[1,2] & Fanny Guyomarc'h[1,2]

[1]*INRA, UMR1253 STLO, 65 rue de Saint Brieuc, 35000 Rennes, France*

[2] *Agrocampus Ouest, UMR1253 STLO, 65 rue de Saint Brieuc, 35000 Rennes, France*

ABSTRACT

The microstructure of milk fat in processed dairy products is poorly known despite its importance in their functional, sensorial and nutritional properties. However, for the last 10 years, several research groups including our laboratory have significantly contributed to increasing knowledge on the organization of lipids in situ in dairy products. This paper provides an overview of recent advances on the organization of lipids in the milk fat globule membrane using microscopy techniques (mainly confocal microscopy and atomic force microscopy). Also, this overview brings structural information about the organization of lipids in situ in commercialized milks, infant milk formulas and various dairy products (cream, butter, buttermilk, butter serum and cheeses). The main mechanical treatment used in the dairy industry, homogenization, decreases the size of milk fat globules, changes the architecture (composition and organization) of the fat/water interface and affects the interactions between lipid droplets and the protein network (concept of inert vs active fillers). The potential impacts of the organization of lipids and of the alteration of the milk fat globule membrane are discussed, and technological strategies are proposed, in priority to design biomimetic lipid droplets in infant milk formulas.

Keywords: Milk Fat Milk fat globule membrane Homogenization Structure

1 INTRODUCTION

Milk is a biological fluid secreted by all female mammals for the growth and development of newborns. Directly transferred from the mother to the newborn, all the constituents of milk (e.g. proteins, lipids, lactose and oligosaccharides) are in their natural state and adapted to bring energy and bioactive molecules. Milk from animals (e.g. cows, buffaloes and goats) is also transformed into various dairy products for infant and adult human consumption.

In particular, milk lipids are secreted in the form of colloidal assemblies called milk fat globules. Technological processes used in the dairy industry, e.g. thermal and mechanical (pumping and homogenization) treatments, can affect the structure and composition of milk components and alter their technological and nutritional properties (Michalski et al. 2013; Garcia et al. 2014). Hence, it is important to have a good knowledge about the organization of lipids in dairy products. The recent development of microscopy techniques has permitted structural investigations in situ in complex food products.

The objectives of this paper are to update information about the organization of lipids in unprocessed milk and to show the role of technological processes on the organization of lipids in processed milk, infant formulas and various dairy products (e.g. cream, butter, buttermilk and cheeses). The major role played by microscopy techniques for the characterization of lipids in situ in milk and processed dairy products is highlighted. This overview compiles the most recent research advances in the field of the organization of milk lipids and brings original structural data in complex dairy products. The potential impacts of the organization of lipids in processed dairy products as compared to unprocessed milk fat globules on the digestibility and metabolic fate of milk lipids are discussed.

For the last 10 years, our laboratory UMR 1253 Science and Technology of Milk and Eggs from the French National Institute for Agricultural Research (INRA) has contributed significantly to the advancement of knowledge on the organization of lipids in the milk fat globule membrane and in situ in various dairy products. Our expertise and research approaches are highlighted throughout the overview.

2 MATERIALS AND METHODS

2.1 Milks, infant milk formulas and processed dairy products

Unprocessed whole bovine milks were provided by a local dairy plant (Triballat, Noyal sur Vilaine, France). Mature human milks were donated by volunteers, as described by Lopez and Ménard (2011). Sterilized UHT, pasteurized and microfiltered fluid milks were bought in supermarkets. Infant milk formulas for 0–6-month babies were provided by industrials or bought in supermarkets (Rennes, France). Sour creams, UHT creams and liquid creams "crème fleurette" were bought in local supermarkets (Rennes, France). Industrial butter, buttermilks and butter serum were provided by a local dairy company (Brittany, France). The cheeses were provided by dairy companies, bought in local supermarkets (Rennes, France) or manufactured in the INRA-STLO pilot plant Plateforme Lait (Lopez et al. 2008).

2.2 Particle size measurements

The size distributions of milk fat globules and processed lipid droplets were determined by laser light scattering using a Mastersizer 2000 (Malvern Instruments, UK) with two laser sources. The refractive indexes used for the laser at 633 and 466 nm were 1.458 and 1.460 for milk fat and 1.462 and 1.460 for blends of vegetable oils in infant formulas, respectively. The refractive index of 1.333 was used for water. About 0.2 mL of sample was diluted in 100 mL of water directly in the measurement cell of the apparatus in order to reach 10% obscuration. The casein micelles were dissociated by adding 1 mL of 35 mM ethylenediaminetetraacetic acid (EDTA)/NaOH, pH 7 buffer to the milks, directly in the apparatus. To determine the size distribution of fat globules in sour creams, samples were diluted 1:10 *v/v* in 1% sodium dodecyl sulfate (SDS) in order to dissociate lipid droplet aggregates and the casein network. Standard parameters were calculated by the software. The experiments were performed in triplicate for each sample.

2.3 Confocal microscopy

The microstructural analysis of the milks, infant milk formulas and dairy products (i.e. creams, buttermilk and cheeses) was performed with an inverted microscope NIKON Eclipse-TE2000-C1si (NIKON, Champigny sur Marne, France) allowing confocal laser scanning microscopy (CLSM) and optical microscopy using differential interferential contrast (DIC, also called Nomarski contrast). Confocal experiments were performed using an argon laser operating at 488 nm excitation wavelength (emission was detected between 500 and 530 nm), a He–Ne laser operating at 543 nm wavelength excitation (emission was detected between 565 and 615 nm) and a diode operating at 633 nm detected with a long pass filter >650 nm. The observations were performed using ×40 (NA 1.33) and ×100 (NA 1.4) oil immersion objectives.

Apolar lipids (i.e. triacylglycerols) were stained either with the lipid-soluble Nile Red fluorescent dye (5H-Benzo α phenoxazine-5-one, 9-diethylamino; $\lambda ex = 543$ nm; Sigma-Aldrich, St Louis, USA) or with LipidTox® ($\lambda ex = 488$ nm; Invitrogen). Milk phospholipids were stained with *N*-(lissamine rhodamine B sulfonyl) dioleoylphosphatidyl ethanolamine (Rhodamine-DOPE, Avanti Polar Lipids Inc., Birmingham, UK) provided at a concentration of 1 $mg.mL^{-1}$ in chloroform (chloroform was evaporated under nitrogen to avoid the possible artefacts caused by this organic solvent). The carbohydrate moities were localized in the milk fat globule membrane by using the lectin wheat germ agglutinin Alexa fluor 488 (WGA; Invitrogen, Cergy Pontoise, France). The staining of proteins was performed by using Fast Green FCF (Sigma-Aldrich, St. Louis, USA) prepared at a concentration of 10 $mg.mL^{-1}$ in water. Acridine Orange fluorescent dye (Aldrich Chemical Company, Inc., Milwaukee, USA) was used to label simultaneously the proteins and the bacteria. The staining protocols followed previously described methods to investigate the microstructure of fat globules in situ in milk (Lopez et al. 2010; Lopez and Ménard 2011), the microstructure of cheeses (Lopez et al. 2007, 2008) and the microstructure of milks during in vitro digestion (Garcia et al. 2014). The two-dimensional images had a resolution of 512 × 512 pixels, and the pixel scale values were converted into micrometers using a scaling factor. In the multiple labeled samples, different colours were used to locate the fluorescent probes (e.g. red colour for phospholipids stained by Rh-DOPE or triacylglycerols stained by Nile red, green colour for proteins stained by Fast green FCF or the carbohydrate moieties stained by WGA-488 and grey levels for DIC images).

2.4 Atomic force microscopy

The preparation of supported lipid bilayers of milk sphingomyelin (MSM), dioleoylphosphatidylcholine (DOPC) and/or cholesterol (chol) has been previously detailed by Guyomarc'h et al. (2014). The preparation of Langmuir-Blodgett monolayers has been detailed by Murthy et al. (2015). AFM imaging of the bilayers and Langmuir-Blodgett monolayers was performed in contact mode using an MFP-3D Bio AFM (Asylum Research, Santa Barbara, CA, USA), silicon MSNL probes (nominal spring constant k ~0.03 N.m^{-1}, Bruker Nano Surfaces, Santa Barbara, CA, USA) calibrated using the thermal noise method and loading forces below ~1 nN. Temperature was 23 ± 0.7°C. The typical scan rate was 0.5 Hz for 256 × 256 pixel images. Sections were drawn across images to measure the height difference H between features of the images (i.e. difference in height between the domains and the fluid phase). Force mapping was performed on MSM/DOPC (50:50 mol%) and MSM/DOPC/cholesterol (40:40:20 mol%) bilayers as described by Guyomarc'h et al. (2014).

2.5 Transmission electron microscopy

For transmission electron microscopy (TEM) analysis, the industrial buttermilk and butter serum samples were concentrated 5-fold at 45°C using ultrafiltration with a scale-laboratory pilot (Pall Centramate) equipped with polymeric membrane (Pall Omega medium screem, cutoff 10 kDa). Then, the concentrated samples were diluted 1:2 *v/v* in agar, fixed for 3 h at room temperature in a solution with final concentration 2.5% glutaraldehyde in 0.15 M Na cacodylate pH 7.2, followed by the post-fixation for 1 h at room temperature in a solution with final concentration 1% osmium in 0.15 M Na cacodylate at pH 7.2. After each fixation, the samples were washed in 0.15 M Na cacodylate at pH 7.2. Samples were dehydrated through an ascending series of ethanol concentrations, infiltrated and embedded in Epon-Araldite-DMP30 resin mixture, then finally polymerized at 60°C for 24 h. Ultra-thin sections (90 nm) were cut in a Leica ultra-microtome and stained with 4% uranyl acetate for 1 h. Sections were investigated with a JEOL 1400 TEM operating at 120 kV; images were recorded on camera Gatan Orius SC 1000 at the Microscopy Rennes Imaging Center (MRic, University Rennes 1, France).

3 OVERVIEW OF THE ORGANIZATION OF LIPIDS IN UNPROCESSED MILK FAT GLOBULES AND CHANGES INDUCED BY INDUSTRIAL TECHNOLOGY: FOCUS ON THE INTERFACE

3.1 Unprocessed milk fat globules: a core of triacylglycerols covered by a biological membrane

Lipids are secreted in milk in the form of spherical entities of about 4 to 5 µm diameter called milk fat globules (Fig. 1a–c), which are enveloped by a biological membrane known as the milk fat globule membrane (MFGM; Fig. 1d). These colloidal assemblies are natural vehicles for triacylglycerols (TAG: ~98% *w/w* of total milk lipids), fat-soluble nutrients (e.g. carotenoids and vitamins) and other biologically active molecules (e.g. phospholipids, sphingolipids, gangliosides, cholesterol and MFGM proteins). Cytoplasmic remnants from the mammary cells were characterized around milk fat globules, particularly in human milk (Fig. 1e; Lopez and Ménard 2011). The amount of fat in the milk, the size of milk fat globules and the chemical composition of milk lipids, both TAG and polar lipids, are sensitive to the diet (Lopez et al. 2014). Several reviews on milk fat globules have been recently published (Lopez et al. 2011; Gallier et al. 2014). In this part of the overview dedicated to unprocessed milk fat globules, we will focus on the most recent advances about the organization of the MFGM obtained by microscopy techniques. The last 10 years have seen an expanding interest in the structure of the MFGM, largely due to its important roles in the physical stability and techno-functional properties of milk fat globules, as well as in the mechanisms of milk fat digestion and metabolism. The MFGM is composed of polar lipids (phosphatidylcholine, PC; phosphatidylethanolamine, PE; phosphatidylinositol, PI; phosphatidylserine, PS and milk sphingomyelin, MSM), gangliosides, cholesterol, proteins and glycoproteins (Dewettinck et al. 2008; Lopez 2011; Gallier et al. 2014). Organization of the MFGM as a trilayer structure is now well accepted (thickness: 10 to 20 nm), with the inner layer composed of proteins and polar lipids from the endoplasmic reticulum and the outer bilayer of polar lipids originating from specialized secretory regions of the apical plasma membrane of the mammary epithelial cells (Heid and Keenan 2005). The distribution of MFGM polar lipids is asymmetric, with PC and MSM largely located in the outer layer of membrane and PE, PI and PS concentrated in the inner surface Deeth (1997). By using freeze-fracture immunochemistry,

Robenek et al. (2006) performed an excellent work that has improved knowledge about the location of proteins in the MFGM. The main research advances regarding the heterogeneities in the outer bilayer of the MFGM have been provided by using confocal microscopy and atomic force microscopy.

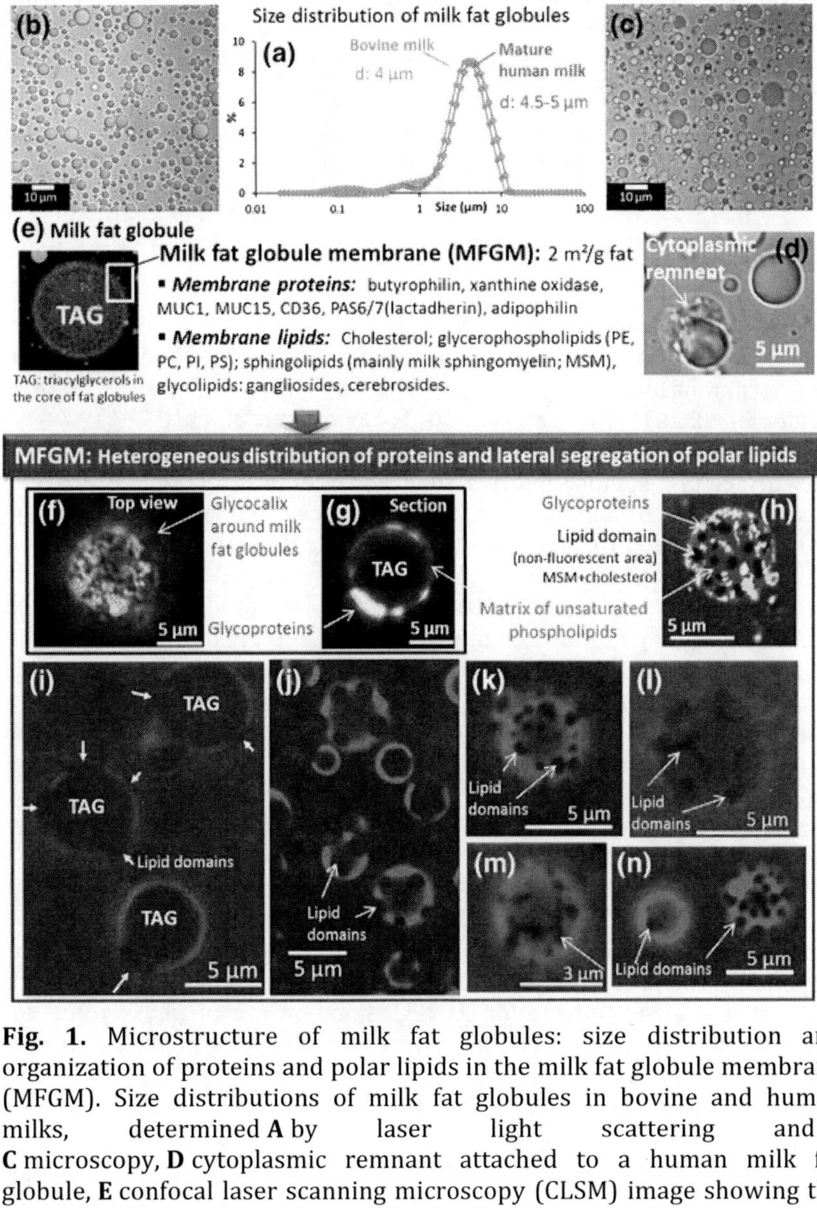

Fig. 1. Microstructure of milk fat globules: size distribution and organization of proteins and polar lipids in the milk fat globule membrane (MFGM). Size distributions of milk fat globules in bovine and human milks, determined **A** by laser light scattering and **B, C** microscopy, **D** cytoplasmic remnant attached to a human milk fat globule, **E** confocal laser scanning microscopy (CLSM) image showing the

triacylglycerol core (TAG; Lipidtox fluorescent dye, *green colour*) and the biological membrane enveloping milk fat globules (Rhodamine-DOPE fluorescent dye, *red colour*). **F-N** CLSM images showing **F–H** the heterogeneous distribution of proteins and glycoproteins in the MFGM, and **H–N** the heterogeneous distribution of polar lipids in the outer bilayer of the MFGM (Rhodamine-DOPE fluorescent dye, *red colour*) with the formation of lipid domain highlighted by *arrows* (non-fluorescent areas)

3.1.1 Lateral segregation of lipids with phase coexistence and heterogeneous distribution of proteins revealed in situ in milk using confocal microscopy

An exciting new aspect of MFGM organization has emerged for the last 5 years, thanks to investigations performed in situ in milk using confocal laser scanning microscopy (CLSM; Evers et al. 2008; Lopez et al. 2010; Gallier et al. 2010; Lopez et al. 2011; Lopez and Ménard 2011; Zou et al. 2012, 2015; Nguyen et al. 2015). CLSM is a powerful tool to penetrate the surface of a sample and to visualize thin optical sections. It provides an opportunity to characterize milk fat globules in situ, without disturbing the internal structure, and to simultaneously localize the main elements of structure: TAG, phospholipids and membrane proteins. Using lipophilic fluorescent probes and lectins, Evers et al. (2008) reported heterogeneities in both the composition and structure of the MFGM within and among fat globules from the same species and also between milks from various mammal species. Using the exogenous fluorescent phospholipid Rh-DOPE, Lopez et al. (2010) observed both fluorescent and non-fluorescent areas at the surface of bovine milk fat globules and interpreted this heterogeneity as the phase separation of polar lipids in the outer bilayer of the MFGM (Fig. 1h–n). Lopez et al. (2010) hypothesized that this could correspond to the lateral segregation of MSM and cholesterol in rigid liquid-ordered (*Lo*) phase microdomains surrounded by the fluid matrix of glycerophospholipids in the liquid-disordered (*Ld*) phase. Milk-SM is 20 to 45% *w/w* of MFGM polar lipids depending on species (Lopez 2011). It contains long chain saturated fatty acids that are responsible for its high temperature of phase transition (Murthy et al. 2015) and is quantitatively the most saturated polar lipid in the MFGM. Moreover, sphingomyelin is known to associate with cholesterol in biological membranes to form domains called "lipid rafts" (Simons and Ikonen 1997). It is therefore hypothesized that the milk SM-rich microdomains revealed in the MFGM could originate from the lipid rafts of the apical plasma membrane of the lactating cells. The heterogeneous distribution of the Rh-DOPE dye in the MFGM has been confirmed by other research groups, *in situ* in bovine milk (Gallier et al. 2010; Zou et al. 2015) and in bovine colostrum (Zou et al. 2015). The lateral segregation of polar lipids in the MFGM was also characterized in human (Lopez and Ménard 2011; Zou et al. 2012) and in buffalo (Nguyen et al. 2015) milks. Differences in the size and pattern of the SM-rich domains have been characterized between fat globules from the same milk and in milks from various mammal species (Lopez et al. 2010; Lopez et al. 2011; Lopez and Ménard 2011) (Fig. 1j–n). A lateral diffusion of the SM-rich domains has been characterized in the plane of the MFGM as a function of time, assimilating them to moving platforms (Lopez et al. 2010; Nguyen et al. 2015). CLSM experiments also revealed that proteins and glycoproteins are heterogeneously distributed in the MFGM and are organized as patches or networks. Notably, the milk SM-rich microdomains are devoid of proteins (Evers et al. 2008; Lopez et al. 2010; Nguyen et al. 2015) (Fig. 1f–h).

On the basis of experimental results obtained by confocal microscopy, our group has proposed a new 3-dimensional and 2-dimensional model for the structure of the MFGM (Lopez et al. 2010; Lopez et al. 2011; Lopez 2011). For the first time, this model proposes the non-random organization of polar lipids, proteins and glycolipids in the plane of the MFGM. It describes the coexistence of at least two lipid phases in the MFGM: (i) the fluid matrix in *Ld* phase composed of the unsaturated glycerophospholipids (PE, PC, PI, PS), proteins, glycoproteins and glycolipids and (ii) the lateral segregation of MSM and cholesterol in *Lo* phase domains (Lopez 2011). The presence of SM-rich domains across the two leaflets of the MFGM bilayer or only on the outer leaflet of the MFGM remains to be elucidated. Recently, Zheng et al. (2014) proposed an updated model, integrating the repartition of the individual polar lipids in the bovine MFGM. The fundamental understanding of molecular arrangements in the MFGM is still in an embryonic stage.

3.1.2 Topography and nanomechanical properties of milk lipid membranes investigated using atomic force microscopy

In the last decade, the emergence of atomic force microscopy (AFM) has opened large perspectives for the investigation of lipid membranes at the nanoscale. AFM imaging allows characterization of the topography of membranes with visualization of phase separations and measurements of the height difference between domains and the fluid phase. The idea of using AFM to investigate the surface nanostructure of interfacial films to better know the organization of the MFGM is recent and roots back into the first statement of interest by Evers (2004). The group of Rafael

Jimenez-Flores (CalPoly, USA) first pioneered the deposition of MFGM components isolated from buttermilk powder to form a monolayer on a mica surface (Langmuir-Blodgett technique) and AFM imaging in air (Jimenez-Flores and Brisson 2008). The same group used this approach to show that polar lipids isolated from milk fractions or processed milk show domains at various surface pressures (Gallier et al. 2010, 2012). Recently, Murthy et al (2015) confirmed the formation of domains in MFGM lipid extract monolayers, protruding by about 1.5 nm at a surface pressure of 30 $mN.m^{-1}$ and 20°C (i.e. below the phase transition temperature of MFGM polar lipids), and revealed the major role played by cholesterol on the topography of the membrane. However, this approach with monolayers may not account neither for the respective roles of the lipid components nor for the hydrated multilayer structure of the MFGM. For these reasons, the use of lipid mixtures with controlled composition (model systems or MFGM lipid extract) to form hydrated bilayers has been employed. Recently, our group has prepared supported lipid bilayers of milk sphingomyelin (MSM; main saturated polar lipid of the MFGM) and dioleoylphosphatidylcholine (DOPC; unsaturated polar lipid) and characterized the phase separation of these lipids for temperature $T < 35°C$ with the formation of gel phase MSM domains protruding by 1 nm from the fluid DOPC phase (Fig. 2a; Guyomarc'h et al. 2014). The investigation of bilayers of MFGM lipid extract also allowed the visualization of domains protruding by ~1 nm from the fluid phase (Fig. 2e). Therefore, the MSM was identified as a major component of the MFGM domains. With the ternary system MSM/DOPC/cholesterol (40:40:20% mol), the authors showed that the MSM-rich domains became dispersed (Fig. 2b; Guyomarc'h et al. 2014). Furthermore, AFM investigation into lipid bilayers allows the quantification of the nanomechanical properties of membranes (Fig. 2c; Garcia-Manyes and Sanz 2010; Li et al. 2011). Using this approach, the authors showed that the membrane's resistance to rupture is significantly higher in the MSM domains as compared to the surrounding fluid DOPC phase (Fig. 2d; Guyomarc'h et al. 2014). Moreover, the authors showed that the addition of cholesterol to MSM/DOPC bilayers decreased the membrane's resistance to rupture, probably through the conversion of the gel phase into liquid-ordered state (Fig. 2d; Guyomarc'h et al. 2015). Ongoing research is detailing the effect of cholesterol and temperature in bilayers of MFGM extract.

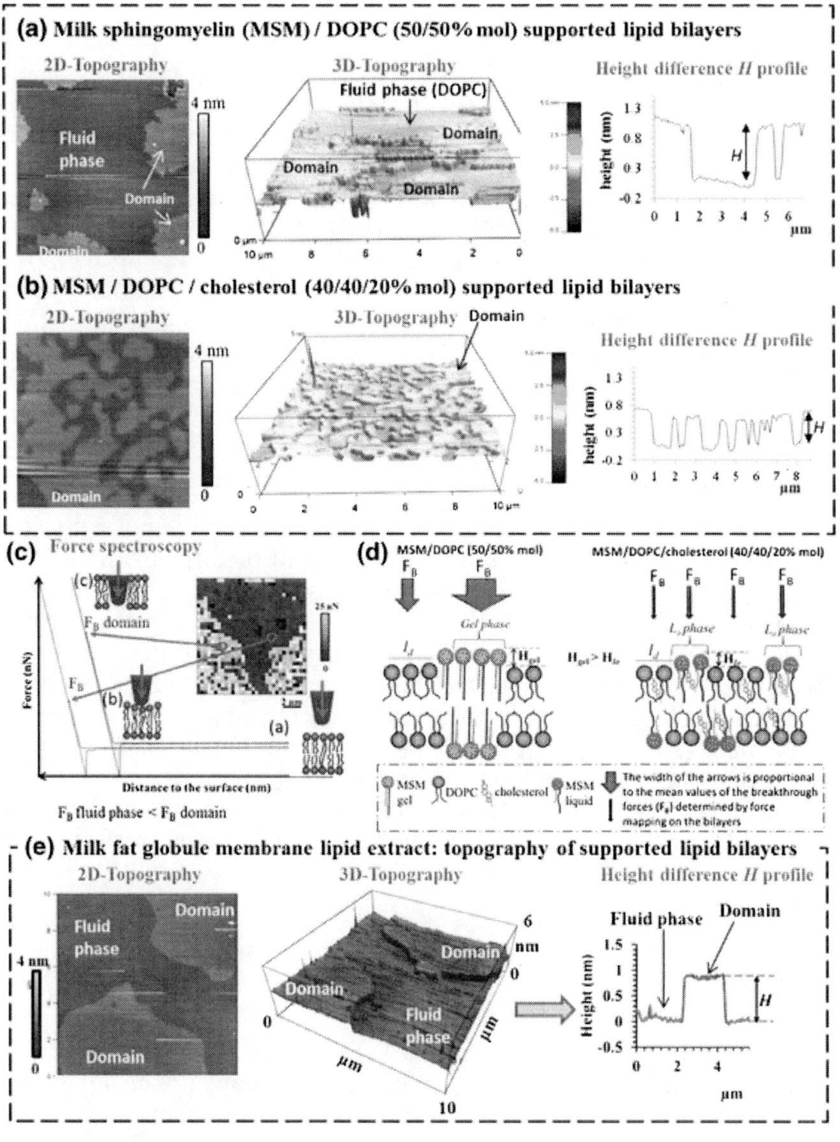

Fig. 2. Topography and nanomechanical properties of supported lipid bilayers investigated by AFM imaging and force spectroscopy. AFM height images of supported lipid bilayers containing **A** MSM/DOPC, **B** MSM/DOPC/cholesterol, **E** MFGM lipid extract with 2-dimensional (*left*) and 3-dimensional (*middle*) representations. The colour scale bars to the right of the images provide height references for the images (*z*-range). *Right*: Cross-sectional line profiles showing height differences *H* between the domains and the fluid phase. **C** Typical force

curves obtained on the different phases of a force map image of a lipid bilayer in fluid conditions (inserted image). **A** Away from the surface, the AFM tip is at rest and shows no applied force **B** when the AFM tip comes in contact with the surface, the applied force increases as the tip pushes against the membrane, **C** the membrane finally ruptures at a value of the force called the "breakthrough force" (F_B) which differs from one phase to another depending on the local stiffness and stability of the membrane. **D** Molecular scheme of the supported lipid bilayers MSM/DOPC and MSM/DOPC/cholesterol with information on differences in the nanomechanical properties between domains (gel or liquid-ordered phases) and the fluid phase. Adapted from Guyomarc'h et al. (2014)

3.1.3 Updated model of the milk fat globule membrane

From the recent knowledge about the lateral segregation of polar lipids in the MFGM, we can propose an updated model of the organization of the MFGM (Fig. 3). Compared with the previous model (Lopez 2011), this model integrates (i) the fact that the domains can be formed by high phase transition temperature polar lipids (e.g. Milk-SM, DPPC) in the gel phase and/or by mixtures of high phase transition temperature polar lipids and cholesterol in the liquid-ordered phase, (ii) that the morphology (size, shape) of the domains and their height difference with the surrounding fluid phase depend on their lipid composition and mainly on the presence of cholesterol and (iii) that the nanomechanical properties are heterogeneous in the MFGM with larger resistance to rupture in the milk-SM rich domains (Fig. 3).

3.2 Processed lipid droplets in milks and infant milk formulas

The fluid milks found in the supermarket for human consumption (both infants and adults) and the infant milk formulas (powders or liquid) are structurally far from the milk produced by female mammals. These structural changes could have processing implications, could affect the digestion and absorption of milk components and have nutritional and health consequences. This part of the paper shows the effect of technological processes, mainly homogenization, on the structure and interface composition of processed lipid droplets found in milks and infant milk formulas.

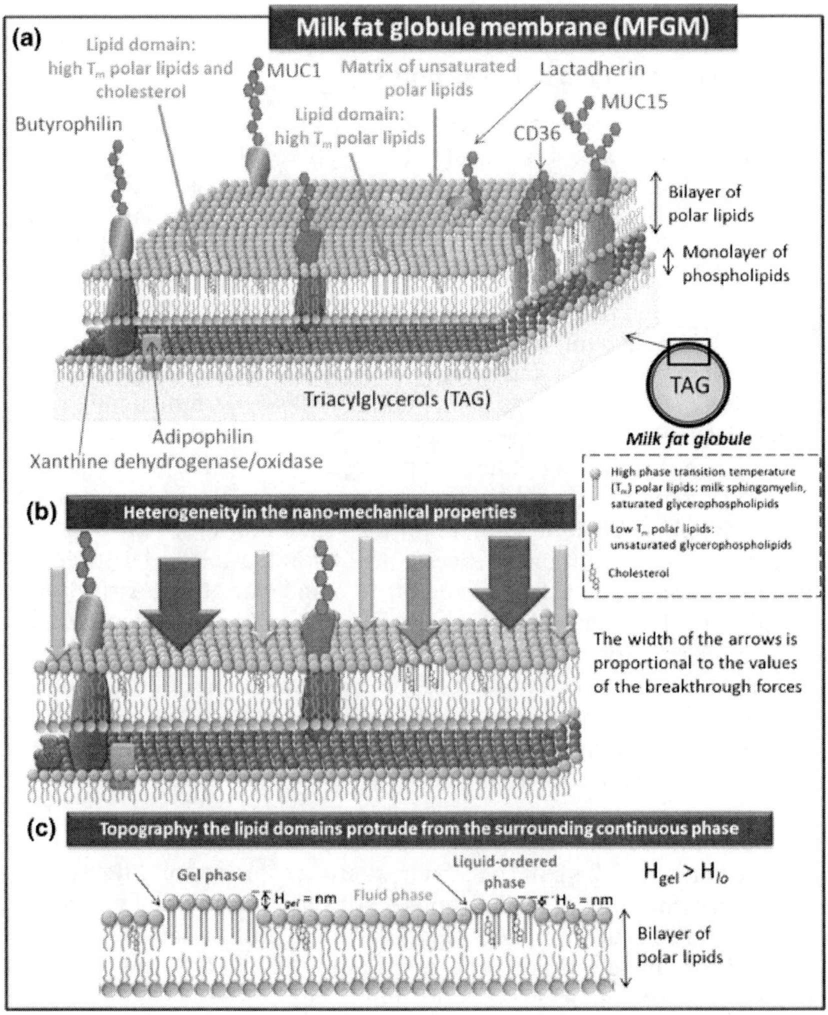

Fig. 3. Updated model of the organization of the biological membrane surrounding fat globules in milk, the milk fat globule membrane (MFGM). **A** 3-Dimensional representation of the trilayered membrane, with the heterogeneous distribution of proteins and the phase coexistence of lipids: lateral segregation of high Tm polar lipids in domains. **B** Heterogeneities in the nano-mechanical properties of the outer bilayer of the MFGM. The width of the arrows is proportional to the mean values of the breakthrough forces determined by atomic force mapping. **C** Schematic representation showing that the MFGM is not flat. Hgel and Hlo are the average height difference between the two lipid phases

3.2.1 Processed lipid droplets in milks

Raw bovine milk is rarely consumed due to its short shelf-life and to fears regarding possible microbiological risks. Commercially available milks are therefore submitted to various technological processes before their consumption, e.g. thermal and mechanical treatments, standardization of fat content. Commercial milks are heat treated (pasteurized or heated at ultra-high temperature; UHT) or microfiltered (microorganisms are physically retained in the retentate by using 1.4 μm pore size membranes) for safe human consumption and to increase storage in the fridge or at room temperature. Fat standardization in commercial milks leads to the preparation of full-fat milks, semi-skimmed milks or skimmed milks. Also, commercial milks undergo a mechanical treatment, e.g. homogenization (pressures applied during industrial process: 5 to 20 MPa). The objective of homogenization is to reduce the size of milk fat globules in order to increase the physical stability of fat for long shelf-storage of the milks and to decrease the perception of excess fat intake due to the fat layer that develops on the milk surface when left to stand (fat globules rise to the top since fat is less dense than the aqueous phase of milk). Figure 4 shows the size distribution of lipid droplets in pasteurized, UHT and microfiltered milks as compared to the size distribution of fat globules in unprocessed bovine milk. The specific surface area of lipid droplets in unprocessed bovine milk is ~2.6 ± 0.2 $m^2.g^{-1}$fat while it is 16.6 ± 3.0 $m^2.g^{-1}$ fat in microfiltered milks (from 13.4 to 20.4 $m^2.g^{-1}$ fat; $n = 7$ milks), 22.3 ± 7.5 $m^2.g^{-1}$ fat in pasteurized milks (from 10.7 to 33.7 $m^2.g^{-1}$ fat; $n = 32$ milks) and 26.4 ± 2.7 $m^2.g^{-1}$ fat in UHT milks (from 23.2 to 34.9 $m^2.g^{-1}$; $n = 21$ milks). In conclusion, there are important differences in the size distribution of lipid droplets between commercially available processed milks and unprocessed bovine milk, with a significant lower size of lipid droplets in processed milks due to homogenization processing. Also, homogenization of milk leads to the disruption of the MFGM and to adsorption of milk proteins (caseins and whey proteins) at the TAG/water interface. After homogenization, the TAG/water interface consists of residual MFGM plus adsorbed milk proteins, of which caseins is the predominant group and whey proteins are mainly represented by β-lactoglobulin (see review Michalski and Januel 2006). Also, heat-denatured whey proteins can interact with MFGM proteins. The sequence of homogenization and heating changes the structure and the composition of the lipid/water interface. The surface increase of the TAG/water interface and changes in the interface composition are both important with respect to chemical reactions (e.g. susceptibility of products to Cu, light-induced flavour deterioration, changes in colour and flavour characteristics) and enzymatic reactions that occur at the interface. The impact on TAG lipolysis and thus the digestibility of fat globules are discussed in the paragraph 4 of this overview.

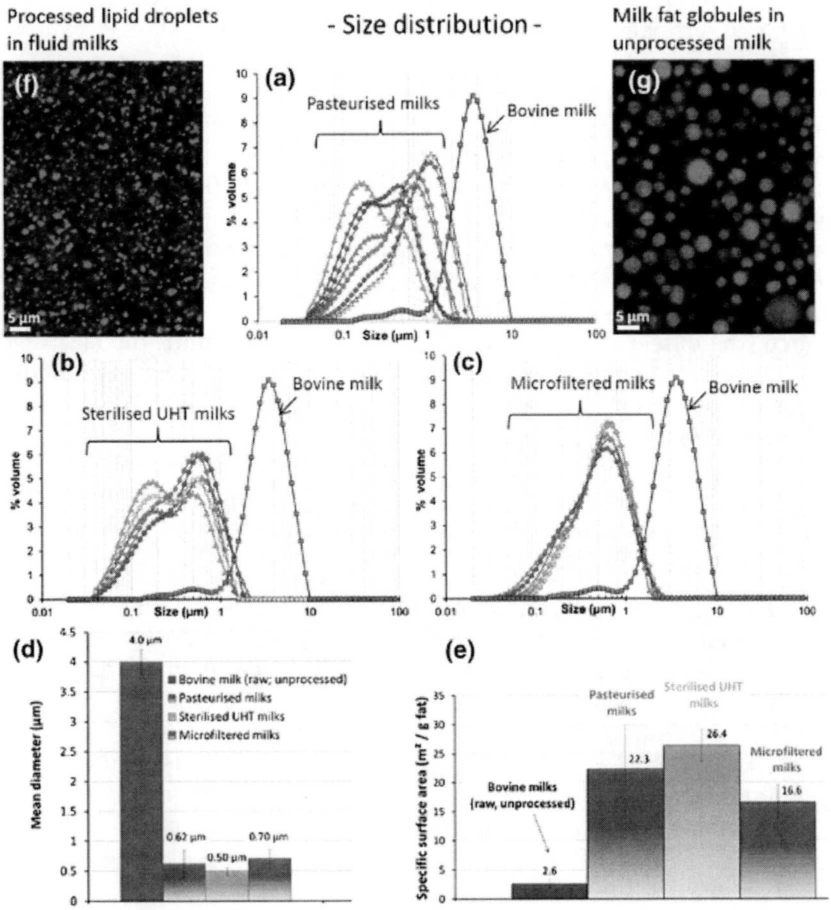

Fig. 4. Microstructure of unprocessed bovine milk vs processed lipid droplets in commercial milks. Laser light scattering experiments showing the size distributions of lipid droplets **A** in pasteurized milks, **B** in sterilized UHT milks and **C** in microfiltrated milks, as compared to unprocessed bovine milk fat globules. Comparison of **D** the mean diameter and **E** the surface area between processed lipid droplets and milk fat globules in unprocessed milk. Confocal laser scattering microscopy images showing the size distribution of **F** processed lipid droplets in commercial processed milk and **G** milk fat globules in unprocessed milk (labeling of total fat by using Nile Red fluorescent dye; *red colour*)

3.2.2 Processed lipid droplets in infant milk formulas

Breastfeeding is highly recommended for infants up to 6 months of age. When mothers cannot or do not want to breastfeed their infant or express their milk, infant milk formula may be used. Infant formulas are manufactured foods designed for the bottle-feeding of babies from powder or liquid. The composition of infant milk formulas is designed to be roughly based on a human mother's milk composition at approximately 1 to 3 months *postpartum*. The most commonly used infant formulas are enriched in cow's milk whey proteins and also contain casein as a protein source (or hydrolyzed proteins to facilitate digestion). They contain a blend of vegetable oils (e.g. palm oil, coconut oil, rapeseed oil and sunflower oil) or mixtures of vegetable oils with bovine milk fat as a fat source (the different sources of fats are combined to mimic the fatty acid composition of human milk), lactose as a carbohydrate source, vitamins, minerals and other ingredients depending on the manufacturer (e.g. prebiotics). Emulsifiers and stabilizers are commonly used to prevent the separation of the oil phase from water in the reconstituted infant milk formula. They include citric acid esters of monoglycerides and diglycerides, lecithins (from vegetable sources such as soya, from fish and krill), gums and maltodextrins.

The microstructure of commercialized infant milk formulas has been investigated in this overview and compared to mature human milk (1 to 6 months *postpartum*). Figure 5 shows the differences in lipid droplet size and interface composition between human milk and infant milk formulas. The size distribution of mature breast milk spans from 0.4 to 13 μm with a mean diameter of 5 μm, corresponding to a surface of 2 $m^2.g^{-1}$ fat (Fig. 5a; Lopez and Ménard 2011). The processed lipid droplets found in the infant milk formulas have a significantly smaller size than breast milk fat globules (mean diameters: 0.3–0.8 μm; Fig. 5a–c) and a higher surface (from about 20 to 40 $m^2.g^{-1}$ fat; Fig. 5a). The small size of the lipid droplets found in infant milk formulas results from the homogenization step involved in the manufacture of these processed products. The objective of homogenization is to create an oil-in-water emulsion by mixing the blend of fats and the other components (proteins, minerals and emulsifiers). The pressure applied upon homogenization is adjusted to form small size droplets (i.e. mainly <1 μm) in order to ensure the physical stability of the processed emulsion during long storage of the powder and after hydration of the powder in the bottle. The differences in the size distributions of processed lipid droplets observed in different commercialized infant milk formulas (Fig. 5a) result from variation in the technological parameters used by different manufacturers or for different

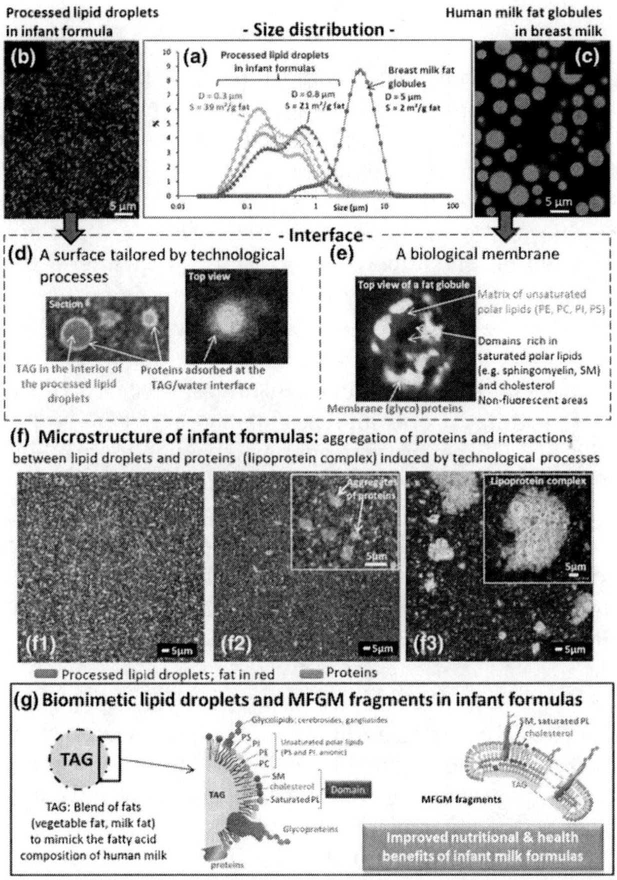

Fig. 5. Comparison of the organization of lipids and interface architecture in human milk vs infant formulas. **A** Size distribution of milk fat globules in mature human milk vs processed lipid droplets in infant formulas, as determined by laser light scattering. Confocal laser scanning microscopy (CLSM) images showing the size distribution, composition and architecture of the interface of **B, D** processed lipid droplets in infant formulas and **C, D** milk fat globules in human milk. In images **B**, **C** and **D**, the labelling of total fat was performed by using Nile red fluorescent dye (*red colour*). In image **E**, the labelling of phospholipids in the biological membrane surrounding milk fat globules was performed by using Rhodamine-DOPE fluorescent dye. In images **D** and **E**, the proteins have been labelled by using Fast Green FCF fluorescent dye. **F** CLSM images showing the microstructure of infant formulas after hydration in a bottle. Total fat was labelled with Nile red fluorescent dye and proteins by Fast green FCF. **G** Schematic representation of biomimetic lipid droplets in infant formulas: presence of MFGM components (polar lipids, PL; membrane proteins) at the surface of processed lipid droplets

markets (i.e. differences in the homogenization process, concentration, evaporation and drying parameters or in the chemical composition). The composition of the TAG/water interface is also different between breast milk fat globules and the processed lipid droplets in infant milk formulas. The CLSM images presented in this overview clearly show the differences in interface composition: Processed lipid droplets in infant milk formulas are mainly covered by milk proteins (Fig. 5d) while breast milk fat globules are enveloped by polar lipids and membrane proteins organized in the MFGM (Fig. 5e), with occasional presence of cytoplasmic remnants (Fig. 1e, d; Lopez and Ménard 2011). Moreover, the structural analysis of various commercialized infant milk formulas revealed the presence of aggregates of proteins or of complexes formed between lipid droplets and proteins (Fig. 5f). These protein aggregates and lipoprotein complexes may result from the thermo-induced denaturation of proteins occurring during the heat treatments performed for the microbial safety of infant formulas. Such lipoprotein complexes induced by the industrial process raise questions about the accessibility of TAG and proteins by the digestive enzymes in the gastro-intestinal tract of newborns and then about the nutritional and health impacts.

The structural analysis presented in this paper clearly shows that the homogenization and the thermal treatments that are used to prepare the emulsion and for microbiological safety, respectively, have a great impact on the structure of fat and on the microstructure of the infant milk formulas. The structure of infant formulas is totally different from human milk, particularly the structure of fat (size of the lipid droplets, interfacial composition and architecture). The potential consequences on the mechanisms of digestion and absorption of lipids as well as on the metabolic programming are discussed in the paragraph 4.2 of this overview. The manufacture of biomimetic processed lipid droplets, with a size of the lipid droplets and a composition and architecture of their surface close to human milk fat globules, could improve the nutritional properties and health impact of infant formulas (Fig. 5g).

3.3 **Microstructure of dairy products showing the effect of technological processes on the organization of lipids**

3.3.1 Cream

Cream is a milk product, defined as an emulsion of milk fat-in-skim milk. In many countries, cream is sold in several grades, depending on the total fat content (heavy cream, light cream and extra-light cream), the viscosity (thickened and fluid) and the heat treatment applied (i.e. pasteurization and UHT) that governs the duration and temperature of storage of the cream. Commercial cream may contain added stabilizers and thickening agents (e.g. sodium alginate, carrageenan, xanthan gum, gelatin, corn starch, sodium bicarbonate, tetrasodium pyrophosphate and alginic acid). In the industrial production of dairy creams, the separation of fat globules from the aqueous phase of milk is accelerated using centrifuges called cream separators. Figure 6a shows the size distribution of milk fat globules in a cream obtained from milk by using a plate separator and pasteurized (fat content ~300 $g.kg^{-1}$). The size distribution of milk fat globules ranges from 1 to 12 μm with a mean volume-weighted diameter of 4.4 ± 0.2 μm (surface area = 1.8 $m^2.g^{-1}$ fat). Figure 6a presents the size distribution of lipid droplets in soured creams containing 30–40% *w/w* fat (thick fermented product; pH ~ 4.5; "crème fraîche"). Figure 6b, c reveals the microstructure of industrial soured creams. Fat can be dispersed as native milk fat globules (traditional process), as shown Fig. 6b, or be at least partially homogenized, as revealed by the lower size of processed fat globules as compared to unprocessed fat globules (Fig. 6c). Microstructural analysis also shows that small processed lipid droplets covered by proteins are connected with the protein network formed by caseins at acidic pH. Bacteria used for culture can be observed in the product, as well as the protein network formed by proteins at pH 4.5 (Fig. 6b). Figure 6d, e shows the size distribution of processed lipid droplets in sterilized UHT creams and liquid pasteurized cream (crème fleurette), respectively. As compared to regular cream, the small size of the processed lipid droplets reveals that milk fat globules have been homogenized. Homogenization of cream is used to increase the physical stability of the cream upon storage (avoid the formation of a cream layer at the top of the product), to limit exudation of whey and to increase the viscosity of the product (and then avoid addition of thickeners). The pressure applied during homogenization, in general from 5 to 25 MPa, depends on the functional properties of the cream. Hence, most of the commercially available creams contain processed lipid droplets to reach functional properties.

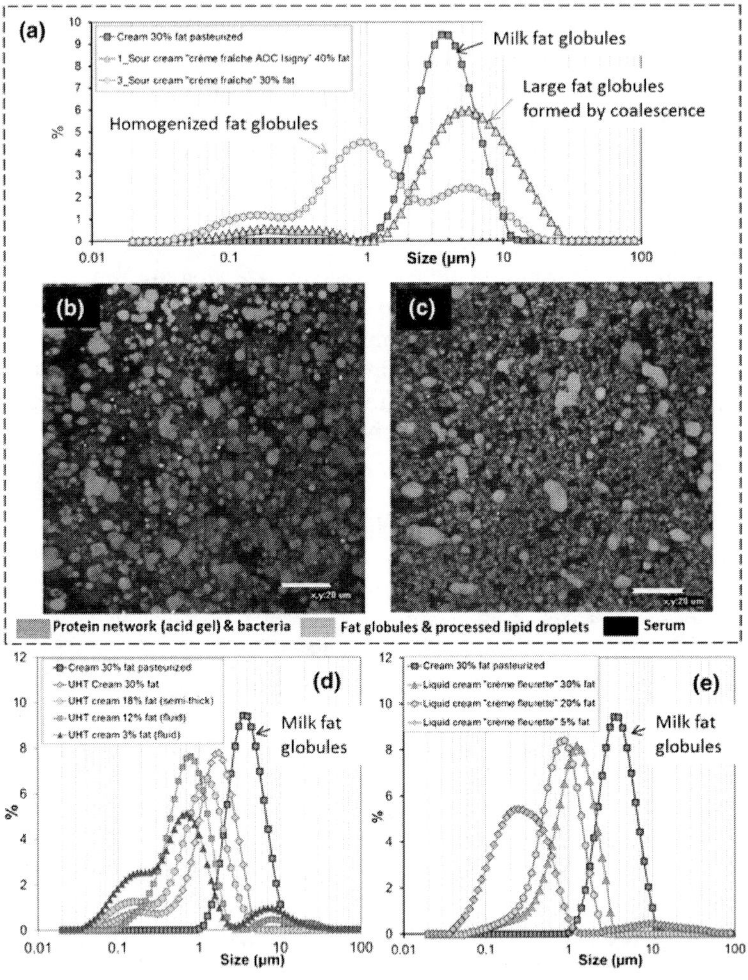

Fig. 6. Size distribution of lipid droplets and microstructure of commercial dairy creams. **A** Size distribution of commercial soured creams vs unprocessed milk fat globules concentrated in a cream determined by laser light scattering. Confocal laser scanning microscopy images showing **B** the microstructure of a non-homogenized soured cream and **C** a homogenized soured cream. Total fat was labeled by using the Nile red fluorescent dye (*red colour*), proteins and bacteria were labeled by using acridine orange fluorescent dye (*green colour*). Size distribution of processed lipid droplets in **D** commercial UHT creams and **E** commercial liquid creams, revealing homogenization of the creams (e.g. processed lipid droplets with a size distribution lower than unprocessed milk fat globules concentrated in a cream)

3.3.2 Microstructure of butter, buttermilk and butter serum

Several technological steps are involved in the manufacture of butter: concentration of fat globules from milk (fat content from 40 $g.kg^{-1}$ in milk to at least 400 $g.kg^{-1}$ in the cream), pasteurization of cream, storage of cream at low temperature to induce TAG crystallization, acidification, churning of cream leading to the disruption of fat globules with phase inversion from oil-in-water to water-in-oil emulsion, production of small butter grains, pressing of the butter grains and then conditioning of butter (820 $g.kg^{-1}$ fat).

Figure 7 shows changes in the structural organization of lipids upon the manufacture of butter. The microstructure of a cream before churning, characterized by CLSM, shows concentrated milk fat globules enveloped by their biological membrane that are dispersed in the aqueous phase, e.g. an oil-in-water emulsion (Fig. 7a, b). The microstructure of butter, characterized by CLSM, shows pockets of serum and residual milk fat globules that are dispersed in the continuous phase formed by partially crystallized TAG (Fig. 7c, d).

Buttermilk and butter serum are the liquid phases released during churning of cream in the process of butter making and during preparation of anhydrous milk fat by melting of butter and then centrifugation to separate anhydrous TAG to the aqueous phase contained in butter, respectively. Buttermilk and butter serum contain about 2 and 11% *w/w* of polar lipids on dry matter, with 19 and 30% *w/w* of sphingolipids on polar lipids, respectively (Rombaut et al. 2006). Figure 7e, f shows the microstructure of buttermilk characterized using CLSM (Fig. 7e) and TEM (Fig. 7f). CLSM images show residual milk fat globules (a core of TAG covered by the MFGM) and butter grains as well as elongated structures that correspond to MFGM fragments (Fig. 7e; Fauquant et al. 2014). The resolution of TEM allowed observation of casein micelles and fragments of the MFGM released from the surface of milk fat globules by the mechanical process applied during churning of the cream. The size of the MFGM fragments released in the buttermilks, ranging from 1 to several micrometers, could be related to the size of the initial fat globule before churning. These TEM images are consistent with previous studies (Corredig and Dalgleish 1997; Morin et al. 2007). Figure 7g shows the microstructure of industrial butter serum composed by casein micelles (black particles with a circular shape in the images), MFGM fragments (elongated black particles) and vesicles of polar lipids (100–400 nm; circular shape with the polar lipid bilayer appearing in black; the interior of the lipid vesicles appears in white colour). These vesicles result from the alteration of the MFGM and self-organization of polar lipids in an aqueous environment. Differences in the microstructure of buttermilk and butter serum and differences in the organization of polar lipids (MFGM fragments vs vesicles of polar lipids, respectively) result from processing (mechanical treatment and temperature).

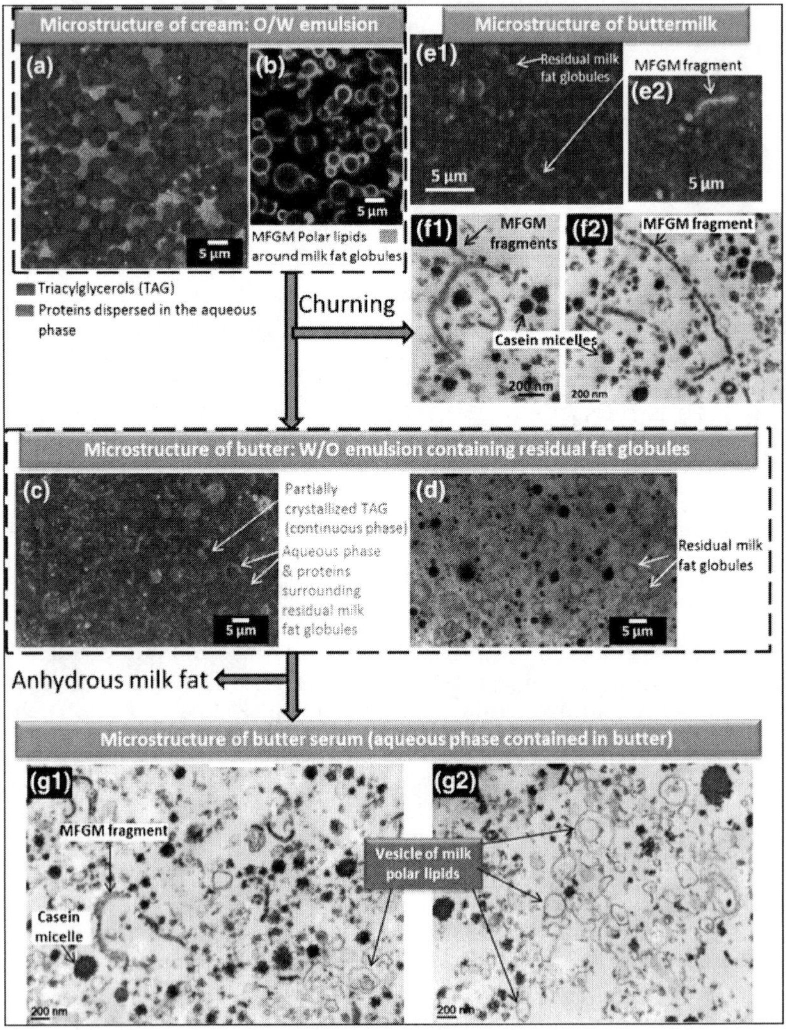

Fig. 7. Changes in the organization of lipids upon the manufacture of butter. **A, B** confocal laser scanning electron microscopy (CLSM) images showing the microstructure of cream. **C, D** CLSM images showing the microstructure of butter. Microstructure of buttermilk characterized by using CLSM (**E**) and transmission electron microscopy (**F**). Microstructure of butter serum revealed by using TEM (**G**)

From this microstructural analysis and from chemical determinations, we learn that the different technology steps involved in the manufacture of butter and anhydrous milk fat lead to the fractionation of milk lipid components according to their polarity. Neutral lipids such as TAG are mainly recovered in butter while the polar lipids organized as vesicles and MFGM fragments are selectively found in the aqueous phases such as buttermilk and butter serum. Among polar lipids, the relative percentage of milk sphingomyelin is higher in butter serum than in buttermilk (Rombaut et al. 2006), which could be related to its specific biophysical properties (i.e. changes in its physical state as a function of temperature, Murthy et al. 2015).

Milk polar lipids have been reported to have interesting functional properties, such as emulsifying and foaming properties, and potentialities to form liposomes for encapsulation (Singh 2006; Dewettinck et al. 2008; Vanderghem et al. 2010). Also, polar lipids have been reported to have interesting nutritional properties and to provide health benefits (Dewettinck et al. 2008). Hence, in the last years, academic and industrials teams have focused on the development of technological processes aiming at purifying and concentrating MFGM fragments and milk polar lipids from buttermilk and butter serum to prepare milk polar lipid or MFGM-enriched ingredients (Dewettinck et al. 2008; Vanderghem et al. 2010; Gassi et al. 2016).

3.3.3 Microstructure of cheeses revealing the major role of processing on the organization of fat

Cheeses are complex dairy products composed of a matrix of protein entrapping fat, aqueous phase containing minerals and bacteria. The structure, texture and mouth feel of cheeses are highly dependent on the fat content within the cheese and the organization of fat. Moreover, the functional properties of cheeses and cheese ingredients (e.g. pizza-style cheeses) are dictated by structure. Hence, analyzing cheese microstructure is of primary importance (Everett and Auty 2008; El-Bakry and Sheehan 2014). Recent research has focused on the organization of fat and the role of the MFGM on cheese structure, texture, functional properties (stretching, melting and free oil formation) and flavour development (Lopez et al. 2006; Lopez et al. 2007; Everett and Auty 2008; Ong et al. 2011; Romeih et al. 2012; Hickey et al. 2015). CLSM associated with pertinent fluorescent probes has permitted (i) to gain information about the spatial organization of individual cheese components in situ in cheeses (TAG, polar lipids, protein network and bacterial colonies) and (ii) the identification of the key steps of the process that can drastically change the properties of fat (Lopez et al. 2006, 2007, 2008; Ong et al. 2011; Ma et al. 2013; El-Bakry and Sheehan 2014). Microscopy techniques showed that the organization of fat in cheeses results from the different steps involved in the technological process, from coagulation to ripening. Examples of different organizations of fat in cheeses, resulting from processing, are presented below.

Regarding hard-type cheeses, very interesting studies on the microstructure of Cheddar cheese have been conducted (Ong et al. 2012), showing destabilization of fat globules within the casein matrix. In Swiss-type cheeses (e.g. Emmental), fat globules are also altered during manufacturing (Fig. 8a–d). Coagulation of milk leads to the formation of a continuous network of casein strands in which fat globules are entrapped (Fig. 8a). Pressing of curd grains and high cooking temperatures up to 55°C results in the disruption of fat globules, coalescence and formation of non-globular free fat (Fig. 8b, c; Lopez et al. 2007). As revealed by CLSM with the fluorescent probe Rhodamine-DOPE, free fat within the Emmental cheese matrix is still covered by polar lipids originating from the MFGM (Fig. 8d; Lopez et al. 2008). The size of the free fat pools formed within the cheese matrix and the oiling-off can be modulated by technological parameters as the heat load applied during pressing and acidification (Richoux et al. 2008). The various structures of fat observed during Swiss-type cheese production and the organization of surface-active material at the interface have been reported (Lopez et al. 2006; 2008) and recently highlighted (Hickey et al. 2015). Studies showed that in hard-type cheeses the bacterial colonies are mainly located at the protein–fat interface, favoring generation of volatile flavour compound from the fat substrate, which influences the final flavour of the ripened cheese (Lopez et al. 2006 ; El-Bakry and Sheehan 2014).

In Mozzarella cheese, fat is located in the aqueous phase entrapped between the elongated protein fibers formed during the pasta filata process (stretching of the curd; Fig. 8e–g). The chemical composition of the cheeses and technological parameters alter the organization of fat (Ma et al. 2013). In high-moisture Mozzarella cheese (e.g. balls found in supermarkets), the structural analysis performed using CLSM showed that fat is organized as globules of about 4 µm in diameter that are mostly aggregated and partially coalesced (Fig. 8e). CLSM performed with the fluorescent probe Rhodamine-DOPE revealed that fat globules are mainly covered by their biological membrane (i.e. the MFGM) (Fig. 8f). In industrial low-moisture Mozzarella cheeses (i.e. pizza cheese ingredient), fat is mainly organized as pools of non-globular fat orientated in the direction of stretching (Fig. 8g), as already reported (Everett and

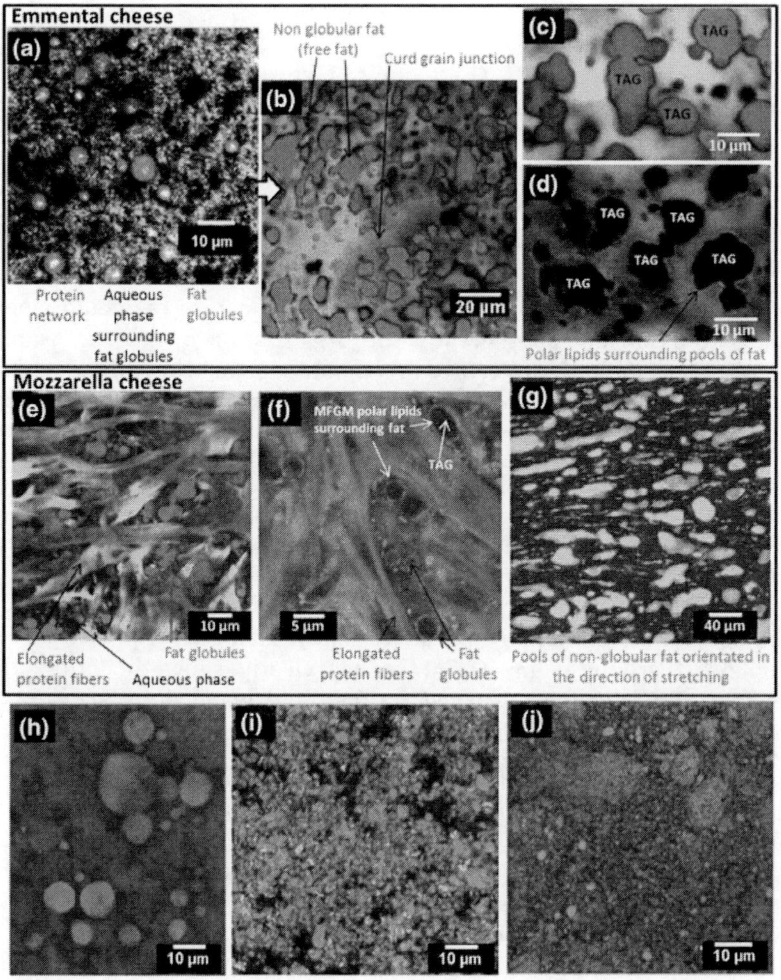

Fig. 8. Microstructure of cheeses showing the effect of dairy technological processes on the organization of lipids. Confocal laser scanning microscopy (CLSM) images of (**A–D**) Emmental cheeses: **A** after coagulation, **B** after pressing of the curd grains, **C**, **D** in the final product, **E, F** high-moisture Mozzarella cheese, **G** low-moisture Mozzarella cheese, **H** mould-ripened soft-type cheese, **I** homogenized cream cheese, **J** homogenized fresh cheese. In all the images, proteins were labelled by using fast green FCF fluorescent dye (*green colour*). Images **A–C, E, G, H–J**: Total fat was labelled using Nile red (*red colour*). Images **D** and **F**: Phospholipids from the milk fat globule membrane were labelled using Rhodamine-DOPE fluorescent dye (*red colour*). Scales bars are in the images

Auty 2008; Ma et al. 2013). These pools of fat result from the disruption of fat globules and their subsequent coalescence due to high temperature cooking, mechanical stretching conditions during manufacturing and ripening. Fat globules covered by the MFGM and pools of fat still covered by polar lipids do not interact with the casein network in the Mozzarella cheese matrix. The size of the fat globules in Mozzarella cheese was positively correlated to meltability and free fat formation (Ma et al. 2013).

The technological steps involved in the manufacture of mould-ripened soft cheeses do not alter greatly the organization of fat. For example, Fig. 8h shows the microstructure of a mould-ripened soft cheese in which milk fat globules are individualized in the casein network (diameter from 4 to 10 μm) or aggregated. Aggregation of milk fat globules facilitates their coalescence and then the increase in the size of milk fat globules (Fig. 8h).

Homogenization of milk is involved in the manufacture of some cheeses, e.g. some fresh cheese or cream cheese. Figure 8i, j shows the microstructure of two commercialized cream cheeses, in which fat globules have been homogenized. The mean diameter of processed lipid droplets is around 1 μm. CLSM images clearly show the interactions between the proteins adsorbed at the lipid/water interface and the proteins in the surrounding phase.

As a conclusion, various organizations of lipids can be present within the cheese matrix, depending mainly on the mechanical treatments (homogenization and/or draining operations) applied during the manufacture of cheeses (Fig. 9). Fat can be dispersed as (i) fat globules of 4 μm in diameter covered by their biological membrane (e.g. high-moisture Mozzarella cheese, mould-ripened soft cheeses, i.e. no homogenization, moderate drainage), (ii) inclusions of non-globular free fat of several μm (e.g. Cheddar cheese, Emmental cheese, low-moisture Mozzarella cheese, i.e. no homogenization and intense drainage), (iii) very small homogenized lipid droplets covered by milk proteins (e.g. fresh cheeses, cream cheese, blue cheeses).

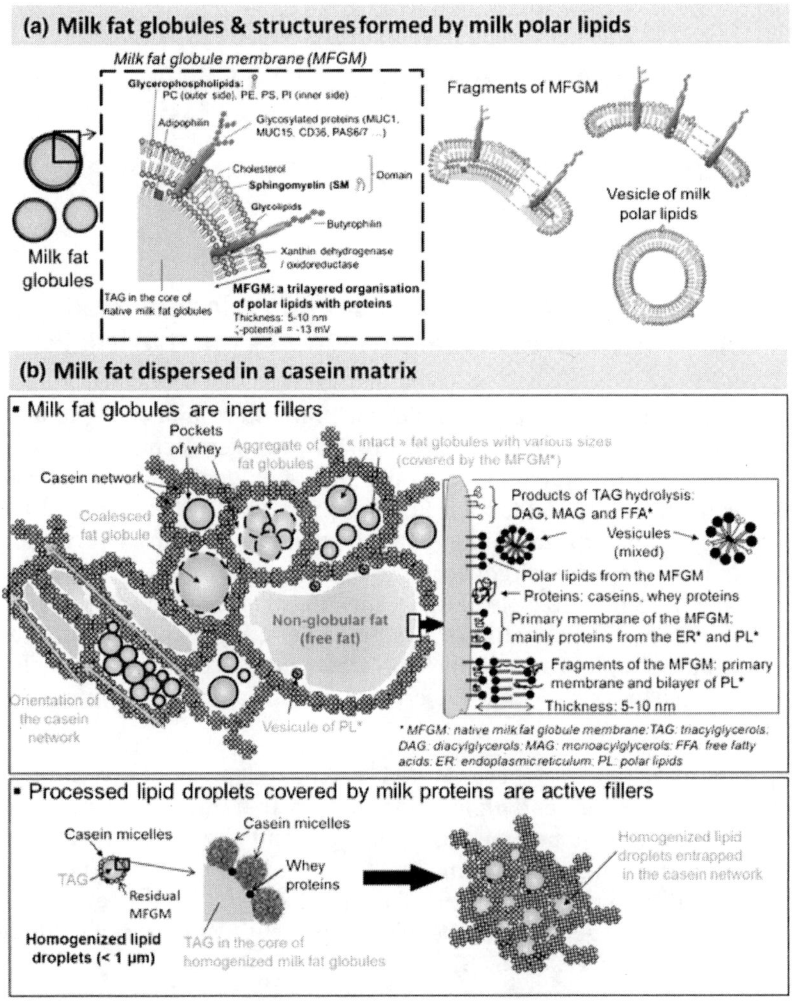

Fig. 9. Microstructure of milk fat components. **A**, *left*: Milk fat globules surrounded by their biological membrane, the milk fat globule membrane (MFGM): a trilayer of polar lipids and proteins. **A**, *right*: Structures derived from the MFGM: fragments of MFGM and vesicles of milk polar lipids characterized in buttermilks and butter serum, respectively. **B** Schematic representation of the organization of milk fat in a casein matrix, as observed in various dairy products such as acidified dairy creams, yoghurt and cheeses

4 THE PROCESSING OF MILK AFFECTS THE MICROSTRUCTURE OF FAT AND ALTERS THE BIOLOGICAL MEMBRANE: DISCUSSION ON THE POTENTIAL IMPACTS

4.1 Technological processes alter milk fat globules and their role in structure and texture: inert vs active fillers

This overview showed that processing of milk alters milk fat globules, changes their size and the architecture of their surface (composition and organization). Also, churning of milk fat globules releases MFGM fragments that can form milk polar lipid vesicles (Fig. 9). The composition of the TAG/water interface (e.g. MFGM, monolayer of phospholipids, caseins and whey proteins) governs the interactions of milk fat with the surrounding protein matrix, leading to specific structural and rheological properties (concept of inert fillers or active fillers; Fig. 9). Changes in the size of the lipid droplets and of the TAG/water interface composition also affect the mechanisms of milk fat crystallization, but this is out of the scope of this overview.

In all cheese matrices where MFGM remnants, polar lipids and/or lipolysis-induced mono- and diglycerides occupy the TAG/water interface, no interaction is seen between the milk fat and the protein network (Fig. 9). Fat is then qualified an "inert filler" in the casein matrix (i.e. have little or no ability to interact; Fig. 9). This is the case for all processes that do not involve homogenization (e.g. hard-type cheese, mozzarella cheese and mould-ripened cheese; Fig. 8).

The structural analysis of various milks and dairy products showed that homogenization is often used in the dairy industry (commercialized milks, Fig. 4; infant formulas, Fig. 5; creams, Fig. 7; cream cheeses, Fig. 8i, i). The main objective of homogenization is (i) to improve the physical stability of the emulsion (example of commercially available milks), (ii) to increase the viscosity of the product, i.e. improve the texture (example of dairy creams), (iii) to create an emulsion (example of the infant formulas) and (iv) to promote lipolysis (blue cheeses). The pressures ranging from 5 to 25 MPa at the industrial scale lead to the formation of small (<1 μm) processed lipid droplets, to the disruption of the MFGM and to the adsorption of milk proteins (mainly caseins) to cover the newly formed interface (Fig. 9). After homogenization, the caseins adsorbed at the TAG/water interface interact with the casein matrix formed (i) by

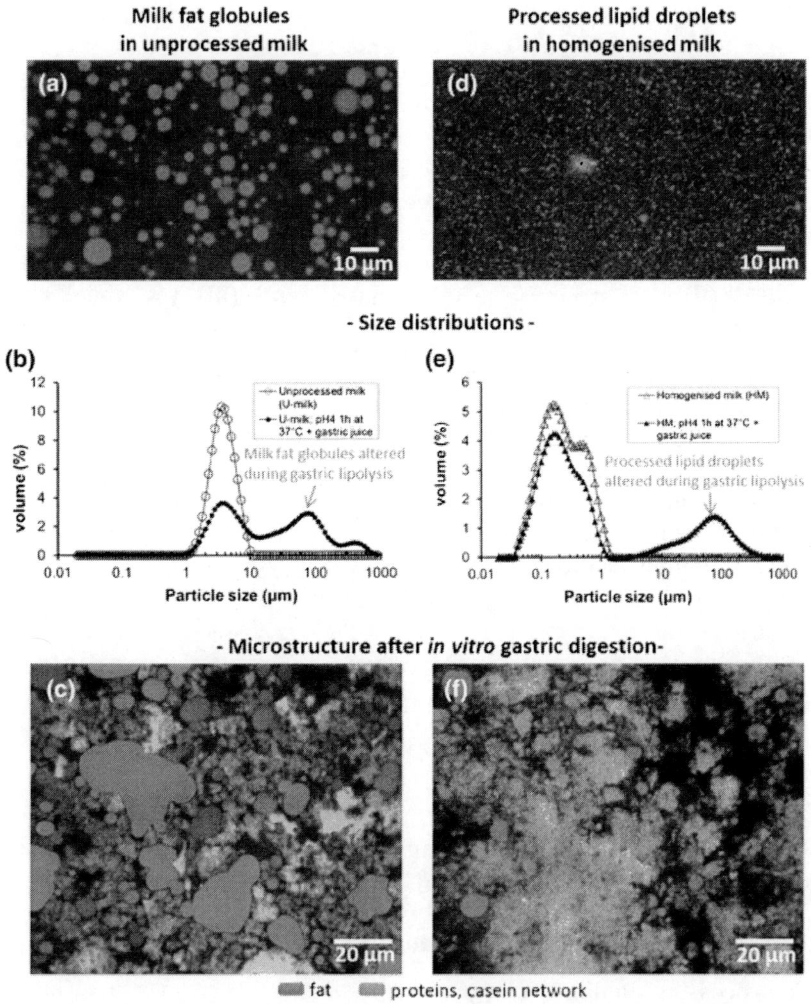

Fig. 10. Structural evolution of milk fat globules in unprocessed milk vs processed lipid droplets in homogenized milk upon simulated human gastric digestion. Confocal laser scanning microscopy (CLSM) images showing the size distribution of the lipids droplets **A** in unprocessed milk, **D** in homogenized milk. Particle size distribution measured in the initial milks and after gastric digestion for **B** unprocessed milk and **E** homogenized milk. CLSM images showing the microstructure upon gastric digestion of **C** unprocessed milk and **F** homogenized milk. Adapted from Garcia et al. (2014)

renneting during cheese-making (Fig. 9i, j) or (ii) at acidic pH for example in the sour dairy creams (Fig. 7c), but also in the stomach upon digestion of milk and dairy products (Fig. 10). Homogenized processed lipid droplets covered by caseins are considered as active fillers (Fig. 9), providing a structural role and increasing firmness of the protein matrix (Everett and Auty 2008). Hence, the composition of the fat globule surface affects (i) the interactions between lipid droplets and the protein network, (ii) the structure of the matrix and (iii) the rheological properties and texture of the product.

For the next decade, the science of the emulsions will bring new opportunities to the dairy industry. The preparation of emulsions with various sizes and specific components located at the TAG/water interface will permit modulation of the interactions with the protein network (inert vs active fillers). Also, the tailoring of emulsions covered by MFGM-originating components (milk polar lipids, cholesterol and MFGM-proteins) should be further considered to prepare small lipid droplets, mimicking the interfacial properties of unprocessed milk fat globules.

4.2 Secreted milk fat globules vs processed lipid droplets in milks and infant milk formulas: industry may learn from biological lipid assemblies to improve the nutritional and health impacts

Milk is a complex arrangement of multiple components that are digested and absorbed though a dynamic and interrelated process, rather than as individual components, highlighting the importance of food microstructure in delivering health benefits.

Milk processing, mainly homogenization, affects the structure of fat as compared to the unprocessed fat globules found in milk: The size decreases, the surface area increases and the architecture of the TAG/water interface changes (bovine or human milk fat globules vs processed lipid droplets; Figs. 4 and 5). Studies showed the importance of the properties of the TAG/water interface, mainly the composition and amount of interface (expressed in m^2 per g of fat; inversely related to the diameter of lipid droplets), in the adsorption and activity of digestive enzymes. It is now well accepted that the native milk fat globules and the processed lipid droplets will evolve differently in the gastrointestinal tract, with consequences in digestive lipolysis and potential impacts on lipid bioaccessibility, absorption and on the postprandial fate of dietary fatty acids.

4.2.1 The accessibility of lipids by the digestive enzymes governs their hydrolysis and metabolic fate: importance of the TAG/water interface amount and architecture and role played by the microstructure upon gastric digestion

Recent studies focused on the digestibility of lipids in milk as a function of their organization, which results from dairy processes (mainly homogenization and heat treatments). Homogenization of milk affects both the size of lipid droplets and the architecture of the TAG/water interface (milk fat globules: 4 µm in diameter, covered by the MFGM vs processed lipid droplets: <1 µm diameter, covered by proteins). It is then difficult to dissociate the effects due to the size of the lipid droplets from those due to the composition of the interface. Garcia et al. (2014) investigated the effect of the droplet size (native milk fat globules of various sizes, homogenized milk lipid droplets) and interface composition (MFGM vs adsorbed milk proteins), in vitro, in conditions close to human physiology. The authors showed that the digestion of TAG is twice as efficient for small native fat globules (1.75 µm; 4.3 $m^2.g^{-1}$ fat) as for large native milk fat globules (6.6 µm; 1.6 $m^2.g^{-1}$ fat) in the gastric phase and is about 30% higher in the intestinal phase. This study provided evidence that, for a similar interface composition (i.e. the MFGM), the amount of the interface directly influences the digestion of milk fat globules (see Lopez et al. 2011 for details about the preparation and composition of size-partitioned milk fat globules). With the same milk (i.e. exactly the same chemical composition), the effect of homogenization on the in vitro digestion of milk lipids has been investigated (Berton et al. 2012; Garcia et al. 2014). The changes in the interface composition (i.e. adsorption of casein micelles and whey proteins) upon decreasing globule size by homogenization had opposite effects to those found when decreasing the diameter of milk fat globules without altering the MFGM, leading to lower levels of lipolysis as expected (Garcia et al. 2014). Heat treatment (UHT sterilization) did not further affect the digestibility of homogenized milk. Structural characterizations performed during in vitro digestion revealed (i) an increase in the size of both native milk fat globules and homogenized lipid droplets upon gastric lipolysis and (ii) differences in the microstructure after coagulation of milk proteins at acidic pH (Fig. 10; adapted from Garcia et al. 2014). It is possible that the interactions between the caseins adsorbed at the surface of homogenized lipid droplets and the protein network have formed upon coagulation of milk in the stomach (Fig. 10f). This would decrease the accessibility for the gastric lipase to TAG and then decrease the levels of lipolysis. Then, the concept of processed lipid droplets as active fillers in the stomach could be responsible for lower levels of lipolysis as compared to inert fillers (i.e. unprocessed milk fat globules). The physical stability of the emulsion in the stomach has also been reported to affect the gastric lipolysis rates and gastric emptying (Golding et al. 2011). Investigations of the effects of milk lipid droplet size and interface composition (MFGM vs. milk proteins) on the activity of the human pancreatic lipase revealed a higher lipolysis for small processed lipid droplets but a lower catalytic efficiency than for milk fat globules covered by the MFGM (twice as much lipolysis despite the 25-fold larger available surface in homogenized milk; Berton et al. 2012). Studies performed in the rat showed that processing of milk (homogenization and heat treatment) affect the microstructure of fat upon digestion (Gallier et al. 2013a, b).

In conclusion, the main parameters affecting the digestibility of milk lipids are the size of the lipid droplets, the composition of the TAG/water interface and the microstructure of the stomach bolus at acidic pH. The role played by the microstructure of milk in the stomach (liquid *vs* coagulated; protein–lipid droplets interactions) could be a means to modulate digestion of lipids.

4.2.2 Processed lipid droplets in infant milk formulas vs breast milk fat globules: when the structure and architecture of the interface really matters for infants

Breast milk or infant milk formulas are the exclusive foods for infants from birth to about 6 months of age. The milk lipids supply 50 to 60% of the calories necessary for newborn growth. Hence, the efficiency of milk lipid digestion and absorption is of particular importance for infants. The differences in TAG composition and internal structure (position of fatty acids on the three glycerol carbons) between human milk and infant milk formulas could affect the digestibility of infant formulas, but it is out of the scope of this overview focused on the organization of lipids at a microscopic level. The pioneering study of Armand et al. (1996) showed in premature newborns that the digestibility of TAG in human milk (native milk fat globules; 4 μm) is more efficient than in infant milk formulas (processed lipid droplets; 0.6 μm). Namely, 25% of TAG in breast milk was hydrolyzed in the gastric phase after 50 min vs 14% for formulas. The absorption of lipids was also higher with human milk as compared to the infant milk formula (Armand et al. 1996). Through this in vivo study, the group of Martine Armand showed that the size of lipid droplets is not the main parameter involved in the digestion of TAG. The unique composition and structure of the biological membrane surrounding fat globules in milk could play a specific role in the mechanisms of adsorption and hydrolysis of TAG by the digestive enzymes, as well as on the metabolic fate of milk lipids. The lateral packing of polar lipids in the outer bilayer of the MFGM (i.e. phase coexistence of SM-rich domains in the gel or liquid-ordered phase in presence of cholesterol and a surrounding fluid phase; Fig. 1) could be involved in the mechanisms of milk fat globule digestion (i.e. adsorption and activity of the digestive enzymes; gastric lipase, bile salt stimulated lipase, pancreatic lipase), as previously discussed (Lopez et al. 2010; Lopez 2011; Lopez and Ménard 2011). Also, the composition of the TAG/water interface can affect the microstructure of milk and infant milk formulas after coagulation of caseins occurring at acidic pH in the stomach (same mechanism as Fig. 10). Breast milk fat globules are inert fillers while processed lipid droplets covered by milk proteins in infant milk formulas are active fillers, which can affect the accessibility of lipids for their digestion and absorption with nutritional impact in infants.

The presence of the MFGM around fat globules or its absence in infant milk formulas is an outstanding question. The microstructure of lipids in infant milk formulas could be closer to breast milk by increasing the size of the processed droplets and by adding MFGM components at the TAG/water interface (Fig. 5g). Infant milk formula manufacturers are developing such new concept of infant formulas with a structural organization of lipids close to breast milk. Oosting et al. (2012) showed that feeding young mice with large lipid droplets coated by milk phospholipids reduced fat accumulation and improved metabolic profile in adulthood as compared to a control infant formula. This study provides a proof of concept that the size and the presence of MFGM components at the surface of processed droplets of early dietary lipids contribute to body composition and metabolic health later in life, then on metabolic programming (Oosting et al. 2012). Moreover, recent studies revealed a direct effect of the presence of MFGM in infant formulas on the cognitive development of infants (Timby et al. 2014). Milk fat globules could supply to the infant the necessary instructions for the development of the intestinal mucosa, of the immune and nervous systems as well as for metabolic activity (Riccio 2004). Moreover, some components of the MFGM are involved with anti-infection or anti-adhesion properties and, hence, with protection of the newborn from various viral and bacterial infections (see reviews: Ward et al. 2006; Dewettinck et al. 2008; Lopez 2011). Therefore, we should consider that both the chemical composition and the structure of biological entities such as milk fat globules components may have evolved over time and may have been optimized through the selective pressure that occurred specifically for each mammal species to assume essential biological functions.

For the next decade, the technological challenge will be to manufacture infant milk formulas containing processed lipid droplets with structural characteristic, interfacial properties and functions close to those of native milk fat globules, i.e. biomimetic lipid droplets (Fig. 5g). To reach this objective, MFGM components concentrated and purified from buttermilk or butter serum (by-products of the bovine dairy industry) could be used. The formulation of milk polar lipids in replacement of non-dairy emulsifiers (e.g. soya lecithin) will bring sphingolipids (milk sphingomyelin, cerebrosides and gangliosides) and bioactive membrane proteins for a positive impact in the gastrointestinal tract of infants. Moreover, the active role played by cholesterol on the physical properties of milk sphingomyelin should be considered in the formulation of biomimetic infant milk formulas (Guyomarc'h et al. 2014; Murthy et al. 2015). These biomimetic lipid droplets will certainly improve the nutritional and health benefits of infant milk formulas.

4.3 A specific role played by the biological membrane enveloping milk fat globules on the nutritional properties and health impact of milk fat?

Milk fat has long been recognized as having cholesterol-raising effects that are caused by the high content of long-chain saturated fatty acids, especially when compared with that of vegetable oils (Iggman et al. 2011). However, epidemiological studies suggested that the association between milk fat intake and cardiovascular disease is partly dependent on the type of dairy food consumed (Goldbohm et al. 2011; Patterson et al. 2013). Controlled studies showed that butter causes higher cholesterol concentrations than do cheeses, even with a similar intake of milk fat (Hjerpsted et al. 2011). The composition (fat, protein and calcium contents), viscosity and microstructure (liquid vs. gel vs. butter) of the food matrix as well as the organization of fat (bulk, emulsified and size of lipid droplets) and its physical state (liquid *vs* crystallized) may possibly modulate the mechanisms of digestion and hypercholesterolemic effects of saturated fat (Michalski et al. 2013). Moreover, recent studies hypothesized that the different effects of various dairy foods on plasma lipids might be caused by the presence of the MFGM, which is rich in bioactive membrane proteins generally considered to exhibit favourable metabolic effects (Spitsberg 2005) and polar lipids (e.g. sphingolipids, mainly MSM). Animal studies showed that MFGM and sphingolipids could lower plasma cholesterol partly by modulating hepatic gene expression (Kamili et al. 2010; Zhou et al. 2012; Chung et al. 2013).

This overview shows that the MFGM is preserved differently in dairy products as a consequence of technological processes that can alter milk fat globules and then alter the physical structure of this biological membrane (Figs. 4–8). The question is: Are the potential beneficial effects due to the whole MFGM surrounding fat globules (i.e. with a trilayered organisation and specific biophysical properties) or to individual components of the MFGM that can be found in MFGM-enriched ingredients? Conflicting results are found in the literature. A recent study showed no effect of MFGM-enriched ingredient on postprandial lipid concentrations (Ohlsson et al. 2010) whereas a study using buttermilk showed significant reductions in both fasting total cholesterol and LDL-cholesterol (Conway et al. 2013). A very recent study compared native milk fat globules covered by their physically intact MFGM with butter oil to determine the effects on lipid profiles (Rosqvist et al. 2015). The results show that the presence of MFGM may counteract the hypercholesterolemic effects of saturated fat. The mechanisms behind the potential effects of the MFGM on plasma lipids are unclear but may involve reduced cholesterol absorption or phospholipid-induced alterations in hepatic gene expression. Whether or not the microdomains of milk SM and cholesterol revealed in the MFGM have some special functions still remains unknown. The roles of the physical structure and single components of the MFGM need additional studies to be understood.

5 CONCLUSION

This overview has documented the current state of the art in relation to the influence of technological processes on the organization of lipids in various dairy products. The relationship between the composition, the structure of lipids and their functions is not fully understood. Recent advanced microscopy studies have permitted to gain important knowledge in situ in complex food products, and their development will progress in the next years. Significant knowledge gaps remain in areas including (i) the role of the organization of fat and microstructure of the acid gel formed in the stomach on the mechanisms of digestion and (ii) the specific role of the MFGM around milk fat globules vs processed lipid droplets even if they are coated by a monolayer of polar lipids. The biofunctionality of the MFGM is not simply a sum of those properties from individual molecules but could result from the specific organization and structure of the polar lipids, cholesterol and membrane proteins. Further nutritional studies are required to investigate the role of milk fat structure on its digestibility and metabolic fate of fatty acids. The role played by specific lipids (e.g. milk sphingomyelin) on the physiology of the intestine and on the microbiota will also improve the knowledge and participate in the improvement of dairy products for beneficial health impacts in infants and adults.

ACKNOWLEDGMENTS

The French National Agency for Research (ANR, Paris, France) is acknowledged for funding the characterization of industrial buttermilks and butter serums using TEM (ALID program, project Valobab; ANR-11-ALID-007). The authors warmly thank the colleagues whose past work has contributed to present knowledge in the organization of lipids in dairy products and its potential impacts, but whose publications could not be cited.

REFERENCES

1. Armand M, Hamosh M, Mehta NR, Angelus PA, Philpott JR, Henderson TR, Dwyer NK, Lairon D, Hamosh D (1996) Effect of human milk or formula on gastric function and fat digestion in the premature infant. Pediatr Res 40:429–437
2. Berton A, Rouvellac S, Robert B, Rousseau F, Lopez C, Crenon I (2012) Effect of the size and interface composition of milk fat globules on their in vitro digestion by the human pancreatic lipase: native versus homogenized milk fat globules. Food Hydrocolloids 29:123–134
3. Chung RW, Kamili A, Tandy S, Weir JM, Gaire R, Wong G, Meikle PJ, Cohn JS, Rye KA (2013) Dietary sphingomyelin lowers hepatic levels and inhibits intestinal cholesterol absorption in high-fat-fed mice. PLoS One 8, e55949
4. Conway V, Couture P, Richard C, Gauthier SF, Pouliot Y, Lamarche B (2013) Impact of buttermilk consumption on plasma lipids and surrogate markers of cholesterol homoeostasis in men and women. Nutr Metab Cardiovasc Dis 23:1255–1256
5. Corredig M, Dalgleish DG (1997) Isolates from industrial buttermilk: emulsifying properties of materials derived from the milk fat globule membrane. J Agric Food Chem 45:4595–4600
6. Deeth HC (1997) The role of phospholipids in the stability of milk fat globules. Aust J Dairy Technol 52:44–46
7. Dewettinck K, Rombaut R, Thienpont N, Le TT, Messens K, van Camp J (2008) Review: nutritional and technological aspects of milk fat globule membrane material. Int Dairy J 18:436–457
8. El-Bakry M, Sheehan J (2014) Analysing cheese microstructure: a review of recent developments. J Food Eng 125:58–96
9. Everett DW, Auty MAE (2008) Cheese structure and current methods of analysis (review). Int Dairy J 18:759–773
10. Evers JM (2004) The milkfat globule membrane—methodologies for measuring milkfat globule (membrane) damage. Int Dairy J 14:747
11. Evers JM, Haverkamp RG, Holroyd SE, Jameson GB, Mackenzie DDS, McCarthy OJ (2008) Heterogeneity of milk fat globule membrane structure and composition as observed using fluorescence microscopy techniques. Intern Dairy J 18:1081–1089
12. Fauquant J, Beaucher E, Sinet C, Robert B, Lopez C (2014) Combination of homogenization and cross-flow microfiltration to remove microorganisms from

industrial buttermilks with an efficient permeation of proteins and lipids. Innovative Food Sci Emerg Technol 21:131–141
13. Gallier S, Gragson D, Jiménez-Flores R, Everett DW (2010) Surface characterization of bovine milk phospholipid monolayers by Langmuir isotherms and microscopic techniques. J Agric Food Chem 58:12275
14. Gallier S, Gragson D, Jiménez-Flores R, Everett DW (2012) beta-Casein-phospholipid monolayers as model systems to understand lipid-protein interactions in the milk fat globule membrane. Int Dairy J 22:58
15. Gallier S, Cui J, Olson TD, Rutherfurd SM, Ye A, Moughan PJ, Singh H (2013a) In vivo digestion of bovine milk fat globules: effect of processing and interfacial structural changes. I. Gastric digestion. Food Chem 141:3273–3281
16. Gallier S, Zhu XQ, Rutherfurd SM, Ye A, Moughan PJ, Singh H (2013b) In vivo digestion of bovine milk fat globules: effect of processing and interfacial structural changes. II. Upper digestive tract digestion. Food Chem 141:3215–3223
17. Gallier S, Laubscher A, Jimenez-Flores R (2014) Chapter 4: the milk fat globule membrane: structure, methodology for its study, and functionality, in food structures, digestion and health, Elsevier Inc., pp 107–142
18. Garcia C, Antona C, Robert B, Lopez C, Armand M (2014) The size and interfacial composition of milk fat globules are key factors controlling triglycerides bioavailability in simulated human gastro-duodenal digestion. Food Hydrocolloids 35:494–504
19. Garcia-Manyes S, Sanz F (2010) Nanomechanics of lipid bilayers by force spectroscopy with AFM: a perspective. Biochimica Et Biophysica Acta 1798:741
20. Gassi JY, Blot M, Beaucher E, Robert B, Leconte N, Camier B, Rousseau F, Bourlieu C, Jardin J, Briard-Bion V, Lambert S, Gésan-Guiziou G, Lopez C, Gaucheron F (2016) Preparation and characterisation of a milk polar lipids enriched ingredient from fresh industrial liquid butter serum: combination of physico-chemical modifications and technological treatments. Int Dairy J 52:34–36
21. Goldbohm RA, Chorus AM, Galindo Garre F, Schouten LJ, van den Brandt PA (2011) Dairy consumption and 10-y total and cardiovascular mortality: a prospective cohort study in the Netherlands. Am J Clin Nutr 93:615–627
22. Golding M, Wooster TJ, Day LXM et al (2011) Impact of gastric structuring on the lipolysis of emulsified lipids. Soft matter 7:3513–3523
23. Guyomarc'h F, Zou S, Chen M, Milhiet PE, Godefroy C, Vié V, Lopez C (2014) Milk sphingomyelin domains in biomimetic membranes and the role of cholesterol: morphology and nanomechanical properties investigated using AFM and force spectroscopy. Langmuir 30:6516
24. Heid HW, Keenan TW (2005) Intracellular origin and secretion of milk fat globules. Eur J Cell Biol 84:245–258
25. Hickey CD, Auty MAE, Wilkinson MG, Sheehan JJ (2015) The influence of cheese manufacture parameters on cheese microstructure, microbial localization and their interactions during ripening: a review. Trends Food Sci Technol 41:135–148
26. Hjerpsted J, Leedo E, Tholstrup T (2011) Cheese intake in large amounts lowers LDL-cholesterol concentrations compared with butter intake of equal fat content. Am J Clin Nutr 94:1479–1484
27. Iggman D, Gustafsson IB, Berglund L, Vessby B, Marckmann P, Riserus U (2011) Replacing dairy fat with rapeseed oil causes rapid improvement of hyperlipidaemia: a randomized controlled study. J Intern Med 270:356–365
28. Jimenez-Flores R, Brisson G (2008) The milk fat globule membrane as an ingredient: why, how, when? Dairy Sci Technol 88:5–18
29. Kamili A, Wat E, Chung RW, Tandy S, Weir JM, Meikle PJ, Cohn JS (2010) Hepatic accumulation of intestinal cholesterol is decreased and fecal cholesterol excretion is increased in mice fed a high-fat diet supplemented with milk phospholipids. Nutr Metab 7:90
30. Li JK, Sullan RMA, Zou S (2011) Atomic force microscopy force mapping in the study of supported lipid bilayers. Langmuir 27:1308

31. Lopez C (2011) Milk fat globules enveloped by their biological membrane: unique colloidal assemblies with a specific composition and structure. Curr Opin Colloid Interface Sci 16:391–404
32. Lopez C, Ménard O (2011) Human milk fat globules: polar lipid composition and in situ structural investigations revealing the heterogeneous distribution of proteins and the lateral segregation of sphingomyelin in the biological membrane. Coll Surf B Biointerface 83:29–41
33. Lopez C, Maillard MB, Briard-Bion V, Camier B, Hannon JA (2006) Lipolysis during ripening of Emmental cheese considering organization of fat and preferential localization of bacteria. J Agric Food Chem 54:5855–5867
34. Lopez C, Camier B, Gassi JY (2007) Development of the milk fat microstructure during the manufacture and ripening of Emmental cheese observed by confocal laser scanning microscopy. Int Dairy J 17:235–247
35. Lopez C, Briard-Bion V, Beaucher E, Ollivon M (2008) Multiscale organization of triglycerides and phospholipids in emmental cheese: from the microscopic to the molecular level. J Agric Food Chem 56:2406–2414
36. Lopez C, Madec MN, Jimenez-Flores R (2010) Lipid rafts in the bovine milk fat globule membrane revealed by the lateral segregation of phospholipids and heterogeneous distribution of glycoproteins. Food Chem 120:22–33
37. Lopez C, Briard-Bion V, Ménard O, Beaucher E, Rousseau F, Fauquant J, Leconte N, Robert B (2011) Fat globules selected from whole milk according to their size: different compositions and structure of the biomembrane, revealing sphingomyelin-rich domains. Food Chem 125:355–368
38. Lopez C, Briard-Bion V, Ménard O (2014) Polar lipids, sphingomyelin and long-chain unsaturated fatty acids from the milk fat globule membrane are increased in milks produced by cows fed fresh pasture based diet during spring. Food Res Int 58:59–68
39. Ma X, Bryony J, Zhang L, Emanuelsson-Patterson EAC (2013) Correlating mozzarella cheese properties to its production processes and microstructure quantification. J Food Eng 115:154–163
40. Michalski MC, Januel C (2006) Does homogenization affect the human health properties of cow's milk? Trends Food Sci Technol 17:423–437
41. Michalski MC, Genot C, Gayet C, Lopez C, Fine F, Joffre F, Vendeuvre JL, Bouvier J, Chardigny JM, Raynal-Ljutovac K (2013) Multiscale structures of lipids in foods as parameters affecting fatty acid bioavailability and lipid metabolism. Prog Lipid Res 52:354–373
42. Morin P, Jimenez-Flores R, Pouliot Y (2007) Effect of processing on the composition and microstructure of buttermilk and its milk fat globule membranes. Int Dairy J 17:1179–1187
43. Murthy AVR, Guyomarc'h F, Paboeuf P, Vié V, Lopez C (2015) Cholesterol strongly affects the organization of lipid monolayers studied as models of the milk fat globule membrane: condensing effect and change in the lipid domain morphology. Biochim Biophys Acta 1848:2308–2316
44. Nguyen HTH, Ong L, Beaucher E, Madec MN, Kentish SE, Gras SL, Lopez C (2015) Buffalo milk fat globules and their biological membrane: in situ structural investigations. Food Res Int 67:35–43
45. Ohlsson L, Burling H, Duan RD, Nilsson A (2010) Effects of a sphingolipid-enriched dairy formulation on postprandial lipid concentrations. Eur J Clin Nutr 64:1344–1349
46. Ong L, Dagastine RR, Kentish SE, Gras SL (2011) Microstructure of milk gal and cheese curd observed using cryo scanning electron microscopy and confocal microscopy. LWT Food Sci Technol 44:1291–1302
47. Ong L, Dagastine RR, Kentish SE, Gras SL (2012) The effect of pH at renneting on the microstructure, composition and texture of Cheddar cheese. Food Res Int 48:119–130

48. Oosting A, Kegler D, Wopereis HJ, Teller IC, van de Heijning BJM, Verkade HJ, van der Beek EM (2012) Size and phospholipid coating of lipid droplets in the diet of young mice modify body fat accumulation in adulthood. Pediatr Res 72:362–369
49. Patterson E, Larsson SC, Wolk A, Akesson A (2013) Association between dairy food consumption and risk of myocardial infarction in women differs by type of dairy food. J Nutr 143:74–79
50. Riccio P (2004) The proteins of the milk fat globule membrane in the balance. Trends Food Sci Technol 15:458–461
51. Richoux R, Aubert L, Roset G, Briard-Bion V, Kerjean JR, Lopez C (2008) Combined temperature-time parameters during the pressing of curd as a tool to modulate the oiling-off of Swiss cheese. Food Res Int 41:1058–1064
52. Robenek HO, Hofnagel et al (2006) Butyrophilin controls milk fat globule secretion. ? Proc Natl Acad Sci U S A 103:10385–10390
53. Rombaut R, Van Camp J, Dewettinck K (2006) Phospho- and sphingolipid distribution during processing of milk, butter and whey. Int J Food Sci Technol 41:435–443
54. Romeih EA, Moe KM, Skeie S (2012) The influence of fat globule membrane material on the microstructure of low-fat cheddar cheese. Int Dairy J 26:66–72
55. Rosqvist F, Smedman A, Lindmark-Mansson H, Paulsson M, Petrus P, Straniero S, Rudling M, Dahlman I, Risérus U (2015) Potential role of milk fat globule membrane in modulating plasma lipoproteins, gene expression, and cholesterol metabolism in humans: a randomized study. Am J Clin Nutr, doi: 10.3945/ajcn.115.107045.
56. Simons K, Ikonen E (1997) Functional rafts in cell membranes. Nature 5:569–572
57. Singh H (2006) The milk fat globule membrane—a biophysical system for food applications. Curr Opin Colloid Interface Sci 11:154–163
58. Spitsberg VL (2005) Invited review: bovine milk fat globule membrane ass a potential nutraceutical. J Dairy Sci 88:2289–2294
59. Timby N, Domellof E, Hernell O et al (2014) Neurodevelopment, nutrition, and growth until 12 mo of age in infants fed a low energy, low protein formula supplemented with bovine milk fat globule membrane: a randomized controlled trial. Am J Clin Nutr 99:860–868
60. Vanderghem C, Bodson P, Danthine S, Paquot M, Deroanne C, Blecker C. (2010) Milk fat globule membrane and buttermilks: from composition to valorization. 14:485-500.
61. Ward RE, German JB, Corredig M (2006) Composition, applications, fractionation, technological and nutritional significance of milk fat globule membrane material. In: Fox PF, McSweeney PLH (eds) Advances dairy chemistry, volume 2: lipids, 3rd edn. Springer, New-York, pp 213–244
62. Zheng H, Jimenez-Flores R, Everett DW (2014) Lateral lipid organization of the bovine milk fat globule membrane is revealed by washing processes. J Dairy Sci 97:5964–5974
63. Zhou AL, Hintze KJ, Jimenez-Flores R, Ward RE (2012) Dietary fat composition influences tissue lipid profile and gene expression in Fischer-344 rats. Lipids 47:1119–1130
64. Zou XQ, Guo Z, Huang JH, Jin QZ, Cheong LZ, Wang XG, Xu XB (2012) Human milk fat globules from different stages of lactation: a lipid composition analysis and microstructure characterization. J Agric Food Chem 60:7158–7167
65. Zou X, Guo Z, Jin Q, Huang J, Cheong L, Xu X, Wang X (2015) Composition and microstructure of colostrum and mature bovine milk fat globule membrane. Food Chem 185:362–370

CHAPTER 3

Role of Somatic Cells on Dairy Processes and Products: A Review

N. Li [1,2,3] R. Richoux [3], M. Boutinaud[4,5] P. Martin [6,7] V. Gagnaire[1,2]

[1].*INRA, UMR 1253Science et Technologie du Lait et de l'ŒufRennesFrance*

[2].*Agrocampus Ouest, UMR 1253Science et Technologie du Lait et de l'ŒufRennesFrance*

[3].*ActaliaRennesFrance*

[4].*INRA, UMR 1348, PhysiologieEnvironnement et Génétique pour l'Animal et les Systèmes d'ÉlevageSaint GillesFrance*

[5].*Agrocampus Ouest, UMR 1348, PhysiologieEnvironnement et Génétique pour l'Animal et les Systèmes d'ÉlevageSaint GillesFrance*

[6].*INRA, UMR 1313Génétique Animale et Biologie IntégrativeJouy-en-JosasFrance*

[7].*AgroParisTech, UMR 1313Génétique Animale et Biologie IntégrativeJouy-en-JosasFrance*

ABSTRACT

Somatic cells are an important component naturally present in milk, and somatic cell count is used as an indicator of udder health and milk quality. The role of somatic cells in dairy processes and products is ill-defined in most studies because the role of these cells combines also the concomitance of physicochemical modifications of milk, bacterial count, and the udder inflammation in the presence of high somatic cell count. The aim of this review is to focus on the role of somatic cells themselves and of endogenous enzymes from somatic cells in milk, in dairy transformation processes, and in characteristics of final products overcoming biases due to other factors. The immune function of somatic cells in the udder defense and their protective role in milk will be primarily considered. Different characteristics of milk induced by various somatic cell counts, types, and their endogenous enzymes influencing directly the technological properties of milk and the final quality of dairy products will be discussed as well. By comparing methods used in other studies and eliminating biases due to other factors not considered in these studies, a new approach has been suggested to evaluate the effective role of somatic cells on dairy processes and products. In addition, this new approach allows the characterization of somatic cells and their endogenous enzymes and, in future research, will allow the clarification of mechanisms involved in the release of these components from somatic cells during dairy processes, particularly in cheese technologies.

Keywords: Milk Somatic cells Enzymes Cathepsin D Cheese

1 INTRODUCTION

Milk is known to be a high-value nutritional biological fluid composed of water, proteins, fat, sugars, minerals, etc. Other important components existing naturally in raw milk are somatic cells (SCs), and the predominant cell type, besides shed epithelial cells, in most species is leucocytes, including macrophages, polymorphonuclear neutrophils cells (PMNs), and lymphocytes (Boutinaud and Jammes 2002).

The amount of SCs, usually called somatic cell count (SCC), in milk is used as an important indicator of udder health since SCs are involved in protecting the mammary gland from infection as part of the innate immune system. SCC in milk is influenced by many factors, such as animal species, milk production level, lactation stage, and also the individual and environmental factors as well as management practices (Rupp et al. 2000). Though SCC is subjected to variation, it is still used as an indicator of milk quality in several species, especially in ruminant and human (Hunt et al. 2013; Sharma et al. 2011). Taking cow milk as an example, when SCC $>2 \times 10^5$ cells.mL^{-1}, the udder is considered to be infected, and when SCC $>4 \times 10^5$cells.mL^{-1}, the milk is deemed unfit for human consumption in the European Union (EU). The legal SCC threshold for milk acceptance in dairy industries varies in different countries, e.g., the values for bovine milk in Germany, Canada, and the USA are 1×10^5, 5×10^5, and 7.5×10^5cells.mL^{-1}, respectively (Olechnowicz and Jaskowski 2012; Schwarz et al. 2011). For caprine and ovine milk, the cutoff value is 1×10^6 in the USA but is not yet defined in the EU (Council Directive 92/46/EEC 1992).

Whether SC is "a friend or a foe" in the dairy field remains a question (Souza et al. 2012). Generally, SCs until now have been considered as negative. High SCC is associated with udder inflammation, which leads to bacteriological problems in milk, an alteration of milk composition, and finally, the major modifications of dairy product characteristics compared to the normal values (Le Maréchal et al. 2011; Lindmark-Mansson et al. 2006; Raynal-Ljutovac et al. 2007; Sharma et al. 2011). However, besides their immune function in the udder and protective functions in milk, SCs have recently been shown to influence, in a positive way, the composition and technological properties of dairy products, thus participating in the final quality of dairy products through their endogenous enzymes (Sanchez-Macias et al. 2013; Souza et al. 2012).

The role of SCs is generally assessed as a global effect, although the influence of the other factors has not been considered separately, and then, includes intrinsic characteristics of milk modified by the inflammation of the mammary gland, consequences on milk biosynthesis and secretion, and bacterial count (Albenzio et al. 2004; O'Farrell et al. 2002; Raynal-Ljutovac et al. 2007). Few research studies until now have given a clear view of the role of SCs without the combination of the other causative factors. This review focuses on the state of our knowledge of SCs in the dairy field and on their effective role in dairy processes and products.

2 SOMATIC CELLS AND UDDER DEFENSE MECHANISM

2.1 The immune function of somatic cells in the udder

SCs are known to be one of the major defense components of the mammary gland against disease or intramammary infections (Paape et al. 1979, 2002; Sharma et al. 2011). The four main cell types composing SC, namely, macrophages, PMNs, lymphocytes, and epithelial cells, are briefly presented in Table 1.

Table 1. Characteristics and composition of somatic cells in healthy milk of different species

	Macrophages	PMNs	Lymphocytes	Epithelial cells	References
Image of bovine somatic cells stained by methylene blue	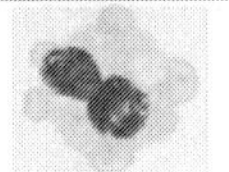		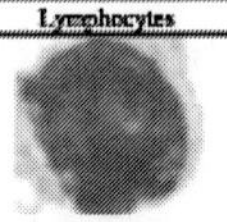	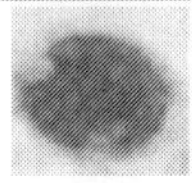	ISO 13366-1/IDF148-1
Morphological characteristics	- Big nucleus	- Plurilobery nucleus	- Dense and round nucleus	- Round nucleus	ISO 13366-1/IDF148-1
Biological functions	- Phagocytosis - Intracellular killing of bacteria - Presentation antigen - Secretion of chemoattractants	- Phagocytosis - Intracellular killing of bacteria - Secretion of antibacterial factors - Detection of chemoattractants	- Specific immune system - Production of immune regulatory cytokines - Lysis of altered or damaged host cells	- Secretion of chemoattractants	Burvenich et al. 2003; Lee et al. 1980; Paape et al. 1979; Paape et al. 2002; Paape et al. 2003; Sordillo et al. 1997; Sordillo and Streicher 2002
Percentage of each cell type in milk of different species					
Bovine	35–79[1]	3–26	16–28	2–15	ISO13366-1/IDF148-1
	26	16	23	36[1]	Benic et al. 2012
	40	10	50[1]	–	Rivas et al. 2006
	60[1]	12	28	–	Marino et al. 2005
	13	28	29	45[1]	Leitner et al. 2000b
Sheep	46–84[1]	2–28	11–20	1–2	Gonzalez-Rodriguez et al. 1995
	57[1]	31	8	2	Morgante et al. 1996
Goat	15–41	45–74[1]	9–20	6	Dulin et al. 1983
	11	79	10	–	Boulaaba et al. 2011

[1] Predominant cell type

Macrophages are generally the predominant cell type in healthy cow milk. They can fight against bacterial invasion quickly by engulfing action. In the case of infection, macrophages release chemical messengers or chemoattractants that are detected by PMNs and direct PMNs in turn towards the infection site. Both macrophages and PMNs can ingest microbial cells by phagocytosis and have an essential role in the innate immune system. Moreover, macrophages participate in the specific immunity as do lymphocytes (Burvenich et al. 2003).

PMNs can be recruited and increase milk SCC when the infection continues. They can be present to a large extent in mastitic milk, even up to 92 % in bovine milk (Paape et al. 1979). When PMNs arrive at the site of infection, they phagocyte microorganisms and kill them by using a combination of oxidative and non-oxidative mechanisms (Pham 2006).

Lymphocytes have a determinant role in the specific immune system. They are the only cells able to recognize the antigens through specific membrane receptors for invading pathogens (Sordillo et al. 1997).

Mammary epithelial cells are the cells that produce milk. They are shed from the mammary epithelium during lactation (Boutinaud and Jammes 2002). Prior to 1980, they were confused with macrophages because of their similar morphological traits and finally discovered in milk by electron microscopy (Lee et al. 1980). The epithelial cells are the first defense line of the mammary glands, and they may participate in the immunity of neonates in different species (Boutinaud and Jammes 2002). This type of cell is often detected below 15 % (Table 1) as also confirmed in other studies (Ben Chedly et al. 2011; Boutinaud et al. 2008; Rankl 2004). The mammary epithelial feature of these cells has been further determined in cows and goats using transcriptomic and proteomic analyses, demonstrating that their content in mRNA and proteins, such as cytokeratin and enzymes involved in milk synthesis, is specific to mammary epithelial cells (Ben Chedly et al. 2011; Boutinaud et al. 2008; Janjanam et al. 2013).

2.2 The protective function of somatic cells in milk

Besides the immune defense role in the udder, SCs can continue their protective function in milk. Additionally, some components identified as being from SCs are present in milk and also help to enhance the host defense. For example, PMNs have bactericidal and respiratory burst activities and they can eliminate the invading bacteria by releasing reactive oxygen species (ROS) and granular enzymes (Paape et al. 2002). Some antibacterial proteins identified in bovine milk also arise from SC such as macrophage scavenger receptor types I and II, PMN peptidoglycan recognition protein and lymphocyte cytosolic protein 1 and cathelicidins. They can continue to exert their protective properties when they are in skim milk, whey, or milk fat globule membranes (Hettinga et al. 2011; Smolenski et al. 2006). Recently, information was obtained on the protein changes of the bovine and goat innate immune system following exposure to lipopolysaccharide (LPS) from the cell walls of gram-negative *Escherichia coli*, leading to an increase in SCC without having the contribution of bacteria themselves (Hinz et al. 2012; Olumee-Shabon et al. 2013). In both ruminant species, an increase in the antimicrobial proteins was observed and more precisely of cathelicidins, which are involved in the degranulation of PMNs. In caprine milk, acute-phase proteins, haptoglobin, and serum amyloid A were shown in milk collected after 18 h after infusion with LPS.

The enzymes initially from SCs can be released or secreted in milk and are therefore considered as endogenous SC enzymes and play an important protective role in milk. The role of the lysozyme, one SC endogenous enzyme, is well known for the ability to destroy bacteria (Paape et al. 2003). Some proteinases from PMNs, such as cathepsin G, elastase, and proteinase 3, have antimicrobial activities during phagocytosis of invading microorganisms. They could also contribute, after release, to the extracellular killing of microorganisms by cleaving their bacterial virulence factors as shown in mice (Pham 2006). Catalase, an endogenous enzyme from PMNs, is one of the main antioxidant enzymes in milk and is suspected of being responsible for changed redox potential of milk that limited the survival capability of microorganisms (Hamed et al. 2008).

Even if the release mechanism of these protective components from SCs is not fully understood yet, the release of these SC endogenous proteins including enzymes has already been mentioned in the last decade, as described in Section 3.1. The contribution of these SC endogenous enzymes to dairy transformation processing and final characteristics of products has been considered previously by several authors, and this issue will be discussed in Section 4.1.

2.3 Identification of somatic cell types

Several methods have been developed to differentiate SCs in milk, and among them, cytology methods are mainly used to visually identify the main cell types by optical microscopy (Baumert et al. 2009; Lindmark-Mansson et al. 2006; Sarikaya et al. 2004). Enzyme-linked immunosorbent assay (ELISA) was also applied for detecting and quantifying SCs, for example, by using the specific antibodies of PMNs. O'Sullivan et al. (1992) suggested to use a direct capture ELISA to diagnose bovine mastitis. In order to separate each type of SCs, immunomagnetic separation has been used for labeling the cell subpopulation. This method was successfully applied to isolate epithelial cells from bovine milk (Boutinaud et al. 2008), macrophages, and PMNs from sheep milk (Albenzio et al. 2009).

Recent advances of SC studies by flow cytometry in medicine diagnostic field give opportunities to study SCs in dairy field at the subpopulation level (Albenzio and Caroprese 2011; Leitner et al. 2000a; Reuter et al. 2014). The combination of immunofluorescence and flow cytometry allows the differentiation of cell types using specific antibodies (Kelly et al. 2000; Park et al. 1991; Riollet et al. 2001). Flow cytometry-cell sorting technologies distinguish and separate differential cells with more precise detection (Dosogne et al. 2003; Piepers et al. 2009). Compared with classical microscopy observation, flow cytometry allows counting SCs and identifying SC types with fewer samples and less time, thus allowing characterizing SCs and determining the roles of SCs in the milk.

2.4 Various somatic cell count and composition

The SC composition (SC types and their percentages in total SCC) in milk varies depending on many factors: animal species, breed, stage of lactation, genetics, parity, day-to-day variation, diurnal variation, milking interval, time of sampling, sampling procedures, stress and trauma, management factors, and seasonal and storage procedures (IDF 466/2013). The macrophage and PMN percentages showed an opposite trend during different stages of lactation in ewes, the highest macrophage level observed in early lactation and the highest PMN level in late lactation (Albenzio et al. 2009). In healthy bovine and ovine milk, macrophages are generally the predominant cells, while in both healthy and unhealthy goat milks, PMNs are predominant (Dulin et al. 1983; Ostensson 1993). Nevertheless, there are several contradictory examples concerning the predominant cell type in the case of healthy bovine milk, macrophages, lymphocytes, or epithelial cells becoming the main cell type according to the authors (Table 1).

Both SC count and composition are related to milk quality, but their relationship is not necessarily associated, except in the case of high SCC corresponding to a high amount of PMNs in bovine milk. SCC gives only the total amount of cells present in milk, but not the distribution of each cell type, making it difficult to have a relevant view of cell composition. Hence, milk with various SCCs and various SC compositions has a particular fingerprint that can lead to different final characteristics of dairy products, and this will be discussed in Section 4.

3 ENDOGENOUS ENZYMES FROM SOMATIC CELLS

Besides protective functions in milk, SCs also provide numerous enzymes mentioned as SC endogenous enzymes in this review. These SCs as well as their endogenous enzymes have different profiles in terms of type, specificity, and activity and can be released into milk and further influence dairy processes and the quality of products. It is worthy to note that these SC endogenous enzymes are only part of all enzymes in milk, usually called milk indigenous enzymes. Actually, these indigenous enzymes in milk are well studied (Kelly et al. 2006; Kelly and Fox 2006; Kelly and McSweeney 2003), in particular the plasmin system (Ismail and Nielsen 2010). In this review, we will focus on SC endogenous enzymes.

3.1 Somatic cells: an important source of enzymes

SCs are an important source of endogenous proteins, including enzymes. A large range of enzymes are released into milk after the lysis of SCs, and among them, lipases (e.g., lipoprotein lipase), oxidases (e.g., catalase and lactoperoxidase), glycosidases (e.g., lysozyme), and proteases (e.g., cathepsins, elastase, and collagenase). The recent advances in proteomic methods have given rise to the identification of numerous proteins from SC (Jethwaney et al. 2007; Lippolis and Reinhardt 2005). Comparison of proteomes from macrophages (Dupont et al. 2004) and milk fat globule membranes (Smolenski et al. 2006) shows that they have many proteins including enzymes in common: annexin, vimentin, and apolipoprotein, α-enolase, heat shock protein, actin, capping protein (involved in cell motility), NADP1-dependent isocitrate dehydrogenase. Even if such similarities could be partly due to the extraction of proteins in the fat layer in which SCs can be trapped and lysed at high-speed centrifugation (Smolenski et al. 2006), we cannot exclude the release of these proteins in milk as a natural phenomenon or during technological processes that can lead to underestimating their role in the final quality of dairy products.

It is only recently that a proteomic approach of the mammary epithelial cells coupled with RT-PCR and Western blotting has been developed, to give an insight into the proteins originated from this cell type in the milk and their expression (Janjanam et al. 2013). Four hundred ninety-seven proteins were identified, and some of them were identical to those originating from milk fat globule membrane (37 %) and in lactating mammary tissue (54 %), while 247 represented new mammary epithelial cell proteins.

The presence of SC endogenous enzymes in milk also suggests the occurrence of leakage or secretion of these endogenous enzymes from SCs (Kelly and Fox 2006). This is the case for cathepsin D, an endogenous enzyme from SCs detected in skim milk or whey (Larsen et al. 2006). It can degrade intracellular proteins and participate in extracellular proteolysis when it is secreted out of SCs (Briozzo et al. 1988). Dupont et al. (2004) confirmed the secretion of some SC endogenous enzymes from macrophages. The main endogenous enzymes from SC in milk are generally from macrophages and PMNs. The location of these endogenous enzymes prior to release into milk is given in Table 2.

Most of the endogenous enzymes from SCs are not identified in milk and, among them, proteolytic enzymes. In fact, due to a high-dynamic range in a concentration of proteins in milk that can vary by at least a factor of 10^6 between caseins and other minor proteins (Gagnaire et al. 2009), most of the proteolytic enzymes from SC are in a concentration that is so low that they are only detected through their activity or immunological studies (Magboul et al. 2001). This part will be discussed below. Additionally, the kinetics of SC death and the dynamics of enzyme release are difficult to measure and still unknown in milk and a fortiori in dairy products.

3.2 Proteolysis of caseins by endogenous enzymes from somatic cells

Among the numerous endogenous enzymes from SCs, proteases are by far the most studied. They are active on caseins in milk as well as during transformation processes. They can therefore modify casein degradation of dairy products, give different textural and organoleptic characteristics

Table 2. Location of main endogenous enzymes in somatic cells

Type of enzymes	Type of somatic cells				
	Macrophages	PMNs	Lymphocytes	Epithelial cells	Cell type unknown
Proteases					
Cathepsin B	+ Guha and Padh (2008)	+ Baggiolini et al. (1978); Magboul et al. (2001); Travis and Fritz (1991)		+ Lah et al. (1996); Guha and Padh (2008)	
Cathepsin C		+ Travis and Fritz (1991)			
Cathepsin D	+ Cohn (1975); Diment et al. (1988); Guha and Padh (2008)	+ Baggiolini et al. (1978); Cohn (1975); Owen and Campbell (1999)		+ Cohn (1975); Lah et al. (1996); Lkhider et al. (2004); Seol et al. (2006); Guha and Padh (2008)	
Cathepsin H	+ Guha and Padh (2008)				
Cathepsin K					+ Moatsou (2010)
Cathepsin L	+ Guha and Padh (2008)	+ Travis and Fritz (1991)		+ Lah et al. (1996)	
Cathepsin G	+ Considine et al. (2002); Campbell et al. (1989)	+ Baggiolini et al. (1978); Considine et al. (2002); Dewald et al. (1975)			
Cathepsin S	+ Guha and Padh (2008)	+ Owen and Campbell (1999)			
Elastase	+ Owen and Campbell (1999); Prin-Mathieu et al. (2002); Campbell et al. (1989)	+ Baggiolini et al. (1978); Considine et al. (2000); Dubin et al. (1994); Owen and Campbell (1999); Travis and Fritz (1991); Prin-Mathieu et al. (2002)	+ Prin-Mathieu et al. (2002)		
Other enzymes					
Catalase					+ Kitchen (1976)
Lipoprotein lipase	+ Azzara and Dimick (1985b)	+ Azzara and Dimick (1985b)			
Collagenase	+ Owen and Campbell (1999); Prin-Mathieu et al. (2002)	+ Verdi and Barbano (1991); Prin-Mathieu et al. (2002)			
Acid phosphatase					+ Kitchen (1976); Andrews and Alichanidis (1975)
Myeloperoxidase	+ Owen and Campbell (1999)	+ Mukherjee et al. (2004)			

to final dairy products, and even reduce the cheese yield (Grappin et al. 1999).

To determine the activity of SC proteases in milk, two strategies are often used: (i) addition of commercially available proteases to milk proteins and study of protein degradation (mainly caseins) and (ii) testing activities of commercial proteases either on caseins or synthetic substrates with or without the addition of inhibitors. Proteolytic activities of commercially available cathepsins B, D, and G and elastase as well as their enzymatic specificity (sites of cleavages) have been determined on α_{s1}- and β-caseins (Considine et al. 1999, 2000, 2002, 2004; Hurley et al. 2000; Kelly and McSweeney 2003; McSweeney et al. 1995). Some other authors (Magboul et al. 2001; Somers et al. 2003) used enzymes extracted from acid whey, or they directly studied the proteolytic activities of endogenous enzymes in milk with different SCCs. Endogenous proteases from SCs, such as cathepsins B and G and elastase, were confirmed as responsible for the hydrolysis of α_{s1}- and β-caseins. Consequences on casein hydrolysis were recently shown after exposure to lipopolysaccharide from the cell walls of gram-negative *E. coli*, inducing an increase in SCC without bacterial damage (Hinz et al. 2012; Olumee-Shabon et al. 2013). Thus, casein proteolysis was higher in bovine (Hinz et al. 2012) than in caprine milk (Olumee-Shabon et al. 2013), and the main casein hydrolyzed was both α_{s1}- and β-caseins in bovine and caprine milk. At least, plasmin, elastase, and cathepsins B and D were shown to participate in the casein degradation.

Cathepsin D can hydrolyze all caseins (α_{s1}-, α_{s2}-, β- and κ-caseins) and was more active on α_{s1}-caseins than on the other caseins with a broad range of cleavage sites (Hurley et al. 2000; McSweeney et al. 1995). It showed cleavage sites similar to chymosin, the main active protease from rennet used in cheese-making process, responsible for the coagulation of milk and causing the milk to separate into curds and whey. Furthermore, cathepsin B has common cleavage sites with cathepsin D and chymosin, notably on Phe_{23}–Phe_{24} bond of α_{s1}-casein (Magboul et al. 2001).

3.3 Various profiles of endogenous enzymes from somatic cells

SC endogenous enzymes show different activities according to their initial location in the cell types (Table 2). The activity of cathepsin D in induced alveolar macrophages was detected to be 60-fold higher than that in PMNs (Cohn 1975), while it was not detectable in lymphocytes (Barabasi and Nassberger 1994). Due to different SC counts and compositions during lactation, the activities of cathepsin D, cysteine proteases (e.g., cathepsin B), and another unidentified milk proteinase measured in milk from the same group of cows fluctuate during lactation (Larsen et al. 2006). Additionally, the presence of some SC endogenous enzymes in milk can change the activity of other enzymes in milk. For example, plasminogen activators, of which one is associated with SC, can modify plasmin activity (Albenzio et al. 2004; Politis and Ng-Kwai-Hang 1988). Such an increase in plasmin activity was observed for SCC below 1×10^5cells.mL^{-1} and represented about 42 % of that observed with high SCC >6×10^5 cells.mL^{-1} (Le Roux et al. 2003). However, as SC types present in milk were generally not determined, enzymatic activities in milk were considered as heterogeneous (Santos et al. 2003).

The profiles of endogenous enzymes in terms of type, quantity, and activity can be influenced by different SC counts and compositions in milk.

Nevertheless, their relationship during milk treatments, cheese processing, and cheese quality is still unknown. The activity of SC proteases was enhanced, while SCC increased (Kelly et al. 2006; O'Farrell et al. 2002). Elastase, one protease mainly present in PMNs, has significantly higher activity when PMNs are massively recruited into milk during infection (Le Roux et al. 2003). The elastase activity was also illustrated to have a positive correlation with SCC in caprine milk (Santillo et al. 2009). Cathepsin G and lipoprotein lipase activities were detected in high-SCC milk (SCC 1.2×10^5–2×10^6 cells.mL^{-1}), correlated with subclinical or clinical mastitis (Azzara and Dimick 1985b). Regarding lipolysis, the lipoprotein lipase is still active after pasteurization and participates in the production of free fatty acid in milk during storage. In addition, pasteurized milk with high SCC is more susceptible to lipolysis than that with low SCC (Santos et al. 2003).

Milk with various SC counts and compositions including different profiles of endogenous enzymes leads to different patterns of proteolysis and lipolysis and finally offers different characteristics of final dairy products. However, the milk in dairy industries, mixing the milk from hundreds or thousands of cows, can minimize the variations between milks in contrast to the milk mixed from several cows in the herd and a fortiori the milk from an individual cow and the quarter milk from the individual's mammary gland. It is still difficult to ascertain a clear relationship between enzyme activities and types of SCs present in milk and to what extent one cell type would be preferable in milk to enhance final dairy product quality.

4 ROLE OF SOMATIC CELLS AND THEIR ENDOGENOUS ENZYMES IN DAIRY PROCESSES AND PRODUCTS

The term "effect of SCs" is mentioned in several studies (Chen et al. 2010; Fernandes et al. 2008; Ma et al. 2000; Politis and Ng-Kwai-Hang 1988; Santos et al. 2002). However, its definition is too imprecise by far. The effect of SCs in these studies includes two aspects: (i) the direct action of SCs themselves and that of their endogenous contents and (ii) the consequence of concomitant high levels of invading bacteria inducing major compositional and physicochemical changes of milk with different modified SCC levels in milk (Albenzio et al. 2011; Somers et al. 2003). As the latter aspect is usually associated with inflammation of the mammary gland, the effect of SC is generally ill-defined leading to confusion with other cofactors, recognized as negative for the researchers and dairy farmers.

In this part, the role of SCs and their impacts on dairy products will be considered. As these literature data concern the effect of SCs in a general way, it is difficult to focus only on the effective role of SCs excluding the influence of other factors associated in the experiments. Hence, these literature data will be discussed here in two parts: in Section 4.1, in which data clearly indicate the participation of some individual endogenous enzymes from SCs on dairy processes and products, and in Section 4.2, in which data concern the general effect of SCs. In addition, different methods of SC preparation used in these studies will be compared in order to bring out a new approach to study the effective role of SCs in the dairy field in the future.

4.1 Participation of endogenous enzymes from somatic cells

The milk indigenous enzymes (including SC endogenous enzymes) were indicated as active in the udder in which the temperature is optimal for most mammalian enzymes. They continued to accumulate their enzyme activities even if these activities could be weak during the refrigerated storage (dairy farm or plant) and modified the characteristics of dairy-processed products (Kelly and Fox 2006). The roles of SC enzymes have been underestimated until now because of the presence of other numerous enzymes in milk: (i) enzymes coming from the blood and present in milk, such as plasmin; (ii) added enzymes during the cheese-making process such as calf rennet (chymosin and pepsin) or fungal rennet; and (iii) enzymes produced by microflora which play an essential role in the final characteristics of cheese, such as texture and flavor. The direct action of SC endogenous enzymes, in particular cathepsin D, in dairy processes and products will be discussed below.

4.1.1 Cathepsin D in cheese

Among the enzymes from SCs, cathepsin D is one of the most studied endogenous enzymes in milk and different types of cheeses. It is a lysosomal acidic protease, which has been greatly discussed in the human medical domain for more than a century because of its multifunctions such as a tumor marker and a cancer indicator (Benes et al. 2008; Minarowska et al. 2008). In the dairy field, studies have increased over the past two decades.

The amount of cathepsin D was measured for the first time by Larsen et al. (1996) in bovine skim milk at 0.4 μg.mL^{-1} level, mainly located in whey at 0.3 μg.mL^{-1} level. It is four- to sevenfold higher in small ruminant milk about 1.8–2.6 μg.mL^{-1} (Albenzio et al. 2009; Santillo et al. 2009). Five molecular forms were identified in bovine milk (Larsen and Petersen 1995): the two inactive forms preprocathepsin D (ppCD) and procathepsin D (pCD) and the three active forms pseudocathepsin D (pdCD), single-chained cathepsin D and two-chained cathepsin D (heavy- and light-chained cathepsin D). In bovine milk, the main form is the inactive form pCD which becomes the active form pdCD under acidic medium conditions (Larsen et al. 1993). The amino acid sequence and three-dimensional structure are close to that of two other aspartic proteinases, chymosin and pepsin (Baldwin et al. 1993). The activator-inhibitor system of cathepsin D is only partly known (Minarowska et al. 2009). Moreover, cathepsin D is able to survive most heat treatments applied during cheese manufacture (Table 3). It is also baroresistant (Moatsou et al. 2008) and stable over a range of acid pH values, from pH 3.5 to 7 (Lee et al. 1998).

Table 3. Variation of cathepsin D activity with different heat treatments

Heat treatment		Residual activity (%) at optimum pH	Dairy matrices	References
Temperature (°C)	Duration			
72	15 s	8	Skim bovine milk	Hayes et al. 2001
		>50	Milk serum/caseins	Larsen et al. 2000
72	60 s	~50	Milk serum	Larsen et al. 2000
		~65	Milk caseins	
65	30 min	<10	Buffer solution	Larsen et al. 2000
55	30 min	45	Skim bovine milk	Hayes et al. 2001

The presence and the activity of cathepsin D have also been illustrated in experimental rennet-free cheese and hard-cooked cheese, in which the rennet is absent or inactivated, respectively (Cooney et al. 2000; Garnot and Mollé 1987; Igoshi and Arima 1993). The presence of intact pCD and an active form derived from pCD was indicated to participate in the proteolysis process of experimental rennet-free UF-Feta cheese (Larsen et al. 2000). Furthermore, the coagulation activity of cathepsin D was confirmed in ewe milk (Albenzio et al. 2009) and in Cheddar cheese (Hurley et al. 1999). According to Santillo et al. (2009), cathepsin D concentration in caprine milk was negatively correlated with casein and protein level, possibly as a consequence of the hydrolysis capability of this enzyme. In caprine milk, cathepsin D activity was not correlated with SCC in contrast to bovine milk in which this link has been shown (Cooney et al. 2000; Kelly 1999; O'Brien et al. 2001).

4.1.2 Other endogenous enzymes from somatic cells in dairy processes and products

Due to the various endogenous enzymes from SCs (shown in Section 4.2) and the various physicochemical conditions encountered during cheese making and ripening, the influence of numerous SC enzymes on the final quality of dairy products could have been underestimated. Cathepsin B has similar cleavage sites to cathepsin D and chymosin; moreover, more than 20 % activity of cysteine protease activity remains after heat treatment at 55 °C for 30 min or 72 °C for 30 s (Magboul et al. 2001). However, to our knowledge, there are few studies concerning the state of SCs and their endogenous enzymes during the dairy process and the contribution of SC endogenous enzymes in varied ranges of dairy products. It will be interesting to elucidate on their amount and forms trapped in the cheese curds, their contribution to proteolysis in cheese varieties, and the mechanism of their release from SCs under different milk treatment conditions such as physical, chemical, and heat treatments.

The role of other SC endogenous enzymes, such as lipoprotein lipase, is not fully understood in dairy products. Lipoprotein lipases are generally considered to be causative of flavor defaults such as "rancid" in some dairy products; they could be ignored or underestimated compared to high lipolytic activities brought by *Penicillium roqueforti* in blue cheeses; they could even be responsible for the desirable flavors (e.g., "picante" flavor) in dairy products for certain consumers. The real contribution of lipase is more difficult to ascertain, due to the numerous origins of lipolytic enzymes that can be also present in cheese. Nevertheless, the desirable but weak lipolysis observed in Swiss-type cheeses is mainly due to the ripening flora *Penicillium freudenreichii* rather than to the lipoprotein lipase being rapidly inactivated during heat treatment of cheese making (Dherbécourt et al. 2010).

4.2 Role of somatic cells

Before discussing the effective role of SCs in the dairy field, it is of great importance to understand the methodologies used in literature, such as types of milk, methods of preparing SCs, and to what extent such approaches can influence final results and interpretation. This part will primarily discuss different approaches used to study the effect of SC as shown in Table 4.

Table 4. Five general approaches for the study on the effect of somatic cells on dairy processes and products

Methods	Description of methods	Advantages	Inconveniences	References
(i)	Using individual milk with different SCC levels	Avoiding mixing different milk	Individual factors involved	Ma et al. (2000); Politis and Ng-Kwai-Hang (1988); Santos et al. (2002)
(ii)	Mixing healthy low-SCC milk with mastitic high-SCC milk	Easy and accurate to obtain different SCCs	Mastitic factors involved	Cooney et al. (2000); O'Farrell et al. (2002); Rogers and Mitchell (1994)
(iii)	Using originally mixed milk and classifying in different SCC categories	Classification of milk SCC is more reasonable	Individual and mastitic factors involved	Albenzio et al. (2004), (2011); Chen et al. (2010); Fernandes et al. (2008); Jaeggi et al. (2003); Somers et al. (2003)
(iv)	Isolating SCs from milk and studying their impact in the dairy products	Avoiding other causative factors in the study of SC roles	Limited quantity of SCs	Albenzio et al. (2009); Azzara and Dimick (1985a, b); Caroprese et al. (2007)
(v)	Concentrating SCs from milk and then adding these SCs with different quantities into milk with low SCC or without SCs	Avoiding other causative factors in the study of SC roles	Limited quantity of SCs	Marino et al. (2005); Sanchez-Macias et al. (2013)

4.2.1 Different approaches to prepare somatic cells and to study their effect

(i)Use of individual milk or quarter milk

This method is based on separately collecting individual milk or quarter milk. There are few studies carried out with this method due to great differences between individual milk samples in terms of milk composition and SC content. In the case of high SCC (>10^6 cells.mL^{-1}), the effect of SCs takes into account other cofactors such as differences between milk samples and the contaminants in milk leading to mastitis. The effect of SCC and the stage of lactation on the quality of various products were studied, such as raw milk, ultrahigh-temperature milk, Cheddar cheese, and full cream milk powder (Auldist et al. 1996a, b, c). One study involving injection of *Streptococcus agalactiae* in a quarter of milk compared the shelf-life of milk containing low and high SCC on the same cows before and after infection (Ma et al. 2000). The low-SCC milk kept a high organoleptic quality during cold storage, even 21 days of shelf-life after pasteurization, in contrast to the high-SCC milk that became rancid, bitter, and astringent. This was correlated with higher lipolysis and proteolysis in high-SCC milk.

(ii)Mix of milk from healthy and mastitic cows

Mixing healthy cow milk containing low SCC with mastitic cow milk containing high SCC in different proportions allows the different SCC levels to be obtained easily and accurately (Cooney et al. 2000; O'Farrell et al. 2002; Rogers and Mitchell 1994). It is difficult to extract information regarding the effective role of the SC from these experiments, since the resulting milk combined the composition of healthy and mastitic milk as well as the various SC counts and compositions. Rogers and Mitchell (1994) found that the organoleptic grade of yoghurt made with low-SCC milk ($<2.5 \times 10^5$ cells.mL^{-1}) was superior to that manufactured from high-SCC milk ($>5 \times 10^5$ cells/mL-1). The quality of Swiss-type cheese, e.g., cheese moisture, salt, pH, and eye production during ripening, was unaffected by SCC, and the patterns of proteolysis in cheese were altered, but this may be due to a number of factors rather than SCC (Cooney et al. 2000).

(iii)Use of originally mixed milk

Collecting the originally mixed milk with different SCCs and classifying it into different SCC categories has often been used in recent years. Nevertheless, it involves both individual differences of milk and other causative factors in the results of the effect of SCs. To have information on the effective role of SCs, interpretation has to be focused on the results of healthy low-SCC experiments. It was shown that Prato cheese made with low-SCC cow milk ($<2 \times 10^5$ cells.mL^{-1}) had a lower number of yeasts and molds, which usually cause deterioration of products (Vianna et al. 2008). Albenzio et al. (2004) have shown that the cheese curds made from low-SCC ewe milk ($<5 \times 10^5$ cells.mL^{-1}) contribute to improve the sensory properties of Canestrato Pugliese cheese during ripening. Additionally, SC count and composition were both parameters associated with the enzyme evolution of milk that contributes to a proteolytic pattern of ewe milk and to cheese sensory quality. For example, macrophages, by producing urokinase-plasminogen activators, were able to regulate the plasminogen-plasmin system that strongly influences cheese ripening.

(iv)Isolating somatic cells from milk

Isolating SCs from milk and then adding them to healthy milk seem perfect to study the effective role of SCs on milk and dairy products as well as selecting one or more cell types and their endogenous enzymes according to the origin of the cells used. The simple method of centrifugation and the advanced methods such as cell sorting flow cytometry and magnetic separation were applied for the cell isolation. Nevertheless, a relatively elevated quantity of milk is required in the case of low-SCC milk from healthy animals to obtain an adequate quantity of SCs. This may be the reason why the study on the effect of SCs until now has been only carried out on milk proteins, rather than on dairy products.

By centrifugation, SCs were isolated as a cell pellet from milk and were prepared in a cell suspension for cell analysis. Nevertheless, only one fourth or less of all SCs in milk was obtained in the precipitated pellet (Prescott and Breed 1910), and whether the small fraction in cell pellet has a fairly constant proportion of whole cells is still a question. By using Ficoll discontinuous density gradient, SCs were isolated from the milk of healthy or mastitic cows. Macrophages were then separately maintained in cell cultures to secrete lipolytic enzymes into the medium. These secreted lipolytic enzymes bind to the fat globule membranes in milk and expose the fat to degradation over storage. The other cell types, PMNs and lymphocytes, did not show lipoprotein lipase activity (Azzara and Dimick 1985a, b). Recently, using a magnetic positive separation method, Caroprese et al. (2007) have isolated macrophages from sheep milk and shown their proteolytic ability on α_s-caseins (~20 %), β-casein (~10 %), and minor degradation on γ-casein (~1 %). Using the same method, Albenzio et al. (2009) isolated not only macrophages but also PMNs from sheep milk. It has been shown that PMNs release lysosomal enzymes that induce a more intense hydrolysis on casein in late rather than in early lactation and positive correlation between these endogenous enzymes (e.g., cathepsin D) and clotting time.

(v)Concentrating somatic cells in milk

Concentrating SCs from milk and then adding different quantities to milk with low SCC or without SCs seem like a more feasible method to study the effective role of SC on milk and on cheese as well. This is a way to collect SCs with a peculiar composition including mainly macrophages, lymphocytes, or PMNs rather than to collect only PMNs from mastitic milk and also to vary the SCC with the same cell type. To our knowledge, Marino et al. (2005) were the first to study the contribution of SC and their proteolytic enzymes in Cheddar-type cheese by adding different quantities of concentrated SCs in a milk containing low SCs (<150,000 cells.mL^{-1}) for cheese making. The milk used to concentrate SCs was collected from milk with >10^6 cells.mL^{-1} to obtain a necessary quantity of SCs. Their results suggested that SCs have a direct influence on cheese moisture content and contribute to proteolysis in milk and Cheddar-type cheese. As mentioned in Section 3.3, the milk from mastitic and healthy cows with different SC counts and compositions has different profiles of SC endogenous enzymes. The cow milk used in this experiment was mastitic so it was mainly composed of PMNs; as a result, the effect of SCs is supposed to be interpreted as the "effect of PMNs." Recently, the effect of SCs concentrated from healthy goat milk was studied on low-fat cheese in order to improve cheese sensorial quality (Sanchez-Macias et al. 2013). Only miniature fresh goat milk cheeses were made in these experiments due to the limited quantity of collected SCs. Even if the ripening period lasted only 7 days, a general reduction in casein fractions was observed in the cheese with added SCs: β-casein (43 vs 26 %), para-κ-casein (37 vs 23 %), α_{s2}-casein (30 vs 20 %), and α_{s1}-casein (25 vs 14 %). Different patterns of casein degradation were found, while SCC was increased in raw or pasteurized goat milk cheeses. Moreover, cheeses with added SC were found to have softer texture and incremental color without any modification of chemical composition (Sanchez-Macias et al. 2013). Even if the SC composition was not determined in this study, the main SCs in healthy goat milk are PMNs (Dulin et al. 1983; Ostensson 1993), which are the same predominant cell type in mastitic cow milk (Azzara and Dimick 1985b). Until now, no study on the effective role of SCs on milk and on cheese has been conducted with healthy cow milk.

4.2.2 A new approach to study the effective role of somatic cells

It is worth noting that different milks and different methods of preparing SCs from milk can both induce different consequences on SCs themselves and then different roles of SCs on dairy processes and products. Adding isolated or concentrated SC with different ratios to cheese milk with low SCC or ideally without SCs will be a good strategy to study the effective role of SCs on dairy products. With the development of membrane filtration techniques in the dairy field (Saboya and Maubois 2000), the selective separation of SCs from raw milk becomes possible and feasible in practical experiments. Comparing other SC preparation technologies to study the role of SCs, the main advantages of membrane filtration techniques are listed below: (i) a great quantity of SCs can be efficiently concentrated at an industrial scale. Thus, collecting sufficient SCs from healthy milk originally poor in SCs is no longer a limiting step, and making diverse dairy products and analyzing the lipolysis and proteolysis in a long term become achievable; (ii) by membrane filtration, the milk deprived in SCs can be obtained by keeping the maximum native characteristics of milk. This SC-free milk can be used as the control in which SCs can be added in different quantities for cheese making. By using the same standardized milk in cheese making, the individual factors between different milk samples in other studies can be excluded; (iii) SCs can be added in different quantities to SC-free milk. Even with a high-added SCC level, the mastitic factors in other studies can be also excluded in this new approach; (iv) the studies on the effect of different SC types can be considered by concentrating SCs from different milk samples, for example, healthy bovine milk enriched in macrophages and mastitic bovine milk enriched in PMNs. However, several aspects of this new approach should be considered: the choice of membrane for optimized filtration, microfiltration, or ultrafiltration; the coseparation of other components in milk such as fat, proteins, and bacteria; and the influence of membrane filtration on the SCs, and to what extent, the final dairy products are affected.

5 CONCLUSION AND PERSPECTIVES

Although different approaches to study the effect of SCs have been conducted, few protocols have been undertaken to highlight the effective role of SCs on dairy products without the other concomitant factors in high-SCC milk. In this review, a new approach to study the effective role of SCs on dairy processes and products is suggested by comparing relevant methods of preparing SC. Isolating or concentrating SCs from milk and adding these SCs at different ratios in SC-free milk seems to be a good strategy. The milk status (mastitis or not mastitis and subclinical, chronic, or acute mastitis), SCC, and SC profiles should be precisely defined. This would allow the control of SC count and composition and, therefore, their endogenous enzyme profiles in milk. Such a fingerprint would help to characterize more precisely the SC impact on technological properties and final quality of dairy products besides the initial microbiological and nutritional characteristics of milk.

By analyzing the literature data, we have pointed out in this review that SCs can have a desirable role, in terms of acceleration of proteolysis in the cheese-ripening process and improvement of cheese sensory quality. Many studies remain to be done to fully understand the effective role of SCs, each cell type, and their extent in the dairy field. In particular, the relationship between SCs and their endogenous enzymes as well as the mechanism by which endogenous enzymes are released from SCs under different conditions and during dairy processes need to be elucidated.

ACKNOWLEDGMENTS

We are grateful to Mary Bret for the English correction of this paper.

REFERENCES

1. Albenzio M, Caroprese M (2011) Differential leucocyte count for ewe milk with low and high somatic cell count. J Dairy Res 78:43–48

2. Albenzio M, Caroprese M, Santillo A, Marino R, Taibi L, Sevi A (2004) Effects of somatic cell count and stage of lactation on the plasmin activity and cheese-making properties of ewe milk. J Dairy Sci 87:533–542
3. Albenzio M, Santillo A, Caroprese M, D'Angelo F, Marino R, Sevi A (2009) Role of endogenous enzymes in proteolysis of sheep milk. J Dairy Sci 92:79–86
4. Albenzio M, Santillo A, Caroprese M, Schena L, Russo DE, Sevi A (2011) Composition, indigenous proteolytic enzymes and coagulating behaviour of ewe milk as affected by somatic cell count. J Dairy Res 78:442–447
5. Andrews AT, Alichanidis E (1975) The acid phosphatases of bovine leucocytes, plasma and the milk of healthy and mastitic cows. J Dairy Res 42:391–400
6. Auldist MJ, Coats ST, Sutherland BJ, Clarke PT, McDowell GH, Rogers GL (1996a) Effect of somatic cell count and stage of lactation on the quality of full cream milk powder. Aust J Dairy Technol 51:94–98
7. Auldist MJ, Coats SJ, Sutherland BJ, Hardham JF, McDowell GH, Rogers GL (1996b) Effect of somatic cell count and stage of lactation on the quality and storage life of ultra high temperature milk. J Dairy Res 63:377–386
8. Auldist MJ, Coats S, Sutherland BJ, Mayes JJ, McDowell GH, Rogers GL (1996c) Effects of somatic cell count and stage of lactation on raw milk composition and the yield and quality of cheddar cheese. J Dairy Res 63:269–280
9. Azzara CD, Dimick PS (1985a) Lipolytic enzyme activity of macrophages in bovine mammary gland secretions. J Dairy Sci 68:1804–1812
10. Azzara CD, Dimick PS (1985b) Lipoprotein lipase activity of milk from cows with prolonged subclinical mastitis. J Dairy Sci 68:3171–3175
11. Baggiolini M, Bretz U, Dewald B, Feigenson ME (1978) The polymorphonuclear leukocyte. Inflamm Res 8:3–10
12. Baldwin ET, Bhat TN, Gulnik S, Hosur MV, Sowder RC, Cachau RE, Collins J, Silva AM, Erickson JW (1993) Crystal structures of native and inhibited forms of human cathepsin-D: implications for lysosomal targeting and drug design. Biochem US 90:6796–6800
13. Barabasi K, Nassberger L (1994) Distribution of cathepsin-D in human polymorphonuclear and mononuclear blood cells. APMIS 102:89–93
14. Baumert A, Bruckmaier RM, Wellnitz O (2009) Cell population, viability, and some key immunomodulatory molecules in different milk somatic cell samples in dairy cows. J Dairy Res 76:356–364
15. Ben Chedly H, Lacasse P, Marnet PG, Komara M, Marion S, Boutinaud M (2011) Use of milk epithelial cells to study regulation of cell activity and apoptosis during once-daily milking in goats. Animal 5:572–579
16. Benes P, Vetvicka V, Fusek M (2008) Cathepsin D—many functions of one aspartic protease. Crit Rev Oncol Hematol 68:12–28
17. Benic M, Habrun B, Kompes G, Mihaljevic Z, Cvetnic Z, Cergolj M, Macesic N (2012) Cell content in milk from cows with S. aureus intramammary infection. Vet Arh 82:411–422
18. Boulaaba A, Grabowski N, Klein G (2011) Differential cell count of caprine milk by flow cytometry and microscopy. Small Rumin Res 97:117–123
19. Boutinaud M, Jammes H (2002) Potential uses of milk epithelial cells: a review. Reprod Nutr Dev 42:133–147
20. Boutinaud M, Ben Chedly MH, Delamaire E, Guinard-Flament J (2008) Milking and feed restriction regulate transcripts of mammary epithelial cells purified from milk. J Dairy Sci 91:988–998
21. Briozzo P, Morisset M, Capony F, Rougeot C, Rochefort H (1988) In vitro degradation of extracellular matrix with Mr 52,000 cathepsin D secreted by breast cancer cells. Cancer Res 48:3688–3692
22. Burvenich C, Van MV, Mehrzad J, Diez-Fraile A, Duchateau L (2003) Severity of E. coli mastitis is mainly determined by cow factors. Vet Res 34:521–564

23. Campbell EJ, Silverman EK, Campbell MA (1989) Elastase and cathepsin G of human monocytes. Quantification of cellular content, release in response to stimuli, and heterogeneity in elastase-mediated proteolytic activity. J Immunol 143:2961–2968
24. Caroprese M, Marzano A, Schena L, Marino R, Santillo A, Albenzio M (2007) Contribution of macrophages to proteolysis and plasmin activity in ewe bulk milk. J Dairy Sci 90:2767–2772
25. Chen SX, Wang JZ, Van Kessel JS, Ren FZ, Zeng SS (2010) Effect of somatic cell count in goat milk on yield, sensory quality, and fatty acid profile of semisoft cheese. J Dairy Sci 93:1345–1354
26. Cohn ZA (1975) The role of proteases in macrophage physiology. In: Reich, Rifkin, Shaw (eds) Proteases and biological control. Cold Spring Harbour Laboratory, New York, pp 483–491
27. Considine T, Healy A, Kelly AL, McSweeney PLH (1999) Proteolytic specificity of elastase on bovine β-casein. Food Chem 66:463–470
28. Considine T, Healy A, Kelly AL, McSweeney PLH (2000) Proteolytic specificity of elastase on bovine αs1-casein. Food Chem 69:19–26
29. Considine T, Geary S, Kelly AL, McSweeney PLH (2002) Proteolytic specificity of cathepsin G on bovine αs1- and β-caseins. Food Chem 76:59–67
30. Considine T, Healy A, Kelly AL, McSweeney PLH (2004) Hydrolysis of bovine caseins by cathepsin B, a cysteine proteinase indigenous to milk. Int Dairy J 14:117–124
31. Cooney S, Tiernan D, Joyce P, Kelly AL (2000) Effect of somatic cell count and polymorphonuclear leucocyte content of milk on composition and proteolysis during ripening of Swiss-type cheese. J Dairy Res 67:301–307
32. Council Directive 92/46/EEC (1992) European Union: Health rules for the production and placing on the market of raw milk, heat-treated milk and milk based products regulations. CELEX-EUR Official Journal L 268:1–31
33. Dewald B, Rindler-Ludwig R, Bretz U, Baggiolini M (1975) Subcellular localization and heterogeneity of neutral proteases in neutrophilic polymorphonuclear leukocytes. J Exp Med 141:709–723
34. Dherbécourt J, Bourlieu C, Maillard MB, Aubert-Frogerais L, Richoux R, Thierry A (2010) Time course and specificity of lipolysis in Swiss cheese. J Agric Food Chem 58:11732–11739
35. Diment S, Leech MS, Stahl PD (1988) Cathepsin D is membrane-associated in macrophage endosomes. J Biol Chem 263:6901–6907
36. Dosogne H, Vangroenweghe F, Mehrzad J, Massart-Leen AM, Burvenich C (2003) Differential leukocyte count method for bovine low somatic cell count milk. J Dairy Sci 86:828–834
37. Dubin A, Potempa J, Travis J (1994) Structural and functional characterization of elastases from horse neutrophils. Biochem J 300:401–406
38. Dulin AM, Paape MJ, Schultze WD, Weinland BT (1983) Effect of parity, stage of lactation, and intramammary infection on concentration of somatic cells and cytoplasmic particles in goat milk. J Dairy Sci 66:2426–2433
39. Dupont A, Tokarski C, Dekeyzer O, Guihot AL, Amouyel P, Rolando C, Pinet F (2004) Two-dimensional maps and databases of the human macrophage proteome and secretome. Proteomics 4:1761–1778
40. Fernandes AM, Moretti TS, Bovo F, Lima CG, Oliveira CAF (2008) Effect of somatic cell counts on lipolysis, proteolysis and apparent viscosity of UHT milk during storage. Int J Dairy Technol 61:327–332
41. Gagnaire V, Jardin J, Jan G, Lortal S (2009) Invited review: proteomics of milk and bacteria used in fermented dairy products: from qualitative to quantitative advances. J Dairy Sci 92:811–825
42. Garnot P, Mollé D (1987) Heat-stability of milk-clotting enzymes in conditions encountered in Swiss cheese making. J Food Sci 52:75–77
43. Gonzalez-Rodriguez MC, Gonzalo C, San Primitivo F, Carmenes P (1995) Relationship between somatic cell count and intramammary infection of the half udder in dairy ewes. J Dairy Sci 78:2753–2759

44. Grappin R, Beuvier E, Bouton Y, Pochet S (1999) Advances in the biochemistry and microbiology of Swiss-type cheeses. Lait 79:3–22
45. Guha S, Padh H (2008) Cathepsins: fundamental effectors of endolysosomal proteolysis. Indian J Biochem Biophys 45:75–90
46. Hamed H, El Feki A, Gargouri A (2008) Total and differential bulk cow milk somatic cell counts and their relation with antioxidant factors. CR Biol 331:144–151
47. Hayes MG, Hurley MJ, Larsen LB, Heegaard CW, Magboul AA, Oliveira JC, McSweeney PLH, Kelly AL (2001) Thermal inactivation kinetics of bovine cathepsin D. J Dairy Res 68:267–276
48. Hettinga K, Van Valenberg H, De Vries S, Boeren S, Van Hooijdonk T, An Arendonk J, Rvoort J (2011) The host defense proteome of human and bovine milk. PLoS One 6:e19433
49. Hinz K, Larsen LB, Wellnitz O, Bruckmaier RM, Kelly AL (2012) Proteolytic and proteomic changes in milk at quarter level following infusion with Escherichia coil lipopolysaccharide. J Dairy Sci 95:1655–1666
50. Hunt KM, Williams JE, Shafii B, Hunt MK, Mehre R, Ting R, McGuire MK, McGuire MA (2013) Mastitis is associated with increased free fatty acids, somatic cell count, and interleukin-8 concentrations in human milk. Breastfeed Med 8:105–110
51. Hurley MJ, O'Driscoll BM, Kelly AL, McSweeney PLH (1999) Novel assay for the determination of residual coagulant activity in cheese. Int Dairy J 9:553–558
52. Hurley MJ, Larsen LB, Kelly AL, McSweeney PLH (2000) The milk acid proteinase cathepsin D: a review. Int Dairy J 10:673–681
53. Igoshi K, Arima S (1993) Acid and semi-alkaline proteinase in Swiss-type cheese. Milchwissenschaft 48:623–626
54. Ismail B, Nielsen SS (2010) Invited review: plasmin protease in milk: current knowledge and relevance to dairy industry. J Dairy Sci 93:4999–5009
55. ISO13366-1/IDF148-1 (2008) Milk—enumeration of somatic cells—part 1: microscopic method (reference method). International Dairy Federation Standard 1–17
56. ISO13366-2/IDF148-2 (2006) Milk—enumeration of somatic cells—part 2: guidance on the operation of fluoro-opto-electronic counters. International Dairy Federation Standard 1–13
57. Jaeggi JJ, Govindasamy-Lucey S, Berger YM, Johnson ME, McKusick BC, Thomas DL, Wendorff WL (2003) Hard ewe's milk cheese manufactured from milk of three different groups of somatic cell counts. J Dairy Sci 86:3082–3089
58. Janjanam J, Jamwal M, Singh S, Kumar S, Panigrahi AK, Hariprasad G, Jena MK, Anand V, Kumar S, Kaushik JK, Dang AK, Mukesh M, Mishra BP, Srinivasan A, Reddy VS, Mohanty AK (2013) Proteome analysis of functionally differentiated bovine (Bos indicus) mammary epithelial cells isolated from milk. Proteomics 13:3189–3204
59. Jethwaney D, Islam MR, Leidal K, De Bernabe D, Campbell K, Nauseef W, Gibson B (2007) Proteomic analysis of plasma membrane and secretory vesicles from human neutrophils. Proteome Sci 5:12–27
60. Kelly AL (1999) Effect of plasmin and somatic cell enzymes on proteolysis in aseptic starter and rennet free model cheeses. Milchwissenschaft 54:249–252
61. Kelly AL, Fox PF (2006) Indigenous enzymes in milk: a synopsis of future research requirements. Int Dairy J 16:707–715
62. Kelly AL, McSweeney PLH (2003) Indigenous proteinases in milk. In: Fox, McSweeney (eds) Advanced dairy chemistry—1 proteins. Springer, US, pp 495–521
63. Kelly AL, Tiernan D, O'Sullivan C, Joyce P (2000) Correlation between bovine milk somatic cell count and polymorphonuclear leukocyte level for samples of bulk milk and milk from individual cows. J Dairy Sci 83:300–304
64. Kelly AL, O'Flaherty F, Fox PF (2006) Indigenous proteolytic enzymes in milk: a brief overview of the present state of knowledge. Int Dairy J 16:563–572
65. Kitchen BJ (1976) Enzymatic methods for estimation of the somatic cell count in bovine milk. 1. Development of assay techniques and a study of their usefulness in evaluating the somatic cell content of milk. J Dairy Res 43:251–258

66. Lah TT, Calaf G, Kalman E, Shinde BG, Somers R, Estrada S, Salero E, Russo J, Daskal I (1996) Cathepsins D, B, and L in transformed human breast epithelial cells. Breast Cancer Res Treat 39:221–233
67. Larsen LB, Petersen TE (1995) Identification of five molecular forms of cathepsin D in bovine milk. Adv Exp Med Biol 362:279–283
68. Larsen LB, Boisen A, Petersen TE (1993) Procathepsin D cannot autoactivate to cathepsin D at acid pH. FEBS Lett 319:54–58
69. Larsen LB, Benfeldt C, Rasmussen LK, Petersen TE (1996) Bovine milk procathepsin D and cathepsin D: coagulation and milk protein degradation. J Dairy Res 63:119–130
70. Larsen LB, Wium H, Benfeldt C, Heegaard CW, Ardo Y, Qvist KB, Petersen TE (2000) Bovine milk procathepsin D: presence and activity in heated milk and in extracts of rennet-free UF-Feta cheese. Int Dairy J 10:67–73
71. Larsen LB, McSweeney PLH, Hayes MG, Andersen JB, Ingvartsen KL, Kelly AL (2006) Variation in activity and heterogeneity of bovine milk proteases with stage of lactation and somatic cell count. Int Dairy J 16:1–8
72. Le Maréchal C, Thiéry R, Vautor E, Le Loir Y (2011) Mastitis impact on technological properties of milk and quality of milk products—a review. Dairy Sci Technol 91:247–282
73. Le Roux Y, François L, Fatima M (2003) Polymorphonuclear proteolytic activity and milk composition change. Vet Res 34:629–645
74. Lee CS, Wooding FBP, Kemp P (1980) Identification, properties, and differential counts of cell-populations using electron-microscopy of dry cows secretions, colostrum and milk from normal cows. J Dairy Res 47:39–50
75. Lee AY, Gulnik SV, Erickson JW (1998) Conformational switching in an aspartic proteinase. Nat Struct Biol 5:866–871
76. Leitner G, Chaffer M, Krifucks O, Glickman A, Ezra E, Saran A (2000a) Milk leucocyte populations in heifers free of udder infection. J Vet Med 47:133–138
77. Leitner G, Shoshani E, Krifucks O, Chaffer M, Saran A (2000b) Milk leucocyte population patterns in bovine udder infection of different aetiology. J Vet Med 47:581–589
78. Lindmark-Mansson H, Bränning C, Aldén G, Paulsson M (2006) Relationship between somatic cell count, individual leukocyte populations and milk components in bovine udder quarter milk. Int Dairy J 16:717–727
79. Lippolis JD, Reinhardt TA (2005) Proteomic survey of bovine neutrophils. Vet Immunol Immunopathol 103:53–65
80. Lkhider M, Castino R, Bouguyon E, Isidoro C, Olivier-Bousquet M (2004) Cathepsin D released by lactating rat mammary epithelial cells is involved in prolactin cleavage under physiological conditions. J Cell Sci 117:5155–5164
81. Ma Y, Ryan C, Barbano DM, Galton DM, Rudan MA, Boor KJ (2000) Effects of somatic cell count on quality and shelf-life of pasteurized fluid milk. J Dairy Sci 83:264–274
82. Magboul AA, Larsen LB, McSweeney PLH, Kelly AL (2001) Cysteine protease activity in bovine milk. Int Dairy J 11:865–872
83. Marino R, Considine T, Sevi A, McSweeney PLH, Kelly AL (2005) Contribution of proteolytic activity associated with somatic cells in milk to cheese ripening. Int Dairy J 15:1026–1033
84. McSweeney PLH, Fox PF, Olson NF (1995) Proteolysis of bovine caseins by cathepsin D: preliminary observations and comparison with chymosin. Int Dairy J 5:321–336
85. Minarowska A, Gacko M, Karwowska A, Minarowski L (2008) Human cathepsin D. Folia Histochem Cytobiol 46:23–38
86. Minarowska A, Karwowska A, Gacko M (2009) Quantitative determination and localization of cathepsin D and its inhibitors. Folia Histochem Cytobiol 47:153–177
87. Moatsou G (2010) Indigenous enzymatic activities in ovine and caprine milks. Int J Dairy Technol 63:16–31

88. Moatsou G, Katsaros G, Bakopanos C, Kandarakis I, Taoukis P, Politis I (2008) Effect of high-pressure treatment at various temperatures on activity of indigenous proteolytic enzymes and denaturation of whey proteins in ovine milk. Int Dairy J 18:1119–1125
89. Morgante M, Ranucci S, Pauselli M, Casoli C, Duranti E (1996) Total and differential cell count in milk of primiparous Comisana ewes without clinical signs of mastitis. Small Rumin Res 21:265–271
90. Mukherjee R, Ram GC, Dash PK, Goswami T (2004) The activity of milk leukocytes in response to a water-soluble fraction of Mycobacterium phlei in bovine subclinical mastitis. Vet Res Commun 28:47–54
91. O'Brien B, Meaney WJ, McDonagh D, Kelly A (2001) Influence of somatic cell count and storage interval on composition and processing characteristics of milk from cows in late lactation. Aust J Dairy Technol 56:213–218
92. O'Farrell IP, Sheehan JJ, Wilkinson MG, Harrington D, Kelly AL (2002) Influence of addition of plasmin or mastitic milk to cheesemilk on quality of smear-ripened cheese. Dairy Sci Technol 82:305–316
93. Olechnowicz J, Jaskowski JM (2012) Somatic cells count in cow's bulk tank milk. J Vet Med Sci 74:681–686
94. Olumee-Shabon Z, Swain T, Smith EA, Tall E, Boehmer JL (2013) Proteomic analysis of differentially expressed proteins in caprine milk during experimentally induced endotoxin mastitis. J Dairy Sci 96:2903–2912
95. Ostensson K (1993) Total and differential leukocyte counts, N-acetyl-β-D-glucosaminidase activity, and serum albumin content in foremilk and residual milk during endotoxin-induced mastitis in cows. Am J Vet Res 54:231–238
96. O'Sullivan CA, Joyce PJ, Sloan T, Shattock AG (1992) Capture immunoassay for the diagnosis of bovine mastitis using a monoclonal antibody to polymorphonuclear granulocytes. J Dairy Res 59:122–133
97. Owen CA, Campbell EJ (1999) The cell biology of leukocyte-mediated proteolysis. J Leukocyte Biol 65:137–150
98. Paape MJ, Wergin WP, Guidry AJ, Pearson RE (1979) Leukocytes—second line of defense against invading mastitis pathogens. J Dairy Sci 62:135–153
99. Paape MJ, Mehrzad J, Zhao X, Detilleux J, Burvenich C (2002) Defense of the bovine mammary gland by polymorphonuclear neutrophil leukocytes. J Mammary Gland Biol 7:109–121
100. Paape MJ, Bannerman DD, Zhao X, Lee JW (2003) The bovine neutrophil: Structure and function in blood and milk. Vet Res 34:597–627
101. Park YH, Fox LK, Hamilton MJ, Davis WC (1991) Bovine mononuclear leukocyte subpopulations in peripheral blood and mammary gland secrections during lactation. J Dairy Sci 75:998–1006
102. Pham CTN (2006) Neutrophil serine proteases: specific regulators of inflammation. Nat Rev Immunol 6:541–550
103. Piepers S, De Vliegher S, Demeyere K, Lambrecht BN, De Kruif A, Meyer E, Opsomer G (2009) Technical note: flow cytometric identification of bovine milk neutrophils and simultaneous quantification of their viability. J Dairy Sci 92:626–631
104. Politis I, Ng-Kwai-Hang KF (1988) Effects of somatic cell counts and milk composition on the coagulating properties of milk. J Dairy Sci 71:1740–1746
105. Prescott SC, Breed RS (1910) The determination of the number of body cells in milk by a direct method. J Infect Dis 7:632–640
106. Prin-Mathieu C, Le Roux Y, Faure GC, Laurent F, Bene MC, Moussaoui F (2002) Enzymatic activities of bovine peripheral blood leukocytes and milk polymorphonuclear neutrophils during intramammary inflammation caused by lipopolysaccharide. Clin Vaccine Immunol 9:812–817
107. Rankl JE (2004) Differentiation of somatic cells in milk of several species with special regard to epithelial cells. Faculty of Veterinary Medicine of the Ludwig Maximilian University, Munich

108. Raynal-Ljutovac K, Pirisi A, De Cremoux R, Gonzalo C (2007) Somatic cells of goat and sheep milk: analytical, sanitary, productive and technological aspects. Small Rumin Res 68:126–144
109. Reuter K, Ehmcke J, Stukenborg JB, Simoni M, Damm OS, Redmann K, Schlatt S, Wistuba J (2014) Reassembly of somatic cells and testicular organogenesis in vitro. Tissue Cell 46:86–96
110. Riollet C, Rainard P, Poutrel B (2001) Cell subpopulations and cytokine expression in cow milk in response to chronic staphylococcus aureus infection. J Dairy Sci 84:1077–1084
111. Rivas AL, Tadevosyan R, Gorewit RC, Anderson KL, Lyman R, Gonzalez RN (2006) Relationships between the phagocytic ability of milk macrophages and polymorphonuclear cells and somatic cell counts in uninfected cows. Can J Vet Res 70:68–74
112. Rogers SA, Mitchell GE (1994) The relationship between somatic cell count, composition and manufacturing properties of bulk milk. 6. Cheddar cheese and skim-milk yogurt. Aust J Dairy Technol 49:70–74
113. Rupp R, Boichard D, Bertrand C, Bazin S (2000) Overview of milk somatic cell counts in the French dairy cattle breeds. Prod Anim 13:257–267
114. Saboya LV, Maubois JL (2000) Current developments of microfiltration technology in the dairy industry. Lait 80:541–553
115. Sanchez-Macias D, Morales-delaNuez A, Torres A, Hernandez-Castellano LE, Jiménez-Flores R, Castro N, Arguello A (2013) Effects of addition of somatic cells to caprine milk on cheese quality. Int Dairy J 29:61–67
116. Santillo A, Kelly AL, Palermo C, Sevi A, Albenzio M (2009) Role of indigenous enzymes in proteolysis of casein in caprine milk. Int Dairy J 19:655–660
117. Santos MV, Ma Y, Barbano DM (2002) Effect of somatic cell count on pasteurized fluid milk quality. National Mastitis Council Regional Meeting Proceedings 27–30
118. Santos MV, Ma Y, Barbano DM (2003) Effect of somatic cell count on proteolysis and lipolysis in pasteurized fluid milk during shelf-life storage. J Dairy Sci 86:2491–2503
119. Sarikaya H, Prgomet C, Pfaffl MW, Bruckmaier RM (2004) Differentiation of leukocytes in bovine milk. Milchwissenschaft 59:586–589
120. Schwarz D, Diesterbeck US, Konig S, Brugemann K, Schlez K, Zschock M, Wolter W, Czerny CP (2011) Flow cytometric differential cell counts in milk for the evaluation of inflammatory reactions in clinically healthy and subclinically infected bovine mammary glands. J Dairy Sci 94:5033–5044
121. Seol MB, Bong JJ, Baik M (2006) Involvement of cathepsin D in apoptosis of mammary epithelial cells. Asian Autral J Anim 19:1100–1105
122. Sharma N, Singh NK, Bhadwal MS (2011) Relationship of somatic cell count and mastitis: an overview. Asian Autral J Anim 24:429–438
123. Smolenski G, Haines S, Kwan FYS, Bond J, Farr V, Davis SR, Stelwagen K, Wheeler TT (2006) Characterisation of host defence proteins in milk using a proteomic approach. J Proteome Res 6:207–215
124. Somers JM, O'Brien B, Meaney WJ, Kelly AL (2003) Heterogeneity of proteolytic enzyme activities in milk samples of different somatic cell count. J Dairy Res 70:45–50
125. Sordillo LM, Streicher KL (2002) Mammary gland immunity and mastitis susceptibility. J Mammary Gland Biol 7:135–146
126. Sordillo LM, Shafer-Weaver K, Dessauge F (1997) Immunobiology of the mammary gland. J Dairy Sci 80:1851–1865
127. Souza FN, Blagitz MG, Penna CFAM, la Libera AMMP, Heinemann MB, Cerqueira MMOP (2012) Somatic cell count in small ruminants: friend or foe? Small Rumin Res 107:65–75
128. Travis J, Fritz H (1991) Potential problems in designing elastase inhibitors for therapy. Am Rev Respir Dis 143:1412–1415
129. Verdi RJ, Barbano DM (1991) Properties of proteases from milk somatic cells and blood leukocytes. J Dairy Sci 74:2077–2081

130. Vianna PCB, Mazal G, Santos MV, Bolini HMA, Gigante ML (2008) Microbial and sensory changes throughout the ripening of prato cheese made from milk with different levels of somatic cells. J Dairy Sci 91:1743–1750

CHAPTER 4

Analytical Assessment of the Intense Heat Load of Whipping Cream, Coffee Cream, and Condensed Milk at Retail in Austria and Germany

Lisa I. Boitz[1] & Helmut K. Mayer [1]

[1] Department of Food Science and Technology, Food Chemistry Laboratory, BOKUUniversity of Natural Resources and Life Sciences Vienna Vienna Austria

ABSTRACT

Time temperature integrators (TTIs) are useful tools in estimating the heat load applied on differently processed dairy products. The objective of this study was to analyze and assess three TTIs – lactulose, furosine, and acid-soluble β-lactoglobulin (β-Lg) – in 70 high heated dairy products at retail in Austria and Germany comprising whipping cream, coffee cream/milk, and condensed milk products. While β-Lg was not appropriate to evaluate the heat load of these products, furosine and especially lactulose increased with rising intensity of heat treatment, and are appropriate to distinguish between several heating categories analyzed. Pasteurized (n = 8) and "heat treated" (n = 5) whipping cream samples showed lowest furosine (48 ± 14/ 45 ± 19 mg.100 g^{-1} protein) and low lactulose (29 ± 10/57 ± 28 $mg.L^{-1}$) concentrations, followed by ESL whipping cream (n = 10), ESL coffee cream

($n = 1$), and UHT whipping cream ($n = 10$) (furosine = 72 ± 37/71 /161 ± 30 mg.100 g^{-1} protein; lactulose = 56 ± 41/161/195 ± 39 mg.L^{-1}), respectively. Sterilized condensed milk samples ($n = 14$) showed the highest concentrations of both TTIs and could be clearly separated from UHT treated samples ($n = 5$) (furosine = 491 ± 196 /216 ± 46 mg. 100 g^{-1} protein; lactulose = 1997 ± 658/409 ± 161 mg.L^{-1}), whereas the so-called heat-treated samples ($n = 9$) had a heat load in between showing an extreme range of variation for both TTIs.

Keywords: Furosine Lactulose β-Lactoglobulin Cream Condensed milk

1 INTRODUCTION

Although bovine liquid milk is the major dairy product consumed with 77 kg per capita and year in Austria, these commodities also include whipping cream, coffee milk or coffee cream, evaporated or condensed milk as well as milk powder (AMA 2014). Moreover, Extended Shelf Life (ESL) products have gained widespread acceptance in Austria and Germany; even "ESL" labeled coffee milk came on the market in Austria recently. The Austrian Food Codex defines "cream" as a product with at least 10% milk fat content, *coffee cream* contains 10–18% milk fat, *whipping cream* at least 30%, and a fat content of 36% is necessary for products labeled as "*extra creamy*" whipping cream (FMH 2011). Condensed or concentrated milk products require an additional evaporation step, commonly implemented in falling film evaporators, thereby increasing its milk solids and viscosity. Due to the same production technologies, and to simplify the following text, the term "condensed milk" includes also concentrated dairy products.

After homogenization and adjustment of fat content, a heat treatment step is necessary to minimize microbial spoilage, inactivate pathogens, and to guarantee a certain shelf life of the product depending on the heat load applied. Table 1 compares processing conditions of pasteurized *milk* to those applied to various types of *whipping cream* and sterilized *coffee cream/milk*. While the label "ESL" for whipping cream is not commonly used in Germany, the label "heat treated" ("wärmebehandelt") could be often found on whipping cream packaging indicating a second heating step using temperatures between 95 and 105 °C of whipping cream prior to filling (indicated with "2x" in Fig.1).

Table 1. Processing conditions of pasteurized milk and various types of whipping cream and sterilized coffee cream/milk as well as their shelf life (FMH 2011; LVBM 2015)

Product	Processing conditions	Shelf life and storage
Pasteurized milk	≤75 °C, few seconds	10 days, chilled
Pasteurized whipping cream	Austria: 85–110 °C Germany: 95–105 °C	10 days, chilled
ESL whipping cream (Austria)	≤135 °C, single flow	21 days, chilled
UHT whipping cream	135–150 °C, few seconds	6 months, ambient conditions
Sterilized coffee cream/milk	In-container sterilized: 121 °C, 20 min	1 year, ambient conditions

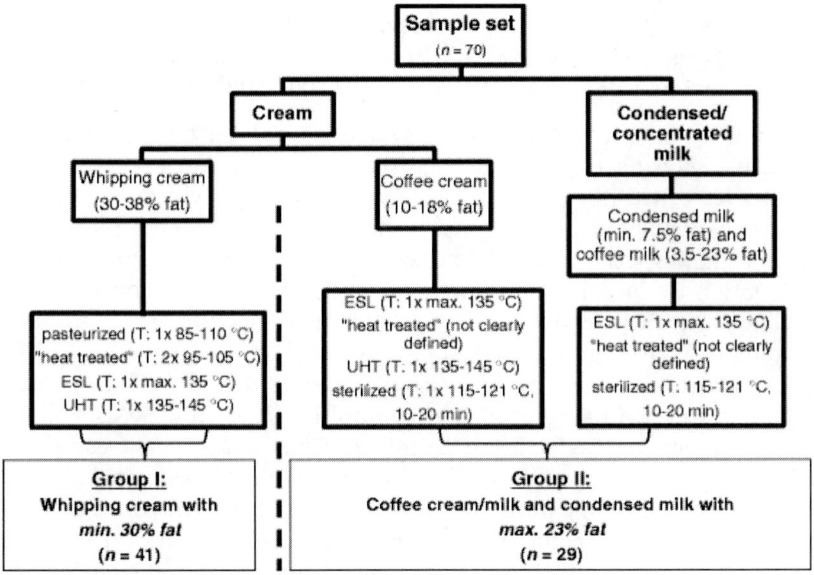

Fig. 1. Classification of sample set according to their production technology, fat content, and applied heat treatment as well as chosen arrangement into "group I" and "group II" for data evaluation. "1x"– single flow heating; "2x"– second heating prior to filling

It is already known and well studied that different time/temperature combinations applied during processing according to the product segment (pasteurization/ESL/UHT or sterilization) can lead to enormous chemical (e.g., Maillard reaction products), nutritional (e.g., vitamin loss), and/or sensorial (e.g., cooked flavor) differences in the final dairy products. To assess the impact of a thermal process on dairy products, intrinsic time temperature integrators (TTIs) provide a useful tool for the quantification of the heat load of liquid milk (Claeys et al. 2002). They are categorized as follows: type I indicators include denaturation, degradation, and inactivation processes of heat labile components, most suitable for the evaluation of low-heat treatments, e.g., enzymes, water- and fat-soluble vitamins, and β-lactoglobulin (β-Lg). Type II indicators describe the formation of substances that are (almost) not present in raw milk, including lactulose, 5-hydroxymethylfurfural, and furosine, and are more appropriate for the assessment of high-temperature processes (Mayer et al. 2010).

Unfolding of the heat sensitive protein β-Lg, comprising 55% of total whey protein in bovine milk, occurs above 60 °C. Thus, the remaining concentration of native acid-soluble β-Lg observed after thermal processing reveals a suitable TTI to assess the heat load of liquid milk (Manzo et al. 2015). For rapid screening, electrophoretic methods are commonly used (Mayer et al. 2010), and for an exact quantification, the IDF recommends a reference HPLC method (ISO/IDF standard 178, 2005).

Furosine (ε-*N*-[2-furoylmethyl]-L-2-lysine), produced by acid hydrolysis of lactulosyllysine, is related to the first stages of Maillard reaction which is associated with the production of every heated dairy product, while its extent depends on the applied time/temperature conditions during processing, especially above 120 °C (Töpel 2007). Due to a positive correlation of furosine concentration and total protein content of the product, the final furosine concentration is given in relation to 100 g protein. Evaluation of furosine in milk is mostly accomplished with HPLC methods, as supposed by the ISO/IDF standard 193 (2004) or other research groups (Mayer et al. 2010). Additionally, lactulose as a further type II indicator was to be determined. This disaccharide comprising fructose and galactose is formed by lactose isomerization, and is a result of the so-called Lobry de Bruyn–Alberda van Ekenstein transformation. Lactulose is not detectable in raw milk, and lowest amounts are found in pasteurized milk, with increasing concentrations with increasing heat load applied (Mayer et al. 2010; Montilla et al. 2005). For quantification, a HPLC method (ISO/IDF standard 147, 2007) and further methods were reported in the literature, e.g., enzymatic determination (Amine et al. 2000).

Intensive research was accomplished considering several TTIs in liquid milk. Furosine and lactulose concentrations increase with rising heat load applied to liquid milk (Gallmann et al. 2001; Mayer et al. 2010; Montilla et al. 2005). However, information on further dairy products besides liquid bovine milk is rare. Some literature considering lactulose contents in high-temperature treated dairy products, such as condensed milk, is available. Nevertheless, these studies are outdated or do not reflect currently used dairy technologies (Adachi and Patton 1961), included only a limited sample set for condensed/sterilized products (Montilla et al. 2005), or did either not consider other TTIs to differentiate the various heating categories of dairy products or were only analyzed in liquid milk or milk powder and not in higher heated dairy products (Sakkas et al. 2014). Hence, the aim of this study was to evaluate appropriate TTIs for dairy products which are generally processed at higher temperatures than liquid milk.

As we have already shown recently (Boitz and Mayer 2015; Mayer et al. 2010), β-Lg as type I indicator is a suitable heat load indicator for liquid milk, but definitely not suitable for the heat load assessment in dairy products with a more intense heat treatment, like whipping cream. Thus, the assumption that β-Lg is also not appropriate as heat load indicator for even higher heated products than whipping cream, such as condensed milk, was to be confirmed in this study. Furthermore, the aim of this study was to determine two different type II indicators, namely lactulose and furosine, in several dairy product segments comprising an overall higher heat load during its production compared to pasteurized liquid milk. Thus, the purpose of our work was to provide novel information on three TTIs for high heated dairy products, which is not available in the present literature. Within the product segments "whipping cream" and "coffee cream/milk and condensed milk", the different heating categories "heat treated", "ESL", "UHT", and "sterilized" were intensively studied. Finally, it was to be proven if lactulose and furosine are suitable TTIs for these dairy product segments as well.

2 MATERIALS AND METHODS

2.1 Chemical references

All solvents used for chromatographic analysis were of HPLC grade, and all chemicals used for sample preparation were of analytical grade. Throughout all experiments, aqueous solutions were prepared with ultrapure water (SG Ultra Clear UV system; Siemens Water 95 Technologies, Warrendale, PA, USA). Acetonitrile (ACN, 100%), disodium hydrogen phosphate (99%), heptanesulfonic acid (>98%), hydrogen peroxide (30%), magnesium sulfate heptahydrate (98%), sodium acetate trihydrate (99.5%), sodium dihydrogen phosphate, octanol (98%), and trifluoroacetic acid (TFA, 99%) were purchased from Sigma-Aldrich (St. Louis, MO, USA). Furosine standard (furosine dihydrochloride; >97%) for calibration was purchased from NeoMPS PolyPeptide Laboratories Group (Strasbourg, France). For native PAGE, ammoniumpersulfate (>99%) and tetramethylethylendiamine were purchased from Serva (Heidelberg, Germany), and bromophenol blue as well as glycine were purchased from Merck (Darmstadt, Germany).

For the determination of lactulose, zinc sulfate (99.5%) and potassium hexacyanoferrate (II) (99.0%), required for Carrez solution I and II, were purchased from Merck (Darmstadt, Germany), and triethanolamine (TEA, 99.0%) buffer was obtained from Acros Organics (Geel, Belgium). Sodium hydroxide (NaOH, 99%), hydrochloric acid (HCl, 37%), and formic acid (FA, 98%) were purchased from Roth (Karslruhe, Germany). Enzymatic determination of lactulose was accomplished with ready-to-use solutions and enzymes included in the commercial enzyme kit (K-Lactul; Megazyme, Bray, Co. Wicklow, Ireland). Solutions and enzymes were stored as prescribed in the manual, and frozen solutions were aliquoted to avoid unnecessary thawing processes.

2.2 Dairy product sample set

The sample set included 70 dairy product samples of different processing categories: whipping cream, coffee cream/milk, and condensed milk. Figure 1 shows a classification of sample set according to the production and the related technology as well as to the respective differentiation into "group I" and "group II." Nutritional values were either taken from the label (carbohydrates) or determined in the laboratory (fat and protein content). All commercial whipping cream samples (group I, $n = 41$) produced by various dairy companies using different heating technologies were obtained from retail outlets either in Austria or in Germany. These samples included previously analyzed samples: 9 pasteurized, 13 ESL (2 lactose-free), and 11 UHT (2 lactose-free) whipping creams as described in Boitz and Mayer (2015), and additionally, 2 UHT samples and 6 German whipping cream samples were labeled as "heat treated." These samples contained 29.50–38.33% fat, 1.76–2.59 g.100 g^{-1} protein, and 3.0–3.4 g.100 g^{-1} carbohydrate. Condensed milk and coffee cream/milk samples (group II, $n = 29$) included 14 samples labeled as "coffee cream" ($n = 13$) or coffee milk ($n = 1$) and 2 samples labeled directly as "concentrated coffee milk". The fat content was between 3.50 and 15.00 g.100 g^{-1}, the protein content was between 2.73 and 6.43 g.100 g^{-1}, and the carbohydrate content was between 3.8 and 10.0 g.100 g^{-1}. Group II also included 13 samples labeled as "condensed milk" (eight unsweetened, five sweetened) with a fat content of 7.50–23.00 g.100 g^{-1}, a protein content of 2.57–7.52 g.100 g^{-1}, and a carbohydrate content of 7.6–55.0 g.100 g^{-1}. All condensed/concentrated products showed an increased fat-free milk dry matter ranging from 12.6 to 20.0%. According to their production process and related heat treatment, samples of group II can be further subdivided into the only available ESL sample, 9 samples labeled as "heat treated" (produced either in Germany or in Austria), 5 UHT, and 14 sterilized samples (including the five sweetened condensed milk samples). All samples were analyzed within their labeled shelf-life period (and stored at 4 °C until first analysis). Long-life products were analyzed within the first quarter of the time period given as "date of expiry". In case of coffee cream samples which were packed in small-sized plastic cups (10 g, respectively), all cups from the sample package were pooled, yielding approximately 200 g of sample. Prior to analysis, all samples were warmed in a water bath set at 40 °C, and this temperature was held for 15 min to get a homogenous emulsion. Samples were aliquoted and kept frozen until further analyses. Unless otherwise specified, all analyses were carried out in duplicate.

2.3 Type II indicator furosine: RP-HPLC analysis

Determination of furosine in all dairy samples was accomplished according to ISO/IDF (2004) with some modifications as described previously (Boitz and Mayer 2015). Due to the high viscosity of some condensed milk samples, a pre-dilution was necessary (*w/w*). Afterwards, 2 g of each sample was mixed with 6 mL 10.6 mol. L^{-1} HCl in screw-cap Pyrex® tubes and exposed to nitrogen for 2 min. For acid hydrolysis, the tubes were closed tightly and heated at 110 °C for 23 h. After heating, the tubes were cooled at ambient conditions, and hydrolysates obtained were filtered (Schleicher & Schuell 595 ½; GE Healthcare, Buckinghamshire, UK) and kept frozen until further analysis. After solid phase extraction of 0.5 mL sample hydrolysate using Waters SepPak® Vac 3cc (500 mg) Certified Silica Cartridges (Waters Corporation, MA, USA), eluates were vacuum dried in a Pico·Tag workstation (Millipore Waters Workstation PTS 3142, pump BOC EDWARDS, model XDS5). Dried samples were diluted in 0.2 mL of freshly prepared diluting solution consisting of 5% ACN and 0.2% FA in ultrapure water. Samples were filtered (0.2 µm, Minisart RC4; Sartorious, Goettingen, Germany) prior to HPLC injection. Furosine detection was performed on a Waters chromatography system using a model 600E multisolvent delivery system, a Rheodyne 7725i injector, guard column (Sentry Guard, Symmetry™ C_{18}, 3.5 µm, 2.1 × 10 mm), and a Symmetry™ 300 C_{18} column (3.5 µm, 2.1 × 150 mm) (Waters Corporation, MA, USA). The isocratic separation (flow rate 0.35 mL.min^{-1}; injection volume 20 µL; column temperature 35 °C) was accomplished with 89% of solvent A consisting of 5 mM heptanesulfonic acid with 0.2% FA and 11% of solvent B (100% ACN). Eluates were detected at 280 nm using a Waters 2489 UV/VIS Detector which was interfaced with a PC running Waters Millennium^{32} chromatography manager. The calibration was performed by plotting peak area versus picomoles of furosine and was linear (R^2 = 0.999) between 5 and 60 pmol per injection. Inter-day repeatability of sample preparation procedure (same analyst and apparatus but different reagents) was verified with eight independent sample work-ups of pasteurized whipping cream (35 mg.100 g^{-1} protein) and UHT whipping cream (155.2 mg.100 g^{-1} protein) samples with resulting RSDs of 2.9 and 1.5%, respectively.

Total nitrogen content required for the final furosine calculation of samples was analyzed in triplicate according to the Kjeldahl method (ISO/IDF 2001). Further, the total fat content of samples was determined according to Gerber's or Roeder's butyrometric method (DIN standard 10284: 1964-03; ISO/IDF 2008).

2.4 Type II indicator lactulose: enzymatic determination

Lactulose concentrations were determined enzymatically because of economical reasons and its easy handling with a modified test kit method (K-Lactul; Megazyme, Bray, Co. Wicklow, Ireland) as described recently (Boitz and Mayer 2015). Chromatographic methods were not necessary due to satisfying precision and sensitivity of used enzyme kit for samples with an expected high concentration of lactulose in the analyzed samples. Briefly, samples were defatted by double centrifugation (first = 20 min, 17,200×*g*, 4 °C; second = 3 min, 15,700×*g*, 4 °C), and after dilution of obtained fat-free phase, the samples were mixed carefully with sodium phosphate buffer (0.4 mol.L^{-1}, pH 7.6) and ultrapure water. After clarification with Carrez solutions I and II, the supernatant was transferred into tightly closing screw-cap microfuge tubes (SC Micro Tube PCR-PT, Sarstedt, Germany) due to intense pressure development during enzymatic reaction in the reaction tube. Each sample required a separate blank, containing water instead of the enzyme β-galactosidase, which was added to the sample hydrolyzing both lactose and lactulose to galactose as well as glucose and fructose, respectively. After the addition of sodium acetate buffer (0.5 mol.L^{-1}, pH 4.5), tubes were incubated 60 min at 40 °C in a covered thermoblock. Afterwards, TEA buffer (1 mol.L^{-1}, pH 7.6), NaOH (0.33 mol.L^{-1}), octanol, and enzyme solution (catalase/glucose-oxidase) as well as hydrogen peroxide were added, followed by another incubation step for 15 min at 40 °C and a subsequent centrifugation (10 min, 15,700×*g*, 20 °C) terminating the enzymatic reactions. After further enzymatic treatment as described in the enzyme kit manual, the absorbance of released NADPH was measured at 340 nm using a UV/VIS spectrophotometer (U-2000; HITACHI, Chiyoda, Japan). The final lactulose concentration (mg.L^{-1}) was calculated from measured lactulose in the fat-free phase considering the fat concentration analyzed for each sample.

Referring to lactose-free or artificially sweetened products, the high concentration of free fructose in the sample led to further enzymatic reactions in the blank as well, making the kit unsuitable for these products. Thus, artificially sweetened and lactose-free samples included in the sample set were not included in further evaluation of results considering lactulose.

2.5 Type I indicator β-Lg: estimation by native PAGE of samples

Due to the assumption of a β-Lg concentration in high-temperature treated dairy products (especially coffee cream/milk and condensed milk samples) lower than the LOQ (20 mg.L^{-1}) of the used RP-HPLC method, a fast and easy electrophoretic technique was chosen to prove this assumption. Native polyacrylamide gel electrophoresis (native PAGE) was used to analyze acid-soluble β-Lg in selected samples separating the whey proteins by their different electrophoretic mobility (according to their electrical charge) during migration towards the positive pole (anode). Precipitation of caseins (including also the heat-denatured whey proteins) was accomplished by acidification of samples. Therefore, the homogenous samples were defatted by centrifugation (15 min, 4 °C, 6450×*g*), and HCl (2 mol.L^{-1}) was added carefully, until a pH of 4.5 was reached. Samples were let stand at ambient conditions and centrifuged again (30 min, 4 °C, 17,200×*g*). The acid whey (supernatant) was aliquoted and kept frozen until further analysis. Sample preparation included a tempering step (10 min, 40 °C) of 100 μL acid whey and the subsequent addition of 400 μL buffer (10 mmol.L^{-1} tris (hydroxymethyl)-aminomethane and 77 mmol.L^{-1} glycine, pH 8.3) as well as 55 μL of bromophenol blue solution. The sample was centrifuged for 5 min (16,100×*g*), and 20 μL of the supernatant was used for electrophoretic separation. Native PAGE (12.5% T) was performed using a dual cooled vertical slab gel electrophoresis unit SE 600 (Hoefer Scientific Instruments, San Francisco, CA, USA) as described in Mayer et al. (2003).

2.6 Type I indicator β-Lg: quantification by RP-HPLC analysis of β-lactoglobulin in whipping cream samples

In order to determine the accurate β-Lg concentration in all whipping cream samples and in some selected group II samples (having visible β-Lg bands in electrophoretograms), these samples were analyzed with RP-HPLC as described previously (Boitz and Mayer 2015).

Briefly, after acid precipitation of caseins (and denatured whey proteins) with HCl (2 $mol.L^{-1}$, pH 4.5), the obtained acid whey was diluted (1:10) with phosphate buffer (0.1 $mol.L^{-1}$, pH 6.7), incubated for 1 h, and samples were filtered prior to HPLC injection (0.2 µm, Minisart RC4; Sartorious, Goettingen, Germany). The same RP-HPLC column and equipment was used as described for the determination of furosine above as well as in Boitz and Mayer (2015). A gradient separation was performed with solvent A consisting of 0.1% TFA in ultrapure water, and solvent B consisting of 0.1% TFA in ACN, using a column temperature of 40 °C and a flow rate of 0.35 $mL.min^{-1}$. UV/VIS detector was set at 205 nm, and injection volume was 10 µL. The calibration was linear ($R^2 = 0.998$) between 0.2 and 3.2 µg per 10 µL injection volume which is equivalent to a β-Lg content of 0.2–3.2 $mg.mL^{-1}$ in liquid milk after the performed dilution (1:10). Considering inter-day repeatability of sample preparation procedure (same analyst and apparatus but different reagents), a RSD of 3.4% was determined for an independent sample work-up ($n = 8$) of a UHT whipping cream sample (130 $mg.L^{-1}$).

2.7 Statistical analysis of results

Calculations of mean values and standard deviations as well as related artworks were accomplished with Microsoft Excel 2010. Moreover, data on lactulose and furosine concentrations were analyzed by one-way ANOVA and post-tested by Tukey test using SPSS 21.0 software, and differences were considered significant at $p < 0.05$.

3 RESULTS

3.1 RP-HPLC analysis of furosine

Seventy retail samples including 41 whipping cream samples (for furosine data of pasteurized and ESL samples; see Boitz and Mayer 2015) and 29 coffee cream/milk and condensed milk samples were analyzed for their furosine concentrations using an existing RP-HPLC method with few modifications in sample preparation because of higher fat content and higher viscosity, especially of condensed milk samples. Figure 2 shows chromatograms of two examples of differently processed whipping cream samples (group I) as well as two examples of coffee cream/milk and condensed milk samples (group II), respectively. While no (or only a slight) dilution was necessary for proper detection of whipping cream samples (Fig. 2a, b), coffee cream/milk samples and condensed milk samples had to be diluted to avoid an overload of the HPLC column

(Fig. 2c, d). Nevertheless, for the undiluted whipping cream samples, peak area increased with increasing intensity of heat treatment. The "ESL" coffee cream sample showed the smallest furosine peak and lowest amount within group II (Fig.2c; 70.9 mg.100 g^{-1}protein) and could be clearly separated from all other samples in group II.

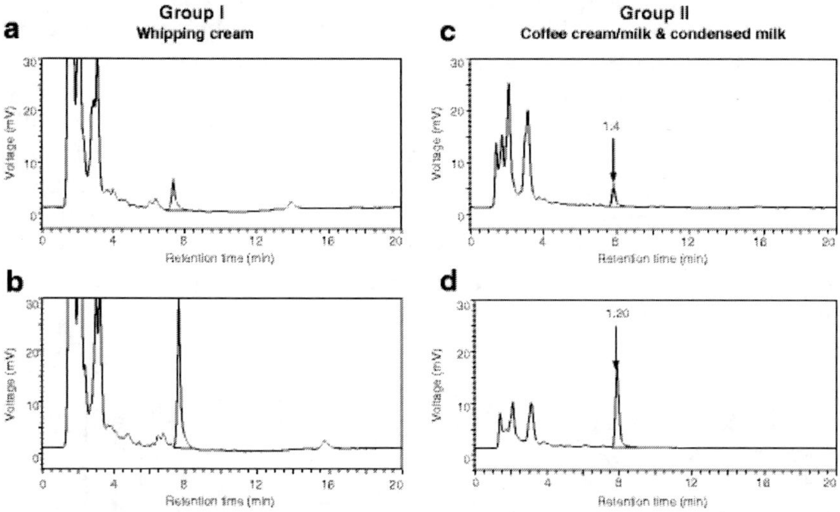

Fig. 2. RP-HPLC chromatograms (λ = 280 nm) of furosine in whipping cream samples (group I) on the left side (**A** pasteurized–52.9 mg. 100 g^{-1} protein, **B** UHT–176.6 mg. 100 g^{-1}protein) and in condensed milk/coffee cream samples (group II) on the right side (**C** ESL–70.9 mg. 100 g^{-1} protein, **D** sterilized–789.1 mg. 100 g^{-1} protein), respectively

Figure 3 shows mean values and standard deviations of the analyzed furosine and lactulose concentrations for pasteurized, ESL, "heat treated", and UHT whipping cream (group I, left side) as well as for ESL, "heat treated", UHT, and sterilized coffee cream/milk and condensed milk samples (group II, right side), respectively. Lowest furosine concentrations were found in pasteurized whipping cream samples (47.8 ± 14.0 mg.100 g^{-1} protein), and increasing concentrations were detected with rising processing temperatures in ESL samples (72.2 ± 36.6 mg.100 g^{-1} protein) and UHT treated samples (161.3 ± 30.2 mg.100 g^{-1} protein), respectively. German whipping cream samples labeled as "heat treated" showed quite similar furosine concentrations (45.2 ± 19.3 mg.100 g^{-1} protein) as pasteurized whipping cream samples, but lower values than ESL whipping cream samples. Referring to group II and when comparing mean values of "heat treated" coffee cream/milk and condensed milk samples

(368.5 ± 248.9 mg.100 g^{-1} protein) to UHT samples (216.2 ± 46.3 mg. 100 g^{-1} protein), "heat treated" samples showed higher furosine concentrations than UHT coffee cream/milk samples. However, "heat treated" coffee cream/milk and condensed milk samples showed more than twice the amount of UHT whipping cream samples of group I (161.3 ± 30.2 mg.100 g^{-1} protein). Interestingly, "heat treated" samples of group II revealed a wide standard deviation of furosine concentrations (368.5 ± 248.9 mg.100 g^{-1} protein). Lowest furosine amount was found in the single ESL sample analyzed with 70.9 mg.100 g^{-1} protein, and highest amounts were found in sterilized coffee cream/milk and condensed milk samples (490.6 ± 196.3 mg.100 g^{-1} protein). As seen in Fig. 3 and confirmed by statistical analysis, sterilized coffee cream/milk samples differed significantly from all other groups ($p < 0.001$), with exception of "heat treated" coffee cream/milk samples. These samples could be clearly distinguished from pasteurized, "heat treated," and ESL whipping cream samples ($p < 0.002$), respectively.

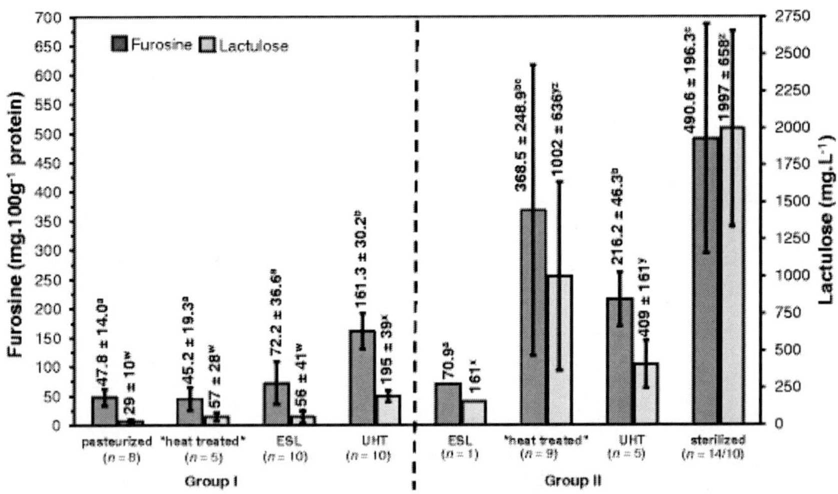

Fig. 3. Heat load indicators furosine and lactulose (mean value ± SD) of whipping cream samples (group I, left side), and coffee milk/cream and condensed milk (group II, right side) according to their heat load category. Lactose-free samples and abnormally high values were not included. Different *superscripts* indicate differing least square means ($p < 0.05$) for furosine (^{abc}) or lactulose (^{wxyz}), respectively. Data for pasteurized and ESL whipping cream samples of group I have been already reported recently (Boitz and Mayer 2015)

3.2 Enzymatic determination of lactulose

All samples were analyzed for their lactulose concentrations in their obtained fat-free phase, and the final lactulose concentration was calculated according to their corresponding fat content (for lactulose data of pasteurized and ESL samples; see Boitz and Mayer 2015). Since the enzymatic determination of lactulose includes a cleavage of disaccharides, the lactulose concentration could not be measured precisely in industrial sweetened products ($n = 5$), and consequently, these samples were not included into the calculation of mean values for lactulose in sterilized samples. Figure 3 shows increasing amounts of lactulose with rising heat load of dairy products. The lowest amounts were found in pasteurized whipping cream samples (29 ± 10 mg.L^{-1}), followed by "heat treated" (57 ± 28 mg.L^{-1}) and ESL labeled whipping cream samples (56 ± 41 mg.L^{-1}). The highest amounts within group I were found in UHT whipping cream samples (195 ± 39 mg.L^{-1}). A slightly lower amount was found in the single ESL coffee milk sample analyzed (group II, 161 mg.L^{-1}). UHT samples of group II showed more than twice the amount (409 ± 161 mg.L^{-1}) of the ESL coffee milk (161 mg.L^{-1}), and interestingly, the "heat treated" samples of group II had approximately a six times higher amount (1002 ± 636 mg.L^{-1}) compared to the ESL sample. As expected, in sterilized samples, even a 12 times higher amount (1997 ± 658 mg.L^{-1}) was found. As for furosine, also lactulose results showed a wide variation especially referring to "heat treated" and sterilized samples of group II (Fig. 3).

3.3 β-Lactoglobulin analysis of samples

Selected whipping cream, coffee cream/milk, and condensed milk samples of different heating categories (pasteurized, "heat treated", ESL, UHT, sterilized) were applied to native PAGE (12.5% T). Acid-soluble whey proteins were separated, and the results of both groups are shown in Fig. 4a and b, where raw milk and pasteurized milk are given as references. Protein patterns of raw milk (Fig. 4a) and pasteurized liquid milk (Fig. 4b) showed intense bands of typical whey protein fractions (e.g., α-La, bovine serum albumin (BSA) and β-Lg), whereas banding patterns of whipping cream samples were weaker (especially β-Lg), thereby showing distinct differences in intensity comparing the different heating categories (Fig. 4a). Obviously, intensity of acid-soluble whey proteins decreased corresponding to an increasing heat load of whipping cream samples analyzed. Thus, pasteurized and ESL whipping cream could be clearly differentiated from UHT whipping cream samples;

corresponding β-Lg concentrations measured by RP-HPLC are given on top of the electrophoretogram (Fig. 4a).

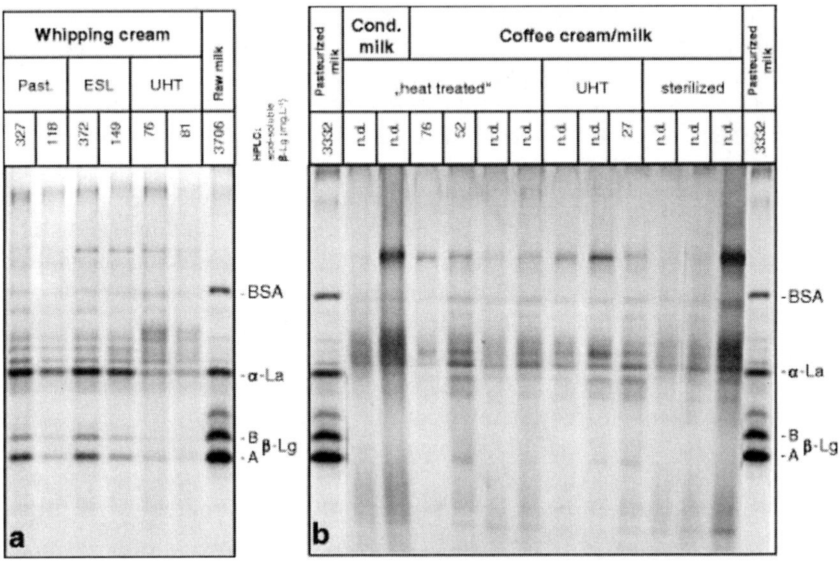

Fig. 4. Native polyacrylamide gel electrophoresis (12.5% T) of whey protein fractions soluble at pH 4.6 of different dairy products. **A** Whipping cream samples (group I, left side): pasteurized, ESL, and UHT samples; **B** selected coffee cream/milk and condensed (cond.) milk samples (group II, right side): "heat treated", UHT, and sterilized samples. Raw milk and pasteurized milk are given as references. BSA (bovine serum albumin), α-lactalbumin (α-La), and β-lactoglobulin (β-Lg) as well as β-Lg concentrations as determined with RP-HPLC in milligrams per liter (if concentration was under LOQ–*n.d.* not determined)

Referring to the acid whey proteins analyzed in group II samples, no β-Lg bands were visible due to excessive heat load, and consequently, no differences in the intensity of β-Lg banding patterns could be observed in these samples (Fig. 4b). Smearing lanes may be a result of concentrated products, although samples were diluted prior to analysis. However, as whipping cream samples could be differentiated by their distinct banding patterns (Fig. 4a), these and some selected samples of group II were analyzed for their accurate β-Lg contents with RP-HPLC. Results for whipping cream samples are shown in Fig. 5, whereas most of the selected samples of group II were below the limit of quantification (LOQ was 20 mg.L^{-1}).

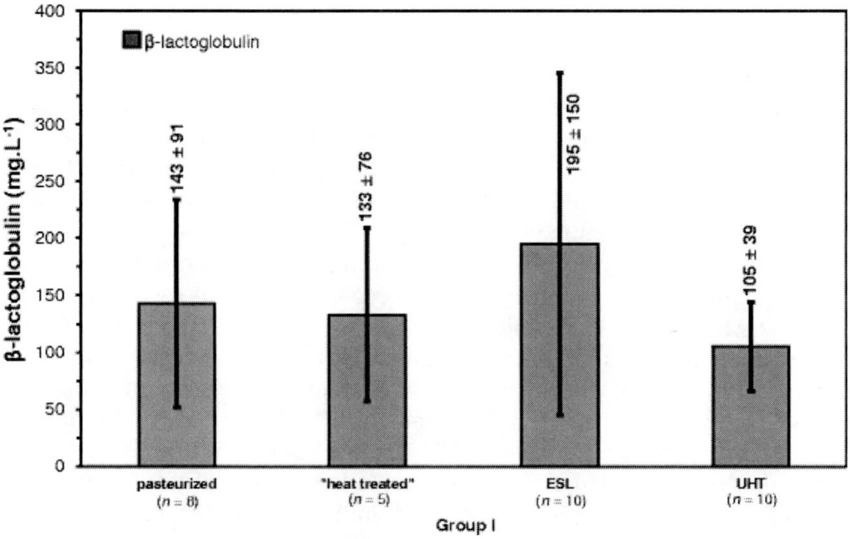

Fig. 5. β-Lg concentrations (mean values ± SD in mg.L^{-1}) of differently processed whipping cream samples purchased from Austrian and German market. Data for pasteurized and ESL samples have been already reported recently (Boitz and Mayer 2015)

4 DISCUSSION

In this study, different dairy products suffering a more intense heat treatment in comparison to liquid milk were analyzed according to several heat load indicators. Results obtained by a rapid screening of acid-soluble β-Lg concentrations of selected samples using native PAGE showed obvious differences in band intensity in group I samples (Fig. 4a), but no remaining bands of acid-soluble β-Lg were obtained for group II samples due to excessive heating (Fig. 4b). Thus, this electrophoretic method is not suitable for coffee cream/milk or condensed milk samples due to low β-Lg concentrations. RP-HPLC analysis revealed concentrations of acid-soluble β-Lg below the LOQ of 20 mg.L^{-1} for most of the selected samples of group II, whereas acid-soluble β-Lg was successfully determined in all whipping cream samples of group I (see Fig. 5). The obtained results for group I samples did not show any correlation between the intensity of applied heat load and the corresponding concentration of acid-soluble β-Lg. Concentrations of German whipping cream products labeled as "heat treated" ("wärmebehandelt"), which are heated twice, were comparable to obtained results of the pasteurized whipping cream samples (heated only once), indicating a mild heat treatment for both heating categories. Nevertheless, ESL whipping cream samples showed highest β-Lg amounts with a great variation of results, and the UHT samples (with the most intense heat treatment) showed lowest β-Lg amounts. As shown in Table 1, processing conditions of ESL whipping cream are situated somewhere between 105 and 135 °C, which is close to those practically applied for UHT whipping cream (135–140 °C). Both the wide range of temperatures applied for ESL processing and the fluent transition of ESL and UHT technology may explain the broad variation of results with the highest values for ESL samples (Fig. 5). However, these results highlight the statement as described previously in Boitz and Mayer (2015) that β-Lg is definitely not appropriate as a heat load indicator for whipping cream. As the concentrations were below the LOQ of RP-HPLC method for the majority of group II samples, β-Lg as type I indicator is not suitable for assessing the heat load of high-temperature treated dairy products such as coffee cream/milk and condensed milk, neither for UHT or "heat treated" nor for sterilized samples.

In contrast to this, type II indicators lactulose and furosine indicated a relationship between overall heat load applied and concentration analyzed. The more intense the head load (including heating time and processing temperature), the higher the lactulose and furosine concentrations. As confirmed by statistical tests, coffee cream/milk and condensed milk samples showed much higher lactulose and furosine values in comparison to whipping cream samples in general. Comparing the lactulose mean values of group I to group II, a significant difference exists ($p < 0.001$) between all heating categories of group I compared to the heating categories "heat treated," UHT, and "sterilized" of group II. This is in contrast to furosine, where UHT whipping cream samples could not be clearly differentiated from "heat treated" samples of group II (compare Fig. 3). Within group II, lactulose and furosine mean values were significantly different between sterilized samples and UHT samples ($p < 0.001$). Thus, the heat load of sterilized coffee cream/milk and condensed milk products is much higher than those of UHT processed samples, whereas the so-called heat-treated coffee cream/milk samples were in between, showing an extreme range of variation for both TTIs. Referring to lactulose and furosine of sterilized samples, variations might be a result of carbohydrate content ranging from 7.6 to 55.0 g.100 g^{-1} due to different evaporation intensities of condensed products in different dairies.

In combination with the acid-soluble β-Lg concentration, lactulose proved to be a useful tool to distinguish between UHT and in-container sterilized milk (European Commission 1992; ISO/IDF 1992, 1993). Thus, the EU and the IDF suggested a maximal lactulose concentration of 600 $mg.L^{-1}$, associated with a minimum β-Lg concentration of 50 $mg.L^{-1}$ for UHT milk (European Commission 1992; ISO/IDF 1992, 1993). In-container sterilized milk shows more than 600 $mg.L^{-1}$ lactulose and less than 50 $mg.L^{-1}$ of acid-soluble β-Lg (Corzo et al. 1996).

In this context, comparing the measured lactulose concentrations, "heat treated" as well as sterilized coffee cream/milk samples showed an amount above 600 $mg.L^{-1}$ and can be clearly distinguished from UHT treated samples showing lactulose concentrations below 600 $mg.L^{-1}$. However, this statement does not apply to the β-Lg concentrations, as UHT coffee cream/milk samples showed values below 50 $mg.L^{-1}$ due to the higher heat load compared to UHT milk. Consequently, the lactulose concentration may indicate a possible differentiation between UHT treated coffee cream/milk and "heat treated" or sterilized products, but β-Lg is not appropriate to distinguish between these categories.

Interestingly, when comparing the UHT treated samples of both groups, the furosine concentration is not significantly ($p > 0.05$) higher in group II samples (216.2 ± 46.3 g.100 g^{-1}protein) compared to group I samples (161.3 ± 30.2 g.100 g^{-1} protein), but more than twice of the lactulose concentration was determined in group II UHT treated samples (409 ± 161 $mg.L^{-1}$) compared to group I samples (195 ± 39 $mg.L^{-1}$). In this case, lactulose appears to be more meaningful than furosine for a successful differentiation of these heating categories. However, although the production technology of whipping cream and coffee cream is quite similar, both TTIs indicate a completely different heat load applied onto these dairy products. The higher concentrations of lactulose and furosine found in group II samples may indicate a much more intense heat load applied to skimmed milk prior to fat adjustment of these products during manufacturing.

Referring to group I, pasteurized and "heat treated" whipping cream samples showed no difference when comparing the furosine concentrations. Once again, lactulose appears to be more meaningful with higher concentrations in "heat treated samples" (57 ± 28 $mg.L^{-1}$) than in pasteurized whipping cream samples (29 ± 10 $mg.L^{-1}$). Higher lactulose concentrations may result from the second heating step during processing of "heat treated" whipping cream. Thus, "heat treated" whipping cream can be placed between pasteurized and ESL samples due to furosine concentrations similar to pasteurized samples and lactulose concentrations comparable to ESL whipping cream. In contrast to these findings, "heat treated" samples of group II showed higher lactulose and furosine mean values than UHT samples. In particular, lactulose concentrations are twice as high in "heat treated" samples compared to UHT samples (1002 ± 636 $mg.L^{-1}$ and 409 ± 161 $mg.L^{-1}$, respectively).

This may show that the heat treatment of coffee cream and condensed cream/milk labeled as "heat treated" (and where no regularly definition exists) can be understood and described as a heat treatment between UHT processing and sterilization. Considering furosine concentrations of "heat treated" samples, a carbohydrate content of 7.6–55.0 g.100 g^{-1} may be one reason for broad variation of results. However, the wide variations of results underline the absence of legal requirements considering time/temperature combinations for coffee cream/milk and condensed milk carrying the label "heat treated." Accordingly, this absence of thresholds or clear regulations (regarding "heat treated") offers a wide margin/tolerance to all dairy companies to choose completely different time-temperature combinations for these products, leading to different qualities of final products having all the same label "heat treated."

As reported recently (Mayer et al. 2010), β-Lg content of raw milk is about 4000 mg.L^{-1}. A minimum β-Lg content of 2600 mg.L^{-1} for pasteurized milk, of 2000 mg.L^{-1} for high pasteurized milk, and of 50 mg.L^{-1} for UHT milk had been proposed by the European Commission, respectively (European Commission 1992). In this context, UHT whipping cream samples fulfill these requirements, but considering group II, only one sample (76 mg.L^{-1}) showed a concentration above this threshold. These results fit to those reported in Feinberg et al. (2006) who found only 44 mg.L^{-1} in indirectly heated UHT milk samples and no β-Lg in sterilized milk, and to those reported in Morales et al. (2000) who found 63.4 mg.L^{-1} in indirectly heated UHT milk and below 10 mg.L^{-1} in sterilized milk samples, respectively.

The obtained results concerning furosine concentrations in condensed milk products are in accordance with Henle et al. (1995) who found concentrations between 340 and 880 mg.100 g^{-1} protein in eight analyzed condensed milk samples, whereas higher furosine concentration (730–836 mg.100 g^{-1} protein) were detected in two samples analyzed by Vallejo-Cordoba et al. (2004). The lower levels obtained in the present study may indicate a more gentle heat treatment of dairy products analyzed as well as advanced production technologies with less heat load applied onto the products. Elliott et al. (2005) reported a furosine concentration of 124 mg. 100 g^{-1} protein in indirectly heated UHT milk, Sakkas et al. (2014) reported 53.3 mg. 100 g^{-1} protein in UHT milk. Obtained results in group I and definitely higher furosine levels of group II reflect an overall higher heat treatment for whipping cream and coffee cream/milk compared to liquid milk.

Considering lactulose, results obtained in this study are different compared to those reported in Montilla et al. (2005), who found lactulose concentrations of 207–313 mg.L^{-1} in condensed milk ($n = 3$) and 623–705 mg.L^{-1} in sterilized milk ($n = 2$). Differences may result from their small sample sets and the gas chromatographic method used for determination. Moreover, the authors did not specify the heat treatment of the analyzed condensed milk samples. Morales et al. (2000) reported a lactulose concentration of 1121 mg.L^{-1} for sterilized milk, applying two heating steps with an overall processing time of approximately 17 min. The obtained lactulose levels in this study were higher, showing a broad standard deviation. Varying carbohydrate content of samples and different industrial conditions and dairy plants may explain these differences. UHT heated milk samples showed lactulose concentrations between 120 and 456 mg.L^{-1}. Obtained results for group I were within in this range, whereas group II samples showed much higher values due to higher heat load applied (time/temperature parameters).

5 CONCLUSION

In conclusion, it could be successfully shown that β-Lg as type I indicator is definitely not appropriate for the assessment of heat load of whipping cream and coffee cream/milk as well as condensed milk samples. In contrast, type II indicators lactulose and furosine were useful tools to distinguish between several heating categories. Lactulose appeared to be more meaningful and showed greater significant differences between several heating categories in comparison to furosine. Finally, UHT treated whipping cream samples could be clearly discriminated from pasteurized, ESL, and "heat treated" whipping cream with both TTIs. Sterilized and "heat treated" coffee cream/milk and condensed milk products could be clearly differentiated from UHT, as well as from ESL products considering lactulose concentrations. Moreover, it could be shown that the label "heat treated" on whipping cream can be interpreted as a heat treatment between pasteurization and ESL heat treatment. However, referring to coffee cream/milk and condensed milk, the heat load of "heat treated" products was higher than an ultra-high-temperature treatment (with high lactulose and furosine concentrations), and just below the excessive heat load of sterilized products.

ACKNOWLEDGMENTS

We are grateful to Veronika Zehetmayer and Christopher Akhgar for their technical assistance in enzymatic and electrophoretic analyses.

CONFLICT OF INTEREST

The authors declare that they have no competing interests.

REFERENCES

1. Adachi S, Patton S (1961) Presence and significance of lactulose in milk products: a review. J Dairy Sci 44:1375–1393
2. AMA-Marketing (2014), Austrian Marketing Agency. Pro-Kopf-Verbrauch. http://www.ama-marketing.at/home/groups/7/Pro_Kopf_Verbrauch_Mopro.pdf. Accessed Jan 2016
3. Amine A, Moscone D, Bernardo RA, Marconi E, Palleschi G (2000) A new enzymatic spectrophotometric assay for the determination of lactulose in milk. Anal Chim Acta 406:217–224
4. Boitz LI, Mayer HK (2015) Evaluation of furosine, lactulose and acid-soluble β-lactoglobulin as time temperature integrators for whipping cream samples at retail in Austria. Int Dairy J 50:24–31
5. Claeys WL, Van Loey AM, Hendrickx ME (2002) Intrinsic time temperature integrators for heat treatment of milk. Trends Food Sci Technol 13:293–311
6. Corzo N, Villamiel M, Martínez-Castro I (1996) Lactulose, monosaccharides and undenatured serum protein contents in commercial UHT creams and their usefulness for thermal treatment assessment. Food Chem 56:429–432
7. DIN, 1964 Standard 10284:1964–03, Cream butyrometer, Roeder type, calibrated. German Institute for Standardization, Berlin, Germany
8. Elliott AJ, Datta N, Amenu B, Deeth HC (2005) Heat-induced and other chemical changes in commercial UHT milks. J Dairy Res 72:442–446
9. European Commission (1992) Proposed draft Commission decision laying down the limits and the methods for distinguishing between the different types of heat treated drinking milk. Dairy Chemists' Group Doc. VI/5726/92.
10. Feinberg M, Dupont D, Efstathiou T, Louâpre V, Guyonnet JP (2006) Evaluation of tracers for the authentication of thermal treatments of milks. Food Chem 98:188–194
11. FMH (2011) Milk and dairy products. Austrian Food Codex, Chapter B32, BMG-75210/0010-II/B/13/2011. Vienna, Austria: Federal Ministry of Health
12. Gallmann P, Eberhard P, Sieber R (2001) Vor- und Nachteile der ESL-Milch (extended shelf life)-Milch Eidg Forschungsanstalt f Milchwirtschaft. FAM-Info 423:1–11
13. Henle T, Zehetner G, Klostermeyer H (1995) Fast and sensitive determination of furosine. Eur Food Res Technol 200:235–237
14. IDF/ISO, 1992 Influence of technology on the quality of heat treated milk and fluid milk products (B-Doc. 222), Int. Dairy Fed., Brussels, Belgium,
15. IDF/ISO, 1993 Influence of technology on the quality of heat treated milk and fluid milk products (B-Doc. 235), Int. Dairy Fed., Brussels, Belgium
16. IDF/ISO, 2001 Standard 20-1, Milk—determination of nitrogen content—part 1: Kjeldahl method. Int. Dairy Fed., Brussels, Belgium,
17. IDF/ISO, 2004 Standard 193, Milk and milk products—determination of furosine content—ion-pair reverse-phase high performance liquid chromatography method. Int. Dairy Fed., Brussels, Belgium,
18. IDF/ISO, 2005 Standard 178, Liquid milk—determination of acid-soluble β-lactoglobulin content—reverse-phased HPLC method. Int. Dairy Fed., Brussels, Belgium

19. IDF/ISO, 2007 Standard 147, Heat-treated milk—determination of lactulose content—method using high-performance liquid chromatography. Int. Dairy Fed., Brussels, Belgium
20. IDF/ISO, 2008 Standard 105, Milk—determination of fat content—Gerber butyrometers. Int. Dairy Fed., Brussels, Belgium
21. LVBM (2015) Was bedeutet "wärmebehandelt" bei Schlagsahne? Landesvereinigung der Bayerischen Milchwirtschaft, Presse-Service Milchland Bayern http:// wwwmilchland-bayern.de Accessed Jan 2016
22. Manzo C, Nicolai MA, Pizzano R (2015) Thermal markers arising from changes in the protein component of milk. Food Control 51:251–255
23. Mayer HK, Bonaparte C, Newart M, Kneifel W (2003) In Authentication of probiotic bifidobacteria using protein and DNA fingerprinting techniques, fermented milk: Proceedings of the IDF Seminar on Aroma and Texture of Fermented Milk held in Kolding, Denmark, June 2002, International Dairy Federation:62–84
24. Mayer HK, Raba B, Meier J, Schmid A (2010) RP-HPLC analysis of furosine and acid-soluble β-lactoglobulin to assess the heat load of extended shelf life milk samples in Austria. Dairy Sci Technol 90:413–428
25. Montilla A, Moreno FJ, Olano A (2005) A reliable gas capillary chromatographic determination of lactulose in dairy samples. Chromatographia 62:311–314
26. Morales FJ, Romero C, Jimenez-Perez S (2000) Characterization of industrial processed milk by analysis of heat-induced changes. Int J Food Sci Technol 35:193–200
27. Sakkas L, Moutafi A, Moschopoulou E, Moatsou G (2014) Assessment of heat treatment of various types of milk. Food Chem 159:293–301
28. Töpel A (2007) Chemie und Physik der Milch: Naturstoff, Rohstoff, Lebensmittel. B. Behr's Verlag GmbH & Co. KG, Hamburg, Germany
29. Vallejo-Cordoba B, Mazorra-Manzano MA, González-Córdova AF (2004) New capillary electrophoresis method for the determination of furosine in dairy products. J Agric Food Chem 52:5787–5790

CHAPTER 5

A Review of The Biology of Calcium Phosphate Sequestration with Special Reference to Milk

Samuel Lenton[1,2], Tommy Nylander[3], Susana C. M. Teixeira[1,2] Carl Holt[4]

[1].EPSAMKeele UniversityKeeleUK

[2].Institut Laue LangevinGrenoble cedex 9France

[3].Department of Physical ChemistryLund UniversityLundSweden

[4].Institute of Molecular, Cell and Systems BiologyUniversity of GlasgowGlasgowUK

ABSTRACT

In milk, a stable fluid is formed in which sequestered nanoclusters of calcium phosphate are substructures in casein micelles. As a result, calcium and phosphate concentrations in milk can be far in excess of their solubility. Variations of calcium, phosphate and casein concentrations in milks, both within and among species, are mainly due to the formation of the nanocluster complexes. Caseins evolved from tooth and bone proteins well before the evolution of lactation. It has therefore been suggested that the role of caseins in milk is an adaptation of an antecedent function in the control of some aspect of biomineralisation. There is new evidence that nanocluster-type complexes are also present in blood serum and, by implication, in many other closely related biofluids. Because such fluids are stable but nevertheless supersaturated with respect to the bone and tooth mineral hydroxyapatite, they allow soft and mineralised tissues to co-exist in the same organism with relative ease. An appreciable concentration of nanocluster complexes exists in fresh saliva. Such saliva may stabilise tooth mineral and help to repair demineralised lesions. In the extracellular matrix of bone, nanocluster complexes may be involved in directing the amorphous calcium phosphate to intrafibrillar spaces in collagen where they can mature into oriented apatite crystals. Thus, evidence is accumulating that calcium phosphate sequestration by phosphopeptides to form equilibrium complexes, first observed in milk, is more generally important in the control of physiological calcification.

Keywords: Amorphous calcium phosphate Blood Saliva Milk Urine Bone Tooth

This paper is part of the special issue dedicated to the 2nd International Symposium on Minerals & Dairy Products (MADP2014) held on 26–28th February in Auckland, New Zealand.

1 INTRODUCTION

Research into the mechanisms of biomineralisation and the chemistry of the mineral phases has been the subject of a number of excellent recent reviews (George and Veis 2008; Wang and Nancollas 2009; Dorozhkin 2011). Hard tissues include bone, cementum, dentine and enamel and contain a basic calcium phosphate mineral called apatite, or hydroxyapatite, named by geologists from the Greek word *apate* meaning deceive because of the variable crystal habit and composition of apatite rocks. Soft tissues and biofluids, by contrast, contain little or no mineral in their normal, physiological state. Soft tissues can become mineralised and hard tissues can become demineralised as a result of degenerative, dysfunctional or diseased conditions, but normally, tissues remain stably mineralised or unmineralised even though they are permeated by the same extracellular fluid. The problem of the easy co-existence of soft and mineralised tissues in the same organism can be stated easily in terms understandable to a chemist. For the hard tissue to stay mineralised, the permeating biofluid must be supersaturated with respect to the hard tissue mineral, and yet, for the soft tissues and biofluids to remain unmineralised, the permeating biofluid must be stable.

A solution to the problem of the stability of biofluids must have been found more than about 500 Ma ago when the first calcium phosphate mineralised tissues appear in the fossil record. These include the cone-shaped denticles or teeth of jawless fishes called conodonts. By about

50 Ma later, homologues of the modern hard tissue types were established. For example, the bone-like head plates of placoderms may have allowed sense organs, mounted on this stable platform, to build up a three-dimensional picture of their environment. Concomitantly, a paralogous group of secreted, calcium (phosphate)-binding phosphoproteins evolved, called SCPPs (Kawasaki et al. 2004; Kawasaki and Weiss 2003). Members of this group of proteins are involved in every aspect of biomineralisation. One of the earliest members of the group of SCPPs is osteopontin which has a very wide occurrence in species, tissues and biofluids (Mazzali et al. 2002).

Caseins are also members of the SCPP group, and it has been known for many years that the colloidal calcium phosphate of casein micelles allows stable milks to be formed, containing much higher calcium and phosphate concentrations than are permitted by the solubility of inorganic calcium phosphates at milk pH. Because of the ready availability of milk and the high concentrations of the caseins responsible for its stability, we have been able to gain a deeper understanding of the reasons for its stability than is available for any other biofluid. These insights have led to an improved understanding of the sources of stability in some other biofluids.

In this review, we summarise some of the evidence of calcium phosphate sequestration dairy chemistry in biofluids and soft and hard tissues. We believe that a dialogue between the dairy chemistry and biomineralisation communities would be to their mutual benefit.

2 BASIC SCIENCE OF AMORPHOUS CALCIUM PHOSPHATE SEQUESTRATION BY PHOSPHOPROTEINS

2.1 Ostwald rule of stages

At physiological pH, calcium phosphate does not precipitate directly from solution as the most thermodynamically stable phase, hydroxyapatite. Instead, it passes through a number of unstable or metastable stages from the most soluble to the least. Typically, an initial and highly unstable amorphous calcium phosphate (ACP) phase, ACP-1, is succeeded by a more stable and less soluble amorphous phase, ACP-2 (Christoffersen et al. 1990). It has been shown that both amorphous phases had the same ratio of calcium to phosphate of 1.35 which is significantly lower than that of a basic tricalcium phosphate (Ca/P = 1.5) but above that of an acidic dicalcium phosphate (Ca/P = 1.0). The chemical formula $Ca(PO_4)_{0.52}(HPO_4)_{0.22}$ is consistent with Ca/P = 1.35 (Christoffersen et al. 1990). In similar studies (Termine and Eanes 1972; Meyer and Eanes 1978; Wuthier et al. 1985; Holt et al. 1989), the Ca/P ratio of ACP has been reported to be as high as 1.5 and as low as 1.18, demonstrating that a range of more-or-less acidic ACP phases are possible. Subsequently, poorly crystalline, impure and non-stoichiometric phases are formed, the best characterised of which is octacalcium phosphate. The mineralised tissues of bone, cementum and dentine all contain poorly crystalline calcium phosphates of these sorts, but enamel comprises highly crystalline prisms of almost pure hydroxyapatite.

2.2 Effect of phosphopeptides

Phosphoproteins and phosphopeptides can affect every aspect of the precipitation process. For example the calcium-sensitive bovine caseins, even at very low concentrations, can extend the lag time before precipitation begins and slow down the rate of maturation of the amorphous state. Phosphopeptides that contain three or more phosphorylated residues (usually seryl) in a short sequence are said to have a phosphate centre (PC). If short, acidic, PC-containing peptides are present in stoichiometric excess over the amount of amorphous calcium phosphate that forms from a supersaturated solution, the solution remains apparently unchanged but contains calcium phosphate nanoclusters with an outer shell of the phosphopeptides. Further investigations have shown that the nanoclusters are equilibrium particles with a defined size and composition. In all other circumstances, a precipitate will develop, although the solution may remain in a metastable state for some time. Certain highly phosphorylated phosphoproteins such as the C-terminal regions of osteopontin or phosphophoryn appear to be able to nucleate hydroxyapatite (Boskey et al. 2010).

2.3 Ion activity product

The ion activity product has played an important part in the science of calcium phosphate sequestration. For a generic chemical formula of an amorphous calcium phosphate, containing both HPO_4^{2-} and PO_4^{3-} (Combes and Rey 2010)

$$Ca^{2+} + yHPO_4^{2-} + \frac{2(1-y)}{3}PO_4^{3-} \leftrightarrow \left[Ca(HPO_4)_y(PO_4)_{\frac{2(1-y)}{3}}\right]_{solid} \quad (1)$$

the corresponding ion activity product is

$$K_S = \{Ca^{2+}\}\{HPO_4^{2-}\}^y\{PO_4^{3-}\}^{2(1-y)/3} \tag{2}$$

where the quantities in curly brackets are activities and the value of y determines how acidic the amorphous phase is. For example, a basic tricalcium phosphate has $y = 0$, the ACP-1 and ACP-2 phases of Christoffersen et al. have $y = 0.22$ and the acidic dicalcium phosphates have $y = 1.0$. We have used the ion activity product to show that sequestered amorphous calcium phosphates can be acidic or basic salts depending on the sequestering phosphopeptide. For a solid phase in equilibrium with the continuous phase, the corresponding ion activity product is invariant with respect to variation in the composition and pH of the fluid. Conversely, if an invariant ion activity is found in a fluid, then the nature of the solid phase (for example, nanoclusters, amorphous or crystalline calcium phosphates) with which it is in equilibrium can be inferred. This is true even if the phase is present at a vanishingly small concentration. J.C. Elliot has written an advanced treatise on calcium phosphates which includes values for the solubility products of all the biologically relevant phases (Elliot 1994).

2.4 Application of a thermodynamic model to physiological fluids

The calculation of the ion equilibria in biological fluids on the assumption that all components are in solution as simple ions and ion complexes is well established (for references, see Holt et al. 2014). A tractable theory of the additional calculation of the partition of salts between sequestered and free forms in a physiological fluid containing phosphopeptides can be formulated on the assumption that there is a further equilibrium reaction in which phosphopeptides bind strongly to amorphous calcium phosphate to form a stoichiometric complex. Experimental evidence confirms that the sequestered complexes have a constant composition and equilibrium size and the solution exhibits an invariant ion activity product (Little and Holt 2004; Holt et al. 2009). A further assumption is that all the phosphate centres have the same capacity to sequester amorphous calcium phosphate and it may sometimes be necessary to know the ion binding properties of free phosphate centres.

The simple ion equilibria calculations require the equilibrium constants for all the ion complexes likely to be present (Holt et al. 1981; Mekmene et al. 2009, 2010; Lyster 1981; Rice et al. 2010; Holt 2004). The additional parameters needed for the partition calculation are listed in Table 1. Each can be determined in independent experiments on well-defined solutions in which there is a known PC-containing phosphopeptide. Because of this and the fact that no fitting process with adjustable parameters is used to describe the unknown fluid, we can describe the calculation as ab initio.

Table 1. Parameters needed for the ab initio calculation of the partition of salts in a fluid

Parameter	Description
C_1, C_2, C_n	Concentrations of the *n* principal salts
R_{C1}, R_{C1}, R_{Cn}	Molar ratio of each salt constituent to PC in the sequestered complex
[PC]	Total concentration of phosphate centres
y	Acidity parameter defined in Eq. (1)
K_S	Solubility constant or invariant ion activity product
$\nu(C_1, C_2, C_n)$	Isotherm describing the simultaneous binding of the *n* principal salts to a free PC

Application of the model to simple nanocluster solutions and to milk has been described (Little and Holt 2004; Holt 2004; Holt et al. 2009).

2.5 Stability diagram of biofluids

A special case of the calculation is for the condition that the concentration of phosphate centres is the minimum needed to sequester the quantity of calcium phosphate formed at equilibrium. The variation of the minimum concentration with pH has been used to calculate the stability diagram for a variety of biofluids (Holt et al. 2014). An example of the stability diagram for a bovine milk-like fluid is shown in Fig. 1.

The stability diagram defines areas of stability and instability. Below a certain pH, called the critical pH, no phosphopeptides are needed to make the solution stable. As pH increases, the minimum concentration of phosphate centres increases in a curvilinear manner up to a plateau value somewhere above pH 8. Close to the boundary line, on the higher pH side, is an ill-defined area of metastability where precipitation may be delayed for some time.

If the solution is a biofluid and if the critical pH is below the physiological pH, then it will be liable to precipitate calcium phosphate. Nevertheless, if it contains a sufficient concentration of a sequestering phosphopeptide, it can be perfectly stable, even at the physiological pH. At equilibrium it will contain nanoclusters of sequestered calcium phosphate. For the example calculation shown in Fig. 1, the physiological pH is about 6.7 where a considerable concentration of phosphate centres is needed to ensure stability.

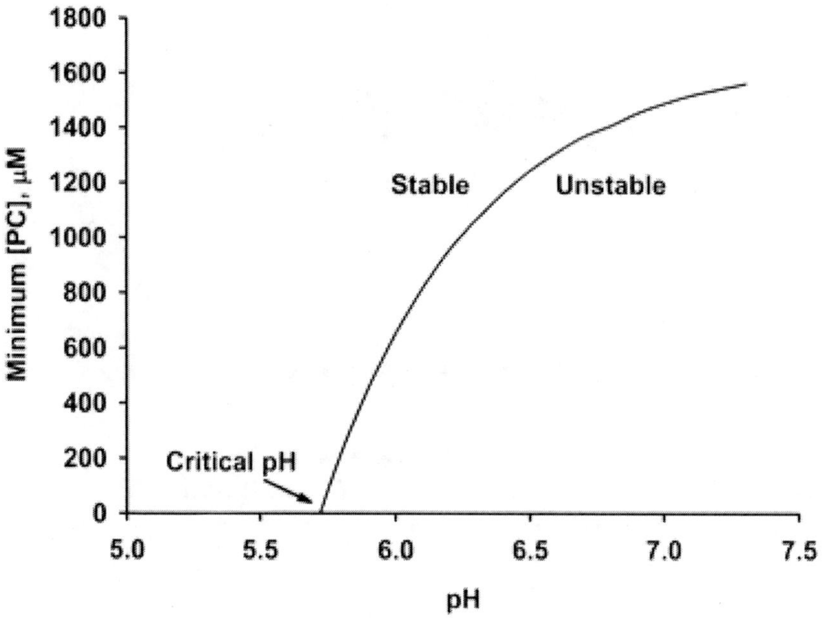

Fig. 1. Stability diagram for a bovine milk-like fluid

3 APPLICATION TO MILK

In milk, the main source of phosphate centres is the calcium-sensitive caseins, but all caseins contain adhesive sequences to promote mutual association. As a result, all milks contain colloidal particles called casein micelles instead of individual nanoclusters. Nevertheless, the casein micelle can be regarded as an assembly of nanoclusters. For example, a bovine milk casein micelle with a radius of 100 nm is an assembly of about 800 nanoclusters and each nanocluster is formed from peptides containing about 50 phosphate centres (Holt et al. 2003). The average spacing between the nanoclusters is about 18 nm which has been mistaken, in the past, for the diameter of protein submicelles (McMahon and Oommen 2012).

Figure 2 shows that cow's milk serum is highly supersaturated with respect to hydroxyapatite and there is no invariant ion activity product for this phase. There is a nearly invariant ion activity product for ACP-2, the more stable phase of amorphous calcium phosphate, but the serum is supersaturated with respect to ACP-2. Milk serum shows virtually the same invariant ion activity product as a solution of casein nanoclusters (Little and Holt 2004).

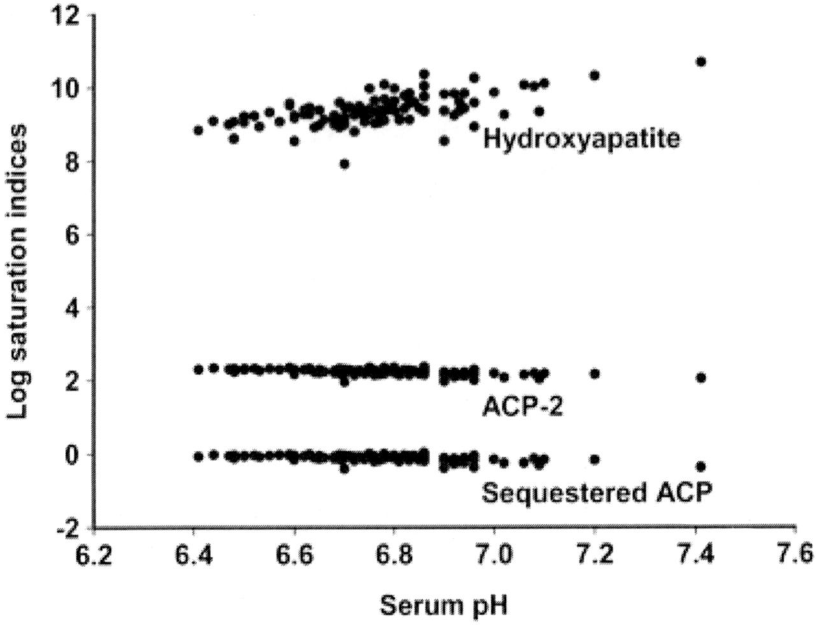

Fig. 2. Saturation indices of bovine milk sera for hydroxyapatite (Holt 1982), ACP-2 (Christoffersen et al. 1990) and ACP sequestered by casein phosphopeptides (Little and Holt 2004). Calculations used the data of individual milk samples in early, middle and late lactation (White and Davies 1958)

When the quantitative predictions of the ab initio calculations of the partition of milk salts are compared to experiment for individual cow milk samples with pH values in the physiological range, the agreement between theory and experiment is very satisfactory (Holt 2004). The extension of the theory to consider some of the non-physiological conditions that are encountered during routine milk processing operations has also been attempted. These include the effects of the adjustment of pH, addition or removal of salts, addition of caseins, concentration and dilution on the partition of salts and the dissociation of individual caseins from the casein micelle. Presently beyond the capabilities of the model are the partition of micronutrients and radionuclides, the effect of temperature and pressure and the consequences of all these manipulations on the aggregation or gelation of milk proteins.

4 APPLICATION TO OTHER PHYSIOLOGICAL FLUIDS

Nearly all physiological fluids are supersaturated with respect to hydroxyapatite and some other crystalline calcium phosphate phases. As a result, artificial biofluids that closely resemble their natural counterparts in salt composition and pH are unstable and will readily precipitate calcium phosphate. They, and natural biofluids, therefore require some means to make them stable. Caseins are largely confined to expression in the mammary gland, but other SCPPs are expressed in physiological fluids and hard and soft tissues. Osteopontin (OPN), in particular, is almost universally expressed in chordate tissues and biofluids (Mazzali et al. 2002). In the extracellular matrix of mammalian bone, for example, it is the major non-collagenous phosphoprotein. It is an unfolded flexible phosphoprotein, like other members of the group of SCPPs (Fedarko et al. 2004; Fisher et al. 2001; Holt et al. 2009). It contains phosphate centre-type sequences in its N-terminal half. We have previously shown that a naturally occurring mixture of OPN phosphopeptides and a purified fraction of the N-terminal phosphopeptides ending at or around residues 145 to 153 of the mature protein (Christensen and Sorensen 2014) can sequester nanoclusters of amorphous calcium phosphate (Holt et al. 2009, 2014).

4.1 Stability diagrams for stimulated saliva, blood serum and urine

The stability diagrams for salt solutions approximating the typical compositions of three very different biofluids, urine, saliva and blood serum, have been calculated, on the assumption that OPN phosphopeptides are responsible for sequestration, and compared to experiment (Holt et al. 2014). For typical urine from non-stone formers, the critical pH is well above the physiological pH so that normally the fluid is not expected to be unstable with respect to calcium phosphate precipitation or to contain calcium phosphate nanoclusters. However, when infected by microorganisms that secrete urease, the pH may rise to above 10 and urolithiasis involving calcium phosphate stones can result.

The stability diagram of typical human saliva is difficult to calculate because the composition and pH are very variable. It is formed by mixing together different secretions from three major and many other minor glands. This biofluid is not very well buffered, so equilibration with air causes the pH to rise following the loss of CO_2. The results for stimulated saliva, however, are more reproducible, and for this fluid, the physiological pH is above the critical pH. According to experiment and calculation, stimulated saliva contains sequestered calcium phosphate (Grøn 1973; Holt et al. 2014) although the nature and amount of the sequestering phosphoprotein or peptide are presently unknown.

The stability diagram has also been calculated for averaged blood serum (Holt et al. 2014) from healthy individuals (Walser 1961, 1962). The physiological pH is close to or a little lower than the critical pH, and because of the tight homeostasis of blood pH, we might not expect it to contain sequestered amorphous calcium phosphate. The point is particularly important for human health because vascular calcification is common in the final stages of kidney failure and is associated with fatal heart attacks. We suggested (Holt et al. 2014) that a background level of sequestering phosphopeptide was desirable to cope with the inevitable fluctuations in composition and ensure that precipitation of calcium phosphate arising from a fluctuation would be fully reversed when homeostasis is restored. The major source of phosphate centres is also uncertain with the principal candidates being OPN and fetuin A (Heiss et al. 2010; Jahnen-Dechent et al. 2008, 2011; Wald et al. 2011a, b), with contributions from other phosphoproteins (Price et al. 2003).

4.2 Ion activity products in stimulated saliva and blood serum

Calculated saturation indices for hydroxyapatite, ACP-2 and nanoclusters sequestered by OPN peptides in fresh saliva and blood serum display a striking similarity to the calculations for milk serum (Fig. 2). The invariant pK_S values for fresh saliva, blood serum and a pure nanocluster solution prepared from OPN peptides are 9.26 ± 0.35, 9.27 ± 0.13 and 9.12 ± 0.05, respectively. These ion activity products are statistically indistinguishable. It is hard to escape the conclusion that milk, blood and saliva all contain similar forms of sequestered amorphous calcium phosphate, albeit at very different concentrations.

5 APPLICATION TO HARD TISSUE MINERALISATION

The discussion so far has been concerned with the role of calcium phosphate nanoclusters in forming stable biofluids, but these structures may be de-stabilised or merely transient intermediates in the mineralisation of hard tissues. A solution of ACP nanoclusters can retain its thermodynamic stability if three conditions are satisfied (Holt et al. 2014). These are that (a) no further ACP can form, (b) the sequestered ACP does not mature into a more crystalline phase and (c) the solution is not in contact with a crystalline calcium phosphate. Condition c is violated in the mouth when saliva comes into contact with the hydroxyapatite of tooth enamel. In contact with teeth, saliva brings about remineralisation of demineralised tooth lesions. In other words, while the nanocluster solution in isolation is thermodynamically stable, when brought into contact with crystals of calcium phosphate, the crystals can grow at the expense of the nanoclusters. In effect, the solution becomes unstable and acts as a reservoir of calcium phosphate to support the growth of the more stable phase. Condition b is the most intriguing one because of the apparently conflicting literature on the roles of phosphoproteins in the inhibition and nucleation of calcium phosphate precipitation. According to the theory of calcium phosphate sequestration, the equilibrium size of the nanocluster increases as the affinity of the peptide for the surface increases. A limited amount of experimental data supports this proposition by showing that sequestration to form a stable complex requires a minimum of three phosphorylated residues in a phosphate centre and that additional, flanking, Asp residues will increase the nanocluster radius (Aoki et al. 1992; Clegg and Holt 2009). The logic is that stronger peptide binding allows bigger clusters to be sequestered, but as the cluster size increases, the probability rises of a more stable phase nucleating and growing within the cluster. This is because there are more molecules that could change in a larger cluster but also because the influence of the surface on the interior of the cluster might only be effective over a limited depth below the surface. Molecules beyond the influence of the surface can therefore behave as if they are in an unstable bulk phase. The minimum size is zero when the free energy of sequestration is zero. The largest known casein phosphate centres, and hence, possibly, the phosphate centres with the highest binding affinities, contain up to eight actual or potential phosphoseryl residues (Holt et al. 2009), but some bone and dental peptides can contain much longer phosphorylated sequences. In this context, it may be relevant that certain highly phosphorylated phosphopeptides such as phosphophoryn appear to be able to nucleate the growth of a solid phase and direct it to specific locations (Boskey et al. 2010; Gericke et al. 2010). The ability of an acidic synthetic polymer, poly-Asp, to promote intrafibrillar mineralisation of collagen has been described (Olszta et al. 2003, 2007). The mechanism involves an amorphous precursor phase which gathers near the gap zone of collagen fibres and then penetrates the fibres to form a parallel array of oriented crystals of hydroxyapatite (Nudelman et al. 2010). It has been suggested that the precursor phase is a liquid-like pre-nucleation cluster for which there is some experimental evidence in calcium carbonate nucleation studies (Gebauer and Coelfen 2011; Gebauer et al. 2008).

However, we consider it more likely that, given the abundance of phosphoproteins, including osteopontin, in the extracellular matrix, the precursor phase is a form of amorphous calcium phosphate stabilised by adsorbed phosphopeptides. Indeed, it has been shown recently that the intrafibrillar mineralisation of collagen can also be accomplished by the nanocluster-forming mixture of OPN peptides (Rodriguez et al. 2014) as well as some other phosphopeptides (Nudelman et al. 2012; Sfeir et al. 2014).

6 CONCLUSIONS AND APPLICATIONS

The formation of thermodynamically stable nanoclusters of sequestered amorphous calcium phosphate is well established in cow's milk. The casein complexes are probably responsible for most of the interspecific variation in calcium and phosphate concentrations in all milks.

The theory of calcium phosphate sequestration is developing to the point where it can make quantitative predictions of the concentrations of ions and nanocluster complexes in quite complex media including physiological fluids other than milk and proteins other than casein.

It is established that the protein OPN, which occurs in a wide distribution of species, tissues and biofluids, can also sequester amorphous calcium phosphate and form somewhat similar nanocluster complexes. The biofluids blood plasma and fresh saliva show an invariant ion activity product for nanoclusters sequestered by OPN phosphopeptides.

Amorphous calcium phosphate complexes with phosphoproteins may be formed transiently in the mineralisation of hard tissues.

There are many possible applications of the science described in this paper. These include, but are not limited to, biomimetic materials for the repair of bone fractures and demineralised teeth, artificial biofluids with superior shelf life that can be used in the irrigation of wounds during and after surgery, the irrigation of joints to help remove erosive crystals from synovial spaces, artificial serum for rapid rehydration of dehydrated patients and a superior kidney dialysis fluid that reproduces more closely the composition of blood serum. Last but not least is a growth medium in tissue engineering applications that can support the growth of mineralised tissues.

ACKNOWLEDGMENTS

TN and ST received funding to support a PhD studentship (held by SL) from the Research Institute for the Environment, Physical Sciences and Applied Mathematics, Keele University, Lund University and the Institut Laue Langevin. CH received an honorarium for delivering a keynote lecture at the Minerals and Dairy Products Symposium, Auckland, 26–27 February 2014.

CONFLICT OF INTEREST

Carl Holt is a named inventor and potential beneficiary of a patent entitled "phosphopeptides and use of the same (2010) WO 2010007441". Samuel Lenton, Tommy Nylander and Susana Teixeira have no conflict of interest.

Compliance with ethical guidelines

This article does not contain any studies with human or animal subjects performed by any of the authors.

REFERENCES

1. Aoki T, Umeda T, Kako Y (1992) The least number of phosphate groups for cross-linking of casein by colloidal calcium phosphate. J Dairy Sci 75(4):971–975
2. Boskey AL, Chiang P, Fermanis A, Brown J, Taleb H, David V, Rowe PSN (2010) MEPE's diverse effects on mineralization. Calcif Tissue Int 86(1):42–46.
3. Christensen B, Sorensen ES (2014) Osteopontin is highly susceptible to cleavage in bovine milk and the proteolytic fragments bind the alpha (v) beta (3)-integrin receptor. J Dairy Sci 97(1):136–146.
4. Christoffersen MR, Christoffersen J, Kibalczyc W (1990) Apparent solubilities of 2 amorphous calcium phosphates and of octacalcium phosphate in the temperature-range 30-42degrees-C. J Crystal Growth 106(2–3):349–354.
5. Clegg RA, Holt C (2009) An E. coli over-expression system for multiply-phosphorylated proteins and its use in a study of calcium phosphate sequestration by novel recombinant phosphopeptides. Protein Expr Purif 67(1):23–34.
6. Combes C, Rey C (2010) Amorphous calcium phosphates: synthesis, properties and uses in biomaterials. Acta Biomaterialia 6 (9):3362–3378.
7. Dorozhkin SV (2011) Calcium orthophosphates: occurrence, properties, biomineralization, pathological calcification and biomimetic applications. Biomatter 1(2):121–164.

8. Elliot JC (1994) Structure and chemistry of the apatites and other calcium orthophosphates. Studies in inorganic chemistry. vol 18. Elsevier, Amsterdam, The Netherlands
9. Fedarko NS, Jain A, Karadag A, Fisher LW (2004) Three small integrin-binding ligand N-linked glycoproteins (SIBLINGs) bind and activate specific matrix metalloproteinases. FASEB J. 18 (2):734 – +.
10. Fisher LW, Torchia DA, Fohr B, Young MF, Fedarko NS (2001) Flexible structures of SIBLING proteins, bone sialoprotein, and osteopontin. Biochem Biophys Res Commun 280(2):460–465
11. Gebauer D, Coelfen H (2011) Prenucleation clusters and non-classical nucleation. Nano Today 6(6):564–584.
12. Gebauer D, Völkel A, Cölfen H (2008) Stable prenucleation calcium carbonate clusters. Science 322(5909):1819–1822.
13. George A, Veis A (2008) Phosphorylated proteins and control over apatite nucleation, crystal growth, and inhibition. Chem Rev 108(11):4670–4693.
14. Gericke A, Qin C, Sun Y, Redfern R, Redfern D, Fujimoto Y, Taleb H, Butler WT, Boskey AL (2010) Different forms of DMP1 play distinct roles in mineralization. J Dent Res 89(4):355–359.
15. Grøn P (1973) Saturation of human saliva with calcium phosphates. Arch Oral Biol 18(11):1385–1392.
16. Heiss A, Pipich V, Jahnen-Dechent W, Schwahn D (2010) Fetuin-A is a mineral carrier protein: small angle neutron scattering provides new insight on fetuin-A controlled calcification inhibition. Biophys J 99(12):3986–3995.
17. Holt C (1982) Inorganic constituents of milk 3. The colloidal calcium phosphate of cows' milk. J Dairy Res 49(1):29–38
18. Holt C (2004) An equilibrium thermodynamic model of the sequestration of calcium phosphate by casein micelles and its application to the calculation of the partition of salts in milk. Eur Biophys J with Biophys Lett 33(5):421–434.
19. Holt C, Dalgleish DG, Jenness R (1981) Inorganic constituents of milk. 2. Calculation of the ion equilibria in milk diffusate and comparison with experiment. Anal Biochem 113(1):154–163. doi:10.1016/0003-2697(81)90059-2
20. Holt C, Vankemenade M, Nelson LS, Hukins DWL, Bailey RT, Harries JE, Hasnain SS, Debruyn PL (1989) Amorphous calcium phosphates prepared at ph 6.5 and 6.0. Mater Res Bull Am Meteorol Soc 24(1):55–62.
21. Holt C, de Kruif CG, Tuinier R, Timmins PA (2003) Substructure of bovine casein micelles by small-angle X-ray and neutron scattering. Coll Surf-Physicocheml Eng A Sp 213(2–3):275–284
22. Holt C, Sorensen ES, Clegg RA (2009) Role of calcium phosphate nanoclusters in the control of calcification. FEBS J 276(8):2308–2323.
23. Holt C, Lenton S, Nylander T, Teixeira SCM (2014) Mineralisation of soft and hard tissues and the stability of biofluids. J. Struct. Biol. 185:383–396.
24. Jahnen-Dechent W, Schafer C, Ketteler M, McKee MD (2008) Mineral chaperones: a role for fetuin-A and osteopontin in the inhibition and regression of pathologic calcification. J Mol Med-Jmm 86(4):379–389.
25. Jahnen-Dechent W, Heiss A, Schaefer C, Ketteler M (2011) Fetuin-A regulation of calcified matrix metabolism. Circulation Res 108(12):1494–1509.
26. Kawasaki K, Weiss KM (2003) Mineralized tissue and vertebrate evolution: the secretory calcium-binding phosphoprotein gene cluster. Proc Natl Acad Sci U S A 100(7):4060–4065.
27. Kawasaki K, Suzuki T, Weiss KM (2004) Genetic basis for the evolution of vertebrate mineralized tissue. Proc Natl Acad Sci U S A 101(31):11356–11361.
28. Little EM, Holt C (2004) An equilibrium thermodynamic model of the sequestration of calcium phosphate by casein phosphopeptides. Eur Biophys J with Biophys Lett 33(5):435–447.

29. Lyster RLJ (1981) Calculation by computer of individual concentrations in a simulated milk salt solution. 2. An extension to the previous model. J Dairy Res 48(1):85–89
30. Mazzali M, Kipari T, Ophascharoensuk V, Wesson JA, Johnson R, Hughes J (2002) Osteopontin—a molecule for all seasons. QJM -an Int J Med 95(1):3–13
31. McMahon DJ, Oommen BS (2012) Casein micelle structure, functions and interactions. In: Fox PF, McSweeney PLH (eds) Advanced dairy chemistry, vol 1A Proteins: basic aspects. 4th edn. Springer, pp 185–210
32. Mekmene O, Le Graet Y, Gaucheron F (2009) A model for predicting salt equilibria in milk and mineral-enriched milks. Food Chem 116(1):233–239.
33. Mekmene O, Le Graet Y, Gaucheron F (2010) Theoretical model for calculating ionic equilibria in milk as a function of pH: comparison to experiment. J Agric Food Chem 58(7):4440–4447.
34. Meyer JL, Eanes ED (1978) Thermodynamic analysis of secondary transition in spontaneous precipitation of calcium-phosphate. Calcif Tissue Res 25(3):209–216.
35. Nudelman F, Pieterse K, George A, Bomans PHH, Friedrich H, Brylka LJ, Hilbers PAJ, de With G, Sommerdijk NAJM (2010) The role of collagen in bone apatite formation in the presence of hydroxyapatite nucleation inhibitors. Nature Mater 9(12):1004–1009.
36. Nudelman F, Bomans PHH, George A, de With G, Sommerdijk NAJM (2012) The role of the amorphous phase on the biomimetic mineralization of collagen. Faraday Discuss 159:357–370.
37. Olszta MJ, Douglas EP, Gower LB (2003) Scanning electron microscopic analysis of the mineralization of type I collagen via a polymer-induced liquid-precursor (PILP) process. Calcif Tissue Int 72(5):583–591.
38. Olszta MJ, Cheng X, Jee SS, Kumar R, Kim Y-Y, Kaufman MJ, Douglas EP, Gower LB (2007) Bone structure and formation: a new perspective. Mater Sci Eng Rep 58(3–5):77–116.
39. Price PA, Nguyen TMT, Williamson MK (2003) Biochemical characterization of the serum fetuin-mineral complex. J Biol Chem 278(24):22153–22160.
40. Rice G, Barber A, O'Connor A, Stevens G, Kentish S (2010) A theoretical and experimental analysis of calcium speciation and precipitation in dairy ultrafiltration permeate. Int Dairy J 20(10):694–706.
41. Rodriguez DE, Thula T, Toro EJ, Yeh Y-W, Holt C, Holliday LS, Gower LB (2014) Multifunctional role of osteopontin in directing intrafibrillar mineralization of collagen and activation of osteoclasts. Acta Biomaterialia 10(1):494–507
42. Sfeir C, Fang P-A, Jayaraman T, Raman A, Xiaoyuan Z, Beniash E (2014) Synthesis of bone-like nanocomposites using multiphosphorylated peptides. Acta Biomaterialia 10(5):2241–2249.
43. Termine JD, Eanes ED (1972) Comparative chemistry of amorphous and apatitic calcium phosphate preparations. Calcif. Tissue Res. 10 (3):171-&.
44. Wald J, Wiese S, Eckert T, Jahnen-Dechent W, Heiss A, Richtering W (2011a) Fetuin-A mediated formation and ripening of colloidal calciprotein particles. Eur Biophys J with Biophys Lett 40:64–65
45. Wald J, Wiese S, Eckert T, Jahnen-Dechent W, Richtering W, Heiss A (2011b) Formation and stability kinetics of calcium phosphate-fetuin-A colloidal particles probed by time-resolved dynamic light scattering. Soft Matter 7(6):2869–2874
46. Walser M (1961) Ion association 6. Interactions between calcium, magnesium, inorganic phosphate, citrate and protein in normal human plasma. J. Clin. Investig. 40 (4):723-&.
47. Walser M (1962) Separate effects of hyperparathyroidism, hypercalcaemia of malignancy, renal failure, and acidosis on state of calcium, phosphate and other ions in plasma. J. Clin. Investig. 41 (7):1454
48. Wang L, Nancollas GH (2009) Pathways to biomineralization and biodemineralization of calcium phosphates: the thermodynamic and kinetic controls. Dalton Trans 15:2665–2672.

49. White JCD, Davies DT (1958) The relation between the chemical composition of milk and the stability of the caseinate complex. 1 General introduction, description of samples, methods and chemical composition of samples. J Dairy Res 25(2):236–255

50. Wuthier RE, Rice GS, Wallace JEB, Weaver RL, Legeros RZ, Eanes ED (1985) In vitro precipitation of calcium-phosphate under intracellular conditions—formation of brushite from an amorphous precursor in the absence of ATP. Calcif Tissue Int 37(4):401–410.

CHAPTER 6

Characterization of the fungal microflora in raw milk and specialty cheeses of the province of Quebec

Karine Lavoie[1,2] Marilyne Touchette[1,2] Daniel St-Gelais[1,3] Steve Labrie[1,2]

[1].*Nutraceuticals and Functional Foods Institute (INAF) and STELA Dairy Research CentreQuébecCanada*

[2].*Department of Food Sciences and NutritionUniversité LavalQuébecCanada*

[3].*Food Research and Development CentreAgriculture and Agri-Food CanadaSaint-HyacintheCanada*

ABSTRACT

The cheese microbial ecosystem is complex, and the presence of non-starter adventitious microorganisms in milk may have an influence on the organoleptic characteristics of cheese. The aim of this study was to analyze the composition and diversity of the fungal flora of raw milk destined for cheesemaking from 19 dairy farms in Quebec and to monitor their evolution throughout ripening. Six hundred ten yeast and mold isolates were collected from raw milk and raw milk cheeses over a 9-month period. Based on the sequences of the rDNA ITS1-5.8S-ITS2 region, 67% of the raw milk isolates were yeasts, which were assigned to 37 species across 11 genera, while 33% were molds, which were assigned to 33 species across 25 genera. A semi-quantitative analysis of the yeasts and molds in the raw milk from four farms was performed over a 5-month period. The composition and diversity of the fungal microflora were totally different for each farm, each of which had a unique species profile. To determine whether adventitious yeast strains from the milk could develop in raw milk cheese, a multilocus-sequence-typing (MLST) analysis was performed on 13 *Issatchenkia orientalis* (syn. *Pichia kudriavzevii*, anamorph: *Candida krusei*) isolates. The same MLST genotypes were identified for strains independently isolated from raw milk and raw milk cheese from a farm processing its own milk. This study contributes to the understanding of the natural fungal microflora of raw milk and suggests that non-starter yeasts and molds can transfer from raw milk to raw milk cheese and may influence cheese ripening.

Keywords: Artisanal cheese Fungi Yeast Molds *Candida krusei Issatchenkia orientalis Pichia kudriavzevii* Multilocus sequence typing MLST Non-starter yeasts and molds

1 INTRODUCTION

Cheeses harbor a complex microbial ecosystem that evolves during ripening. The coexistence of microorganisms with different activities determines the organoleptic characteristics of cheeses (Arfi et al. 2005). This ecosystem is essential since aseptic cheeses lack the aromatic compounds produced during ripening (Perry and McGillivray 1964). Controlling the ripening process requires in-depth knowledge of the succession of microbial communities and their activities in cheese. In addition to the starter culture, uninoculated (adventitious) microorganisms can participate in the cheese ripening (Swearingen et al. 2001). Non-starter lactic acid bacteria (NSLAB) in cheddar cheese are a good example of this (Beresford and Williams 2004). The characterization of NSLAB, mainly lactobacilli, has revealed that they make a positive contribution to the ripening of a number of cheeses (Law 2010; Swearingen et al. 2001), which has led to the development of commercial adjunct cultures to accelerate cheese ripening (Settanni and Moschetti 2010).

Studies have also shown that adventitious yeasts and molds can contaminate milk and cheese and contribute to the ripening of specialty cheeses (Mounier et al. 2006; Petersen et al. 2002; Vacheyrou et al. 2011). Potential sources of contamination include ambient air, farm workers, cheesemakers, brine, and cheesemaking equipment (Borelli et al. 2006; Mounier et al. 2006; Pereira-Dias et al. 2000; Seiler and Busse 1990; Simov et al. 2006; Vacheyrou et al. 2011; Viljoen et al. 2003). Over 40 non-starter yeasts and molds (NSYM) species have been isolated from milk, brines and cheeses, the most common being *Candida catenulata*, *Candida intermedia*, *Candida lusitaniae*, *Candida parapsilosis*, *Candida rugosa*, *Candida zeylanoides*, *Cryptococcus curvatus*, *Issatchenkia orientalis* (syn. *Pichia kudriavzevii*, anamorph: *Candida krusei*), *Kluyveromyces marxianus*, *Pichia fermentans*, *Pichia guillermondii*, *Saccharomyces cerevisiae*, *Torulaspora delbrueckii* (anamorph: *Candida colliculosa*), and *Yarrowia lipolytica* (Baroiller and Schmidt 1990; Borelli et al. 2006; Callon et al. 2007; El-Sharoud et al. 2009; Fadda et al. 2004; Jacques and Casaregola 2008; Lopandic et al. 2006; Roostita and Fleet 1995; Rohm et al. 1992; Seiler and Busse 1990; Seker 2010; Sohier et al. 2009). The main non-starter mold species in dairy products are *Cladosporium cladosporioides*, *Epicoccum nigrum*, *Mucor racemosus*, *Penicillium brevicompactum*, *Penicillium chrysogenum*, and *Penicillium commune* (Hocking and Faedo 1992; Montagna et al. 2004; Samson et al. 2004). While some of these species are known to be undesirable (e.g., *Cladosporium cladosporioides*, *M. racemosus*, *Penicillium commune*), the impact of most NSYM in cheeses is unknown. It has, however, been suggested that they may make important contributions to the flavor and texture of specialty cheeses (Fleet 1990; Pereira-Dias et al. 2000).

While milk may be a source of NSYM, little is known about the composition and diversity of the fungal microflora of milk. A study of Chinese milk reported the presence of 11 yeast species but did not address mold diversity (Chen et al. 2010). Little is known about fungal diversity in raw cow milk. No molecular analyses of fungal species in cheese or studies on changes in yeast and mold diversity over time of milk from dairy farms that produce their own cheeses have been published to our knowledge.

Over 300 cheese varieties are produced in Canada, 53% of which come from the province of Quebec (Statistics Canada 2010). Most varieties are produced by artisan cheesemakers, some of whom process the milk from their own dairy cows to better control the quality of the milk and cheese. The aim of the present study was to analyze the fungal composition and diversity of milk from 19 dairy farms in the province of Quebec. Yeasts and molds in 111 raw milk samples were isolated and characterized. A more detailed analysis of the raw milk and cheese from four dairy farms over a 5-month period was also conducted to characterize their specific microflora and how it varied over time. A multilocus sequence typing (MLST) analysis of adventitious yeast *I. orientalis* isolates (anamorph: *Candida krusei*) was also performed to determine whether NSYM from raw milk could be transferred to raw cheese.

2 MATERIALS AND METHODS

2.1 Selection of dairy farms and sampling of cow milk and cheeses

Nineteen dairy farms in six different administrative regions of the province of Quebec were selected. The farms produced specialty cheeses from the cow milk they produced. The names and locations of the farms are being withheld for confidentiality purposes. One hundred eleven raw milk samples were collected between January and November 2009. The major yeast and mold species in each sample were isolated and identified. The samples were collected monthly in each farm directly from bulk tanks using sterile 1-L bottles. Depending from the size of the farm, milk obtained could be the result of one or more milking. The bottles were placed on ice and were sent to the laboratory as quickly as possible. While most samples were analyzed within 24 h, some took 48 h due to travel time constraints. Others were discarded because they took longer than 48 h to arrive or the temperature of the sample was too high on arrival. The farms also provided specialty cheeses for analysis (Table 1). The cheeses were allowed to ripen for 7–10 days before the samples were sent to the laboratory where NSYM flora was analyzed. Samples were aseptically removed from the rind (9 cm^2 on each face) and core (11 g).

Table 1. Characteristics of the cheeses analyzed for non-starter yeasts and molds

	Milk		
Cheese type	Raw	Thermized	Pasteurized
Soft ripened	3	6	–
Washed rind	10	4	2
Natural rind	–	–	1
Ripened in the mass	–	1	–
Mixed (soft ripened and washed)	–	1	–
Total	13	12	3

2.2 Isolation of yeasts and molds from milk and specialty cheese samples

The milk samples were serially diluted in sterile 0.1% (*w*/*v*) peptone solution and were plated on YEGC agar medium (10 $g.L^{-1}$ yeast extract, 10 $g.L^{-1}$ glucose, 0.1 $g.L^{-1}$ chloramphenicol, 15 $g.L^{-1}$ agar) (Guichard and Bonnarme 2005). The rind and core samples (11 g) were placed in 99 mL of sterile 2% (*w*/*v*) sodium citrate preheated to 45 °C and were homogenized using a Stomacher lab blender (Seward Stomacher® 400 Circulator) for 90 s at 250 rpm (Lopandic et al. 2006). Serial dilutions in peptone solution were prepared and spread on YEGC agar plates. The plates were incubated at 25 °C for 3 days. Individual colonies were selected based on shape and color, assuming that colonies with the same morphology belonged to the same genus or species. The colonies were re-streaked on YEGC agar plates, which were incubated at 25 °C for 3 days and then kept at 4 °C until the colonies were identified. Samples of the isolates were also suspended in YEGC broth containing 10% (*v*/*v*) glycerol and stored at −80 °C. A total of 608 strains, 505 from the milk samples and 103 from the cheese samples, were isolated.

2.3 Semi-quantitative analysis of fungal microflora of milk from four dairy farms

Four dairy farms (A, B, C, and D) were selected for a semiquantitative analysis of the fungal microflora of the milk. Milk samples from each farm were collected once a month from June to October 2009. The samples were diluted, plated, and incubated as described above. Since selecting colonies based on morphology alone can introduce bias, every colony was systematically purified and identified. Five to 25 colonies were isolated, depending on the sample, for a total of 272 isolates. The purified isolates were grown on YEGC agar plates and stored at 4 °C until the isolates were characterized. For long-term storage, samples were suspended in YEGC broth containing 10% (*v*/*v*) glycerol and were stored at −80 °C. The semiquantitative assessment was based on the proportion of each species identified and the total yeast and mold counts in each sample.

2.4 DNA isolation and sequencing of the ITS1-5.8S-ITS2 rDNA region

Genomic DNA was isolated according to the method of Al-Samarrai and Schmid (2000). A ~550-pb fragment from each isolate was PCR amplified using the universal ITS1 (5′- TCC GTA GGT GAA CCT GCG G - 3′) and ITS4

(5′- TCC TCC GCT TAT TGA TAT GC -3′) primers flanking the ITS1-5.8S-ITS2 ribosomal DNA (rDNA) region. The 50-μL PCR reaction mixtures contained 2 U of *Taq* DNA polymerase (New England Biolab), 1× *Taq* buffer (10 mM Tris–HCl, 50 mM KCl, 1.5 mM $MgCl_2$, pH 8.3), 100 μM dNTP (GE Healthcare), 50 μM of each primer (ITS1 and ITS4), and 50 ng of DNA template. The PCR assays were performed using a PTC-200 DNA Engine Thermal Cycler (BioRad). Samples were incubated for 2 min at 94 °C followed by 35 cycles using the following conditions: 30 s at 94 °C, 2 min at 55 °C, 2 min at 72 °C, and a final 5-min extension step at 72 °C. The PCR products were sequenced using the ITS1 and ITS4 primers and a 3130xl genetic analyzer (Applied Biosystems). Sequences were assembled using the Staden software package and compared to public databases using the nucleotide–nucleotide basic local alignment search tool (BLAST).

2.5 Multilocus sequence typing of *I. orientalis* isolates

Thirteen *I. orientalis* isolates (anamorph *Candida krusei*) were analyzed by multilocus sequence typing (MLST) as described previously by Jacobsen et al. (2007). Briefly, six loci (*ADE*2, *HIS*3, *LEU*2, *LYS*2, *NMT*1, and *TRP*1) were sequenced and were compared to the *Candida krusei* scheme at http://pubmlst.org/ckrusei/. The locus and sequence type (ST) numbering system of the scheme was used when possible. New STs were identified and deposited in the pubmlst.org database.

3 RESULTS AND DISCUSSION

3.1 Yeast and mold diversity in milk

Colonies (505) from 111 cow milk samples were isolated and selected based on differences in shape and color. This approach was selected to increase the probability of isolating a large number of species. The molecular identification approach using the ITS1-5.8S-ITS2 rDNA region revealed that 339 (67%) of the isolates were yeasts that could be assigned to 37 species belonging to 11 genera (Table 2) and that the remaining isolates were molds that could be assigned to 33 species belonging to 25 genera (Table 3).

Table 2. Phylogenetic affiliation of 339 yeasts isolated from raw cow milk from 19 dairy farms between February and October 2009

Genus	Total	Species	Total	Number of farms
Candida	148	*apicola*	2	1
		catenulata	1	1
		glaebosa	48	2
		glabrata	1	1
		intermedia	2	2
		parapsilosis	41	12
		pararugosa	4	4
		rugosa	10	4
		silvae	2	1
		tropicalis	14	4
		zeylanoides	23	5
Cryptococcus	37	*carnescens*	2	2
		chernovii	1	1
		curvatus	17	7
		infirmominiatum	1	1
		macerans	7	4
		magnus	5	2
		victoriae	3	2
		sp.	1	1
Debaryomyces	71	*hansenii*	70	14
		nepalensis	1	1
Galactomyces	6	*candidus*	6	4
Issatchenkia	25	*orientalis*	25	8
Nakazawaea		*holstii*	1	1
Pichia	27	*anomala*	10	4
		burtonii	1	1
		fermentans	15	8
		jadinii	1	1
Rhodotorula	11	*glutinis*	1	1
		mucilaginosa	9	5
		slooffiae	1	1
Trichosporon	10	*aquatile*	1	1
		cutaneum	3	2
		guehoae	1	1
		inkin	3	2
		mycotoxinivorans	1	1
		sp.	1	1
Trichothecium		*roseum*	1	1
Yarrowia		*lipolytica*	2	2

The total number of isolates (total) for each genus and species is indicated as well as the number of farms whose milk contained these isolates

Table 3. Phylogenetic affiliation of 166 molds isolated from raw cow milk from 19 dairy farms between February and October 2009

Acremonium	15	*strictum*	14	7
		sp.	1	1
Alternaria		*alternata*	4	3
Apiospora		*montagnei*	1	1
Aspergillus		*versicolor*	5	2
Aureobasidium		*pullulans*	1	1
Cladosporium	28	*cladosporioides*	9	8
		herbarum	8	6
		sphaerospermum	9	4
		sp.	2	2
Emericella		sp.	6	5
Epicoccum		*nigrum*	1	1
Eurotium	39	*amstelodami*	6	5
		niveoglaucum	3	2
		rubrum	14	8
		sp.	16	4
Fusarium		*oxysporum*	1	1
Geosmithia		*pallida*	1	1
Gibberella	3	*avenacea*	2	2
		moniliformis	1	1
Lecytophora		*hoffmannii*	5	2
Lichtheimia		*corymbifera*	22	9
Microdochium		*bolleyi*	1	1
Mucor		*racemosus*	1	1
Nectria		*mauritiicaula*	1	1
Ophiostoma		sp.	1	1
Penicillium	11	*commune*	2	1
		expansum	2	2
		glabrum	1	1
		sp.	4	2
Peyronellaea		*glomerata*	2	2
Phaeosphaeria	2	*avenaria*	1	1
		nodorum	1	1
Phoma	11	*herbarum*	8	6
		macrostoma	2	2
		sp.	1	1
Plectosphaerella		*cucumerina*	2	2
Scopulariopsis		*brevicaulis*	1	1
Verticillium		*nigrescens*	1	1

The total number of isolates (total) for each genus and species is indicated as well as the number of farms whose milk contained these isolates

Of the 339 yeast isolates, *Debaryomyces hansenii* was the most abundant species (21%) and was detected in the milk from 14 of the 19 dairy farms. *Candida* was the most abundant genus (44%) and the most diverse, with 148 isolates and 11 species. *Cryptoccocus* spp. and *Pichia* spp. made up 10.9% and 8.0% of the yeast isolates, respectively. Of the 37 yeast species identified, 14 were isolated only once. *Eurotium* was the most abundant mold genus (39 isolates or 23.5%) and was detected in the milk from 13 of the 19 farms. *Lichtheimia* (*Lichtheimia corymbifera*) was the second most abundant mold genus, with 22 isolates (13.3%) from nine farms. Fifteen mold species were isolated only once.

Candida glaebosa was the second most frequent yeast isolate and was found in the raw milk from two farms (Table 2). However, 47 of the *Candida glaebosa* 48 isolates were found in the milk of a single dairy farm (dairy farm A). This species has been identified in cheeses affected by thread mould spoilage but has not been clearly associated with this phenomenon (Hocking and Faedo 1992). These findings indicate that the fungal profile of milk differs from one farm to another. While other investigators have reported that most of the species identified in our study are commonly found in dairy products (Borelli et al. 2006; El-Sharoud et al. 2009; Fadda et al. 2004; Lopandic et al. 2006; Montagna et al. 2004; Pereira-Dias et al. 2000; Prillinger et al. 1999; Roostita and Fleet 1995; Vasdinyei and Deák 2003), only a few have been shown to be part of the native fungal ecosystem of milk.

We also isolated *Prototheca zopfii*, a yeast-like algal species, from two milk samples. While this non-photosynthetic algal species is thought to be a putative causal agent of bovine mastitis, it has not been possible to make a definite correlation between the presence of *Prototheca zopfii* and the sub-clinical and clinical symptoms of the disease (Kishimoto et al. 2009; Marques et al. 2010).

Despite the large number of isolates, it is likely that only a fraction of the fungal species in the samples were actually identified. Some species may have had difficulties growing on the synthetic medium or were present in such small numbers that they were masked by the dominant species. However, this overview of the fungal microflora of the milk samples provides an inkling of the complexity of this ecosystem.

3.2 Semiquantitative analysis of fungi in milk from four dairy farms over a 5-month period

To determine whether the fungal ecosystem differed from farm to farm or changed over time for a given farm, the yeast and mold populations in milk samples from four dairy farms were analyzed over a 5-month period.

However, rather than selecting colonies based on morphology, every colony was purified and identified by sequencing the ITS1-5.8S-ITS2 rDNA region. This approach was used to eliminate any bias that may have been introduced by selecting colonies based on morphology.

Dairy farm A produced milk with the highest fungal count, with an average of 1.8×10^3 $CFU.mL^{-1}$ (Online resource 1), while farm c produced milk with the lowest fungal count, with an average of 3.6 $CFU.mL^{-1}$ (Online resource 2). Yeasts dominated the microflora of the milk from farms A–C, accounting for 91%, 86%, and 82% of the isolates, respectively. Molds dominated the microflora of the milk from farm D, accounting for 67% of the isolates (Online resource 3). While this systematic approach gave a more precise, quantitative portrait of the fungal ecosystem, less species diversity was observed. It also revealed that the dominant species and the composition of the ecosystem differed from one milk sample to another. The samples analyzed over the 5-month period revealed that the milk from farm A was dominated by the fungal species *Candida glaebosa*, which made up 59% of the isolates. This species was also present in every sample obtained from this farm. *Candida zeylanoides* and *Candida parapsilosis* were the dominant species in the milk of farm b, accounting for 31% and 25% of the isolates, respectively (Online resource 4). However, they were only found in three of the five milk samples.

Interestingly, *D. hansenii* was the dominant yeast species in the milk from farm C. On average, it made up 49% of the total fungal population and was found in four of the five samples. This species is commonly inoculated in dairy products. Some strains have a positive effect on cheese ripening through their proteolytic and lipolytic activities and are able to protect against contaminants (Breuer and Harms 2006; Corsetti et al. 2001). Farm D was unique in that, while *D. hansenii* (16%) was the predominant yeast species, the microflora of the milk was dominated by a diverse population of mold species (67% of isolates), including *Lichtheimia corymbifera* (found in all samples), *Eurotium* spp., *Acremonium* spp., *Aspergillus* spp., *Cladosporium* spp., and *Penicillium* spp. (Online resource 3).

The composition of the NSYM from milk differs from farm to farm but may also differ along the months. This study also corroborates other studies reporting that the period had a significant effect on the global yeast profile of milk. This could be due to the combined effects of the feeding regime, physiological state of the animal, the weather, and the indoor and outdoor environments (Baroiller and Schmidt 1990; Callon et al. 2007; Zheng et al. 2006). Moreover, cheesemaking process and equipment could be sources of NSYM, especially brine and ripening chambers (Seiler and Busse 1990; Prillinger et al. 1999; Zheng et al. 2006).

3.3 Non-starter yeasts and molds isolated from cheese samples

Non-starter yeasts and molds (103 isolates) were isolated from 28 surface and core samples from 1-week-old specialty cheeses produced from raw and pasteurized milk (Table 1, Table 4). The NSYM colonies were selected based on morphology in order to isolate as many NSYM species as possible. Given that some species are used as ripening cultures, including *Penicillium camemberti*, *Geotrichum candidum* (teleomorph *Galactomyces candidus*), *Kluyveromyces lactis*, and *D. hansenii*, isolates corresponding to these species were not included in the analysis since it was impossible to determine whether they were natural contaminants or members of the ripening culture.

Thirty different NSYM species were isolated from the rind and core samples. Of these, 12 species were isolated only once. Yeasts were isolated more frequently than molds, and accounted for 73% of the isolates. Many NSYM isolated from the cheese samples were not found in the milk, including *Aspergillus fumigatus*, *Cochliobolus sativus*, *Fusarium domesticum*, *Mucor circinelloides*, *Cryptococcus diffluens*, *Pichia membranifaciens*, *Saccharomyces servazii*, *Trichosporon asahii*, *Trichosporon domesticus*, and *Trichosporon jirovecii*. The proportion of these species in the milk samples may have been too low for them to be detected, or they may have been contaminants from the environment, workers, the ripening chamber, or other sources.

The most frequent yeast species isolated was *I. orientalis* (syn. *P. kudriavzevii*), which accounted for 25% of the yeast isolates. However, this frequency may have been due to its colonial morphology, which is different from the common white smooth regular colonies often observed with other species such as *Candida zeylanoides*, *D. hansenii*, and *Candida glaebosa*. *I. orientalis* was reported as a common species found in Graukäse cheese and was also isolated from brines (Prillinger et al. 1999; Seiler and Busse 1990). However, the contribution of this species to cheese ripening process needs to be investigated. *M. racemosus* was the most frequently isolated mold species. This species is frequently isolated as NSYM and is undesired in most specialty cheeses, such as Camembert (Bekada et al. 2008).

Table 4. NSYM isolated from 28 cheeses after 7–10 days of ripening: isolates were recovered either from the rind or core of the cheese

Genus	Total	Species	Total	Rind	Core
Molds					
Aspergillus		*fumigatus*	1	0	1
Aureobasidium		*pulullans*	1	0	1
Cladosporium	5	*cladosporioides*	4	1	3
		sp.	1	1	0
Cochliobolus		*sativus*	1	0	1
Eurotium	2	*niveoglaucum*	1	0	1
		sp.	1	0	1
Fusarium		*domesticum*	2	2	0
Lecythophora		*hoffmannii*	2	0	2
Lichtheimia		*corymbifera*	4	0	4
Mucor	8	*circinelloides*	1	0	1
		racemosus	7	4	3
Peyronellaea		*glomerata*	2	1	1
Total molds			28	9	19
Yeasts					
Candida	26	*catenulata*	1	0	1
		parapsilosis	11	6	5
		pararugosa	4	1	3
		tropicalis	5	4	1
		zeylanoides	5	0	5
Cryptococcus	2	*diffluens*	1	0	1
		curvatus	1	1	0
Issatchenkia		*orientalis*	19	7	12
Kluyveromyces		*marxianus*	1	0	1
Pichia	5	*jadinii*	1	0	1
		fermentans	2	1	1
		membranifaciens	2	2	0
Rhodotorula	11	*glutinis*	1	1	0
		mucilaginosa	10	6	4
Saccharomyces		*servazzii*	2	0	2
Trichosporon	9	*aquatile*	1	1	0
		asahii	4	2	2
		domesticum	2	1	1
		jirovecii	2	0	2
Total yeasts			75	33	42

Most of the isolates (59%) came from the core samples. Seven of the 19 yeast species and seven of the 11 mold species from the core samples were not detected in the rind samples (Table 4). Because most cheeses analyzed in this study are surface ripened, the use of ripening culture may have reduced the number of NSYM isolated at the surface. *F. domesticum, C. curvatus, Pichia membranifaciens,* and *Trichosporon aquatile* were only isolated from the rind samples. Fewer species were isolated from the rind samples than from the core samples. However, the microflora of the ripening culture may have masked the NSYM species to some extent.

3.4 MLST genotyping of *I. orientalis* isolates

MLST genotyping was performed on 13 *I. orientalis* isolates (Table 5) to assess the distribution and persistence of non-starter yeasts in cheese. Leaving aside LMA-694, LMA-696, and LMA-726, two main groups were identified. The first group consisted of LMA-250, LMA-519, LMA-545, LMA-652, and LMA-805, which belong to the St-67 sequence type. The second group was more heterogeneous and was comprised of LMA-422, LMA-783, LMA-546, LMA-503, and LMA-666. The strains in this group shared many loci and all harbored the same *ADE*1 locus.

Eight MLST profiles were identified in the nine isolates from the milk samples, indicating that at least eight different strains were present. Four of these isolates came from the milk from farm A. LMA-545 and LMA-546 came from the same milk sample but had different MLST profiles.

The allelic St-67 profile encompassed all the isolates from cheese 1 (farm A), despite the fact that the cheese samples were collected from three different lots (April 3, June 12, and August 12), suggesting that this strain or close relatives persisted in the processing plant. Interestingly, the same MLST genotype was found in the raw milk samples from this farm, suggesting that *I. orientalis* in milk may have survive the cheesemaking process and grow in raw milk cheese. An isolate from cheese sample 2 from dairy farm A had a totally different genotype profile (LMA-503), confirming that several *I. orientalis* strains can coexist in the same cheesemaking environment and develop in cheeses.

It is noteworthy that LMA-250 and LMA-545 had the same MLST genotype but were isolated from different farms. This suggested that closely related strains could be found in widely separated locations. While the MLST analysis could not guarantee that these two isolates were identical strains, it did suggest a strong genetic affiliation.

Table 5. MLST performed on 13 *Issatchenkia orientalis* strains isolated from milk and cheese samples

Isolates	Source	St	*ADE2*	*HIS3*	*LEU2*	*LYS2D*	*NMT1*	*TRP1*
LMA-519	Cheese 1[a] (milk A), April 3	67	2	4	9	9	8	8
LMA-652	Cheese 1[a] (milk A), June 12	67	2	4	9	9	8	8
LMA-805	Cheese 1[a] (milk A), Aug. 12	67	2	4	9	9	8	8
LMA-545	Milk A, April 2	67	2	4	9	9	8	8
LMA-250	Milk E, February 11	67	2	4	9	9	8	8
LMA-422	Milk A, March 5	100	22	2	1	14	2	25
LMA-783	Milk A, July 14	149	22	2	5	14	2	25
LMA-546	Milk A, April 2	102	22	2	1	14	13	26
LMA-503	Cheese 2[b] (milk A), April 2	101	22	4	5	10	2	25
LMA-666	Milk C, May 20	103	22	1	5	4	2	25
LMA-694	Milk G, April 15	19	5	1	2	10	3	5
LMA-696	Milk H, April 15	104	6	1	18	10	11	1
LMA-726	Milk F, April 15	105	2	1	12	13	3	10

Sequence type (St) and allelic numbers correspond to the pubMLST database for *Candida krusei* (http://pubmlst.org/ckrusei/)

[a] Raw milk cheese, washed rind, semi-hard

[b] Raw milk cheese, Brie type

4 CONCLUSION

The NYSM population in milk is very diverse, and the importance of this microflora is often underestimated. The present study provided a clearer picture of the fungal microbiota of milk and confirmed that the diversity and composition of the microbiota can vary from farm to farm. In most cases, yeasts predominated, but some ecosystems were dominated by molds, indicating that the relative proportions of yeasts and molds can vary considerably. The present study also revealed that the fungal ecosystem of milk can remain stable over time. In addition, the MLST genotyping of non-starter yeast isolates strongly suggested that *I. orientalis* in milk could survive the cheesemaking process and develop in raw milk cheese. Our results also suggested that NSYM can transfer from milk to cheese and influence the raw cheese ripening. Additional studies will be required to determine whether NSYM play a positive or negative role in the development of the cheese flavor.

ACKNOWLEDGMENTS

This work was funded by the Novalait Inc. Fonds Québécois de la Recherche sur la Nature et les Technologies (FQRNT), the Ministère de l'Agriculture, des Pêcheries et de l'Alimentation du Québec (MAPAQ), and Agriculture and Agri-Food Canada (AAFC).

Open Access

This article is distributed under the terms of the Creative Commons Attribution Noncommercial License which permits any noncommercial use, distribution, and reproduction in any medium, provided the original author(s) and source are credited.

REFERENCES

1. Al-Samarrai TH, Schmid J (2000) A Simple Method For Extraction Of Fungal Genomic DNA. Lett Appl Microbiol 30:53–56
2. Arfi K, Leclerq-Perlat MN, Spinnler HE, Bonnarme P (2005) Importance Of Curd-Neutralising Yeasts On The Aromatic Potential Of Brevibacterium Linens During Cheese Ripening. Int Dairy J 15:883–891
3. Baroiller C, Schmidt JL (1990) Contribution À L'étude De L'origine Des Levures Du Fromage De Camembert. Lait 70:67–84
4. Bekada AMA, Benakriche B, Hamadi K, Bensoltane A (2008) Modelling Of Effects Of Water Activity, Ph And Temperature On The Growth Rate Of Mucor Racemosus Isolated From Soft Camembert Cheese. World J Agr Sci 4:790–794
5. Beresford T, Williams A (2004) The Microbiology Of Cheese Ripening. In: Fox PF (Ed) Cheese: Chemistry, Physics And Microbiology. Vol. 1, 3rd Edn. Elsevier Academic Press, Amsterdam
6. Borelli BM, Ferreira EG, Lacerda ICA, Franco GR, Rosa CA (2006) Yeast Populations Associated With The Artisanal Cheese Produced In The Region Of Serra Da Canastra, Brazil. World J Microbiol Biotechnol 22:1115–1119
7. Breuer U, Harms H (2006) Debaryomyces Hansenii—An Extremophilic Yeast With Biotechnological Potential. Yeast 23:415–437
8. Callon C, Duthoit F, Delbes C, Ferrand M, Le Frileux Y, De Crémoux R, Montel M-C (2007) Stability Of Microbial Communities In Goat Milk During A Lactation Year: Molecular Approaches. Syst Appl Microbiol 30:547–560
9. Chen L-S, Ma Y, Maubois J-L, Chen L-J, Liu Q-H, Guo J-P (2010) Identification Of Yeasts From Raw Milk And Selection For Some Specific Antioxidant Properties. Int J Dairy Tech 63:47–54
10. Corsetti A, Rossi J, Gobbetti M (2001) Interactions Between Yeasts And Bacteria In The Smear Surface-Ripened Cheeses. Int J Food Microbiol 69:1–10
11. El-Sharoud WM, Belloch C, Peris D, Querol A (2009) Molecular Identification Of Yeasts Associated With Traditional Egyptian Dairy Products. J Food Sci 74:341–346

12. Fadda ME, Mossa V, Pisano MB, Deplano M, Cosentino S (2004) Occurrence And Characterization Of Yeasts Isolated From Artisanal Fiore Sardo Cheese. Int J Food Microbiol 95:51–59
13. Fleet GH (1990) Yeasts In Dairy Products. J Appl Bacteriol 68:199–211
14. Guichard H, Bonnarme P (2005) Development And Validation Of A Plate Technique For Screening Of Microorganisms That Produce Volatile Sulfur Compounds. Anal Biochem 338:299–305
15. Hocking AD, Faedo M (1992) Fungi Causing Thread Mould Spoilage Of Vacuum Packaged Cheddar Cheese During Maturation. Int J Food Microbiol 16:123–130
16. Jacobsen MD, Gow AR, Maiden MCJ, Shaw DJ, Odds FC (2007) Strain Typing And Determination Of Population Structure Of Candida Krusei By Multilocus Sequence Typing. J Clin Microbiol 45:317–323
17. Jacques N, Casaregola S (2008) Safety Assessment Of Dairy Microorganisms: The Hemiascomycetous Yeasts. Int J Food Microbiol 126:321–326
18. Kishimoto Y, Kano R, Maruyama H, Onozaki M, Makimura K, Ito T, Matsubara K, Hasegawa A, Kamata H (2009) 26S Rdna-Based Phylogenetic Investigation Of Japanese Cattle-Associated Prototheca Zopfii Isolates. J Vet Med Sci 72:123–126
19. Law B (2010) Cheese Adjunct Cultures. Aust J Dairy Technol 65:45–49
20. Lopandic K, Zelger S, Banszky LK, Eliskases FL, Prillinger H (2006) Identification Of Yeasts Associated With Milk Products Using Traditional And Molecular Techniques. Food Microbiol 23:341–350
21. Marques S, Silva E, Thompson G (2010) Phenotypic Characterisation Of Mastitic Prototheca Spp. Isolates. Res Vet Sci 89:5–9
22. Montagna MT, Santacroce MP, Spilotros G, Napoli C, Papa A, Dragoni I (2004) Investigation Of Fungal Contamination In Sheep And Goat Cheese In Southern Italy. Mycopathologia 158:245–249
23. Mounier J, Goerges S, Gelsomino R, Vancanneyt M, Vandemeulebroecke K, Hoste B, Brennan NM, Scherer S, Swings J, Fitzgerald GF, Cogan TM (2006) Sources Of The Adventitious Microflora Of A Smear-Ripened Cheese. J Appl Microbiol 101:668–681
24. Pereira-Dias S, Potes ME, Marinho A, Malfeito-Ferreira M, Loureiro V (2000) Characterisation Of Yeast Flora Isolated From An Artisanal Portuguese Ewes' Cheese. Int J Food Microbiol 60:55–63
25. Perry KD, Mcgillivray WA (1964) The Manufacture Of 'Normal' And 'Starter-Free' Cheddar Cheese Under Controlled Bacteriological Conditions. J Dairy Res 31:155–165
26. Petersen KM, Westall S, Jespersen L (2002) Microbial Succession Of Debaryomyces Hansenii Strains During The Production Of Danish Surfaced-Ripened Cheeses. J Dairy Sci 85:478–486
27. Prillinger H, Molnar O, Eliskases-Lechner F, Lopandic K (1999) Phenotypic And Genotypic Identification Of Yeasts From Cheese. Antonie Van Leeuwenhoek 75:267–283
28. Rohm H, Eliskases-Lechner F, Bräuer M (1992) Diversity Of Yeasts In Selected Dairy Products. J Appl Microbiol 72:370–376
29. Roostita R, Fleet GH (1995) The Occurrence And Growth Of Yeasts In Camembert And Blue-Veined Chesses. Int J Food Microbiol 28:393–404
30. Samson RA, Hoekstra ES, Frisvad JC (2004) Introduction To Food- And Airborne Fungi, 7th Edn. Centralbureau Voor Scchimmelcultures, Utrecht
31. Seiler H, Busse M (1990) The Yeast Of Cheese Brines. Int J Food Microbiol 11:289–304
32. Seker E (2010) Identification Of Candida Species Isolated From Bovine Mastitic Milk And Their In Vitro Haemolytic Activity In Western Turkey. Mycopathologia 169:303–308
33. Settanni L, Moschetti G (2010) Non-Starter Lactic Acid Bacteria Used To Improve Cheese Quality And Provide Health Benefits. Food Microbiol 27:691–697
34. Simov ZI, Simova ED, Beshkola DM (2006) Impact Of Two Starter Cultures On Proteolysis On Kashkaval Cheese. World J Microbiol Biotechnol 22:147–156

35. Sohier D, Le Dizes A-S, Thuault D, Neuveglise C, Coton E, Casaregola S (2009) Important Genetic Diversity Revealed By Inter-LTR PCR Fingerprinting Of Kluyveromyces Marxianus And Debaryomyces Hansenii Strains From French Traditional Cheeses. Dairy Sci Technol 89:569–581
36. Statistics Canada, Agriculture Division, Livestock And Animal Products Section (2010) Dairy Statistics, Fourth Quarter 2009. Statistics Canada, Ottawa, P 44
37. Swearingen PA, O'Sullivan DJ, Warthesen JJ (2001) Isolation, Characterization, And Influence Of Native, Nonstarter Lactic Acid Bacteria On Cheddar Cheese Quality. J Dairy Sci 84:50–59
38. Vacheyrou M, Normand A-C, Guyot P, Cassagne C, Piarroux R, Bouton Y (2011) Cultivable Microbial Communities In Raw Cow Milk And Potential Transfers From Stables Of Sixteen French Farms. Int J Food Microbiol 146:253–262
39. Vasdinyei R, Deák T (2003) Characterization Of Yeast Isolates Originating From Hungarian Dairy Products Using Traditional And Molecular Identification Techniques. Int J Food Microbiol 86:123–130
40. Viljoen BC, Knox AM, De Jager PH, Lourens-Hattingh A (2003) Development Of Yeast Populations During Processing And Ripening Of Blue Veined Cheese. Food Technol Biotechnol 41:291–297
41. Zheng Q-Y, Westermark S-O, Rasmuson-Lestander Â, Wang X-R (2006) Detection And Quantification Of Cladosporium In Aerosols By Real-Time PCR. J Environ Monit 8:153–160

CHAPTER 7

Milk and Dairy Products: Vectors to Create Probiotic Products

Gabriel-Danut Mocanu[1] and Elisabeta Botez[1]

[1] *Department of Food Science, Food Engineering and Applied Biotechnology, Faculty of Food Science and Engineering, „Dunarea de Jos" University of Galati, Romania*

1. INTRODUCTION

The most important function of alimentation is represented by the assurance of human metabolic needs as well as wellbeing and satisfaction induced by sensorial characteristics of food. In the same time, by modulating some target functions of the body, the food components might have benefic psychological and physiological effects, beside the nutritional ones, already accepted.

In fact, food must contribute to health improving/protection and sustain systems of defence against different aggressions. We are situated at a new frontier of nutrition, in which the foods are evaluated by their biological potential and by their ability to reduce the risk of developing certain diseases. We can talk today about the fact that food for health represent an expanding field: *probiotic functional food.*

In essence, probiotic functional food are products that, by their biological active compounds and consumed in current diets, contribute to optimal human physical and psihycal health.

The appearance and development of functional probiotic food are the response of production field to the results of cellular and molecular biology field research, which demonstrates the implication of food components in proper functioning cellules and subcelular structures. The importance of these studies is essential in contemporaneous context in which the environment assaults by many ways the human body, fully stressing it's protection, adaption and equilibrium maintenance systems. By their specific action, the food components might contribute to the maintain the normal parameters of cellular edificium and of the human body equilibrium.

Nowadays we are assisting to an intensification of research in food – alimentation – health relationship field. The ideea that food might increase/defend health due to active biological components from it's composition conquers more and more acceptability in the scientific community and there are many publication in this field. Unlike the last years, the customers from many countries become more and more interested in health beneficial determined by alimentation, including probiotic functional food. In Romania, even before the adherence to UE, there were registered studies concerning manufacturing of probiotic functional foods, especially in dairy industry and explaining the induced benefits for health.

In this trend of food science are included some of the studies developed over the years by researchers from Galati Food Science and Engineering Faculty.

2. PROBIOTICS: WHAT ARE THEY?

2.1. Definitions

The name probiotic comes from the Greek *„pro bios"* which means *„for life"*. The history of probiotics began with the history of man; cheese and fermented milk were well known to the Greeks and Romans, who recommended their consumption, especially for children and convalescents. Probiotics are defined as the living microorganisms administered in a sufficient number to survive in the intestinal ecosystem. They must have a positive effect on the host [1].

The term „probiotic" was first used by [2] in 1965 to describe the „substances secreted by one microorganism that stimulate the growth of another". A powerful evolution of this definition was coined by [3] in 1974, who proposed that probiotics are „organisms and substances which contribute to intestinal microbial balance" [4]. In more modern definitions, the concept of an action on the gut microflora, and even that of live microorganisms disappeared [5] in 1998 defined probiotics as the „food which contains live bacteria beneficial to health", whereas [6] in 2001 defined them as „microbial cell preparations or components of microbial cells that have a beneficial effect on the health and well-being".

Some modern definitions include more precisely a preventive or therapeutic action of probiotics. [7] in 1997 for example, defined probiotics as „microorganisms which, when ingested, may have a positive effect in the prevention and treatment of a specific pathologic condition". Finally, since probiotics have been found to be effective in the treatment of some gastrointestinal diseases [6], they can be considered to be therapeutic agents. It is clear that a number of definitions of the term „probiotic" have been used over the years but the one derived by the Food and Agriculture Organization of the United Nations/World Health Organization [8] and endorsed by the International Scientific Association for Probiotics and Prebiotics [9] best exemplifies the breadth and scope of probiotics as they are known today:"live microorganisms which, when administered in adequate amounts, confer a health benefit on the host".

This definition retains historical elements of the use of living organisms for health purposes but does not restrict the application of the term only to oral probiotics with intestinal outcomes [10]. Despite these numerous theoretical definitions, however, the practical question arises whether a given microorganism can be considered to be a probiotic or not.

Some strict criteria have been proposed. [11] in 1992, for example, proposed the following parameters to select a probiotic: total safety for the host, resistance to gastric acidity and pancreatic secretions, adhesion to epithelial cells, antimicrobial activity, inhibition of adhesion of pathogenic bacteria, evaluation of resistance to antibiotics, tolerance to food additives and stability in the food matrix.

The probiotics in use today have not been selected on the basis of all these criteria, but the most commonly used probiotics are the strains of lactic acid bacteria such as *Lactobacillus, Bifidobacterium* and *Streptococcus* (*S. thermophilus*); the first two are known to resist gastric acid, bile salts and pancreatic enzymes, to adhere to colonic mucosa and readily colonize the intestinal tract [4, 12].

2.2. Properties of Lactic Acid Bacteria

The lactic acid bacteria are generally defined as a cluster of lactic acid-producing, low %G+C, non-spore-forming, Gram-positive rods and cocci that share many biochemical, physiological, and genetic properties. They are distinguished from other Gram positive bacteria that also produce lactic acid (e.g., *Bacillus, Listeria,* and *Bifidobacterium*) by virtue of numerous phenotypic and genotypic differences. According to current taxonomy, the lactic acid bacteria group consists of twelve genera (table 1). All are in the phylum *Firmicutes,* Order, *Lactobacillales.* Based on 16S rRNA sequencing and other molecular techniques, the lactic acid bacteria can be grouped into a broad phylogenetic cluster, positioned not far from other low G +C Gram positive bacteria.

Five sub-clusters are evident from this tree, including: (1) a *Streptococcus-Lactococcus* branch (Family *Streptococcaceae*), (2) a *Lactobacillus* branch (Family *Lactobacillaceae*), (3) a separate *Lactobacillus-Pediococcus* branch (Family *Lactobacillaceae*); (4) an *Oenococcus-Leuconostoc-Weisella* branch (Family *Leuconostocaceae*), and (5) a *Carnobacterium-Aerococcus-Enterococcus-Tetragenococcus-Vagococcus* branch (Families *Carnobacteriaceae, Aerococcaceae,* and *Enterococcaceae*).

Seven of the twelve genera of lactic acid bacteria, *Lactobacillus, Lactococcus, Leuconostoc, Oenococcus, Pediococcus, Streptococcus,* and *Tetragenococcus,* are used directly in food fermentations. Although *Enterococcus* sp. Are often found in fermented foods (e.g., cheese, sausage, fermented vegetables), except for a few occasions, they are not added directly. In fact, their presence is often undesirable, in part, because they are sometimes used as indicators of fecal contamination and also because some strains may harbor mobile antibioticresistance genes.

Importantly, some strains of *Enterococcus* are capable of causing infections in humans. Likewise, *Carnobacterium* are also undesirable, mainly because they are considered as spoilage organisms in fermented meat products. Finally, species of *Aerococcus, Vagococcus,* and *Weisella* are not widely found in foods, and their overall significance in food is unclear.

2.3. Probiotics as Functional Foods

In the last decades consumer demands in the field of food production has changed considerably. Consumers more and more believe that foods contribute directly to their health [13, 14]. Today foods are not intended to only satisfy hunger and to provide necessary nutrients for humans but also to prevent nutrition-related diseases and improve physical and mental well-being of the consumers [15, 16].

[1]Adapted from [17]; [2]Adapted from [18]; [3]Refers to the general properties of the genus; some exceptions may exist; [4]Species of *Lactobacillus* may be homofermentative, heterofermentative, or both; [5]This phenotype is variable, depending on the species; [6]Some species produce D-, L-, or a mixture of D- and L-lactic acid.

Table 1. Genera of lactic acid bacteria and their properties [1, 2, 3]

Genus	Cell morphology	Fermentation route	Growth at		Growth in NaCl at		Growth at pH		Lactic acid isomer
			10ºC	45ºC	6.5%	18%	4.4	9.6	
Lactobacillus	rods	homo/hetero[4]	±[5]	±	±	-	±	-	D, L, DL[6]
Lactococcus	cocci	homo	+	-	-	-	±	-	L
Leuconostoc	cocci	hetero	+	-	±	-	±	-	D
Oenococcus	cocci	hetero	+	+	±	-	±	-	D
Pediococcus	cocci (tetrads)	homo	±	±	±	-	+	-	D, L, DL
Streptococcus	cocci	homo	-	+	-	-	-	-	L
Tetragenococcus	cocci (tetrads)	homo	+	-	+	+	-	+	L
Aerococcus	cocci (tetrads)	homo	+	-	+	-	-	+	L
Carnobacterium	rods	hetero	+	-	-	-	-	-	L
Enterococcus	cocci	homo	+	+	+	-	+	+	L
Vagococcus	cocci	homo	+	-	-	-	±	-	L
Weisella	coccoid	hetero	+	-	±	-	±	-	D, L, DL

[1]Adapted from [17]
[2]Adapted from [18]
[3]Refers to the general properties of the genus; some exceptions may exist
[4]Species of *Lactobacillus* may be homofermentative, heterofermentative, or both
[5]This phenotype is variable, depending on the species
[6]Some species produce D-, L-, or a mixture of D- and L-lactic acid.

In this regard, functional foods play an outstanding role. The increasing demand on such foods can be explained by the increasing cost of healthcare, the steady increase in life expectancy, and the desire of older people for improved quality of their later years [19, 15, 20].

The term "functional food" itself was first used in Japan, in the 1980s, for food products fortified with special constituents that possess advantageous physiological effects [21, 22]. Functional foods may improve the general conditions of the body (e.g. pre- and probiotics), decrease the risk of some diseases (e.g. cholesterol-lowering products), and could even be used for curing some illnesses.

The European Commission's Concerted Action on Functional Food Science in Europe (FuFoSE), coordinated by International Life Science Institute (ILSI) Europe defined functional food as follows: "a food product can only be considered functional if together with the basic nutritional impact it has beneficial effects on one or more functions of the human organism thus either improving the general and physical conditions or/and decreasing the risk of the evolution of diseases. The amount of intake and form of the functional food should be as it is normally expected for dietary purposes. Therefore, it could not be in the form of pill or capsule just as normal food form" [23].

European legislation however, does not consider functional foods as specific food categories, but rather a concept [22, 24]. Therefore, the rules to be applied are numerous and depend on the nature of the foodstuff. Functional foods have been developed in virtually all food categories. From a product point of view, the functional property can be included in numerous different ways as it can be seen in table 2.

Table 2. Able 2.prominent types of functional food [20, 25, 26]

Type of functional food	Definition	Example
Fortified product	A food fortified with additional nutrients	Fruit juices fortified with vitamin C
Enriched products	A food with added new nutrients or components not normally found in a particular food	Margarine with plant sterol ester, probiotics, prebiotics
Altered products	A food from which a deleterious component has been removed, reduced or replaced with another substance with beneficial effects	Fibers as fat releasers in meat or ice cream products
Enhanced commodities	A food in which one of the components has been naturally enhanced through special growing conditions, new feed composition, genetic manipulation, or otherwise	Eggs with increased omega-3 content achieved by altered chicken feed

It should be emphasized however, that this is just one of the possible classifications. According to alternative classification, some functional products are (1) "add good to your life", e.g. improve the regular stomach and colon functions (pre- and probiotics) or "improve children's life" by supporting their learning capability and behaviour. It is difficult, however to find good biomarkers for cognitive, behavioural and psychological, functions. Other group (2) of functional food is designed for reducing an existing health risk problem such as high cholesterol or high blood pressure. A third group (3) consists of those products, which "makes your life easier" (e.g. lactose-free, gluten-free products) [27].

These products have been mainly launched in the dairy-, confectionery-, soft-drinks-, bakery- and baby-food market [16, 20, 26].

3. HEALTH BENEFITS OF PROBIOTICS

Since Metchnikoff's era, a number of health benefits have been contributed to products containing probiotic organisms. While some of these benefits have been well documented and established, others have shown a promising potential in animal models, with human studies required to substantiate these claims. More importantly, health benefits imparted by probiotic bacteria are very strain specific; therefore, there is no universal strain that would provide all proposed benefits, not even strains of the same species. Moreover, not all the strains of the same species are effective against defined health conditions. Some of these strain specific health effects are presented in figure 1.

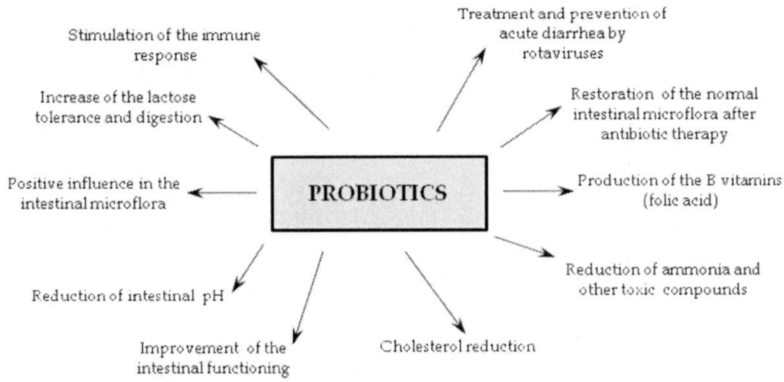

Figure 1. Probiotic beneficial effects on human health [**28**, **29**]

4. PROBIOTIC DAIRY PRODUCTS

Foods that affect specific functions or systems in the human body, providing health benefits beyond energy and nutrients—functional foods—have experienced rapid market growth in recent years. This growth is fueled by technological innovations, development of new products, and the increasing number of health-conscious consumers interested in products that improve life quality. Since the global market of functional foods is increasing annually, food product development is a key research priority and a challenge for both the industry and science sectors. Probiotics show considerable promise for the expansion of the dairy industry, especially in such specific sectors as yogurts, cheeses, beverages, ice creams, and other desserts. This book chapter presents an overview of functional foods and strategies for their development, with particular attention to probiotic dairy products.

4.1. Types of Probiotic Dairy Product

The most common probiotic dairy products worldwide are various types of yogurt, other fermented dairy product, various lactic acid bacteria drinks and mixture of probiotic (fermented) milks and fruit juice. Probiotic cheese, both fresh and ripened, have also been launched recently. In table 3 are listed some dairy functional food products that have been developed recently in Faculty of Food Science and Engineering.

4.1.1. Fermented milks and beverages

Fermented milks and beverages make up an important contribution to the human diet in many countries because fermentation is an inexpensive technology, which preserves the food, improves its nutritional value and enhances its sensory properties.

Table 3. able 3.Some examples of dairy probiotic products developed

Type of dairy functional food products	Description/Name
Fermented dairy products	Drink yogurts with La-5 and carrot juice/ BIOCOV
	Yogurt with La-5 and biomass of *Spirulina platensis*/YLaSP
	Yogurt with BB 12 and biomass of *Spirulina platensis*/YBbSP
	Yogurt with ABY 3/ABT 5 and medicinal plant extracts/ AFINOLACT
	Yogurt with ABY 3/ABT 5 and medicinal plant extracts/ CATINOLACT
	Yogurt with ABY 3/ABT 5 and medicinal plant extracts/ ROSALACT
Cheeses	Dessert based on fresh cheese and some fruit pulp
	Appetizer – type fresh cheese
	Probiotic Telemea cheese

[30, 31] in 2011, proposed the realization of a probiotic dairy drink with added carrot juice. This probiotic product was obtained using goat milk (fat = 3.63%, proteins = 3.05%, lactose = 4.55%, dry matter = 12.05% and density = 1.030 g mL^{-1}) which has been pasteurized at a temperature of 72ºC, for 20 minutes, a probiotic culture type Nutrish containing *Lactobacillus acidophilus* La-5 and carrot juice (dry matter = 9.35%, pH = 6.23, titratable acidity = 0.14 malic acid/100g, ash = 0.7%). After pasteurization, milk was quickly cooled to inoculation temperature at

37°C. The incubation of obtained fermented dairy drink was made at 37°C for 5 hours.

The addition of carrot juice (at a percentage of 10%) had a positive effect on physical – chemical and microbiological parameters of fermented dairy drink. Combining goat milk with carrot juice can get some food with potential therapeutic role.

As a result of the lactose fermentation, the titratable acidity increased fast during the incubation period. At the end of the storage period (after 5 days), the highest value of titratable acidity was 61 ºT. The pH of the obtained new product decreased during incubation period, and will stabilize during storage period, pH = 5.1 after 5 days of storage. The evolution of the number of microorganisms was analyzed for each sample during incubation and storage period. It was observed that the fermented dairy drink with added carrot juice product had been preserving its functional properties during storage (over 10^8cfu mL^{-1} probiotic bacteria).

The products were analyzed in terms of fluid flow thus establishing their rheological behavior. The literature shows that the rheological properties of fermented dairy products depend on the development of lactic bacteria as a consequence of metabolic changes leading physicochemical substrate in milk.

In figure 2 is presented the variation of shearing stress (τ, Pa) according to the shearing rate (*[Math Processing Error]*ϑ, s^{-1}). There was determined that samples have a rheological behavior similar with the one of the non-Newtonian fluids, time independent, therefore a pseudoplastic behavior. Specific for a fluid with this type of behavior is the flow resistance decrease as a result of the fluid shearing rate increase.

For all samples, it was noted that for low values of shear rate, tangential shear stress variation depending on shear rate was increasing (regression coefficient R^2 values varies from 0.962 and 0.995).

To obtain yoghurt with *Spirulina platensis* biomass was used pasteurized cow milk (non fat dry matter = 9.08%, fat = 1.5%, proteins = 3.52%, lactose = 4.32%, mineral salts = 0.72%). Pasteurization of milk is achieved by maintaining standardized milk at 95 °C for 5 minutes. After pasteurization, milk was cooled to inoculation temperature at 42°C.

The inoculation of milk for obtaining these fermented dairy products is with a probiotic culture containing *Lactobacillus acidophilus* La-5 respectively *Bifidobacterium lactis* BB 12, at this time was added and biomass of *Spirulina platensis* (0.5 – 1% according to [32]).

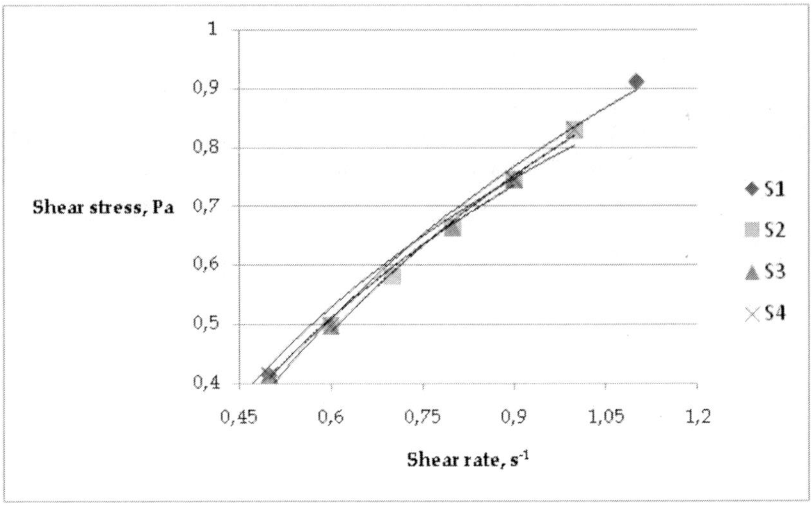

Figure 2. The shearing stress variation according to the shearing rate

After inoculation follows the distribution and packaging and incubation was made at 42°C for 6 hours in the thermostats set at the optimal temperature for the development of these bacteria. Meanwhile yoghurt gel gets a specific consistency. Cooling and storage of obtained yoghurts is performed at 6 °C for 15 days. In this storage period, coagulum is more compact, the flavor and taste become more pleasant. As a result of the lactose fermentation, the titratable acidity increased. This is slightly higher for the samples with La 5 from those with BB12.

All products with Spirulina platensis biomass have titratable acidity higher than control sample (1.1 times higher for samples with BB 12 and 1.2 times higher for samples with La 5).

The evolution of pH is correlated with lactose fermentation intensity and increased with titratable acidity, but in the same time it is influenced by the buffer substances that are found in *Spirulina platensis* biomass or formed during the manufacture of yoghurt. The pH of fermented dairy products fall between the values 4.11 and 4.53, values considered normal for such products.

The addition of *Spirulina platensis* biomass (figure 3) has positively influenced the number of viable probiotic microorganisms.

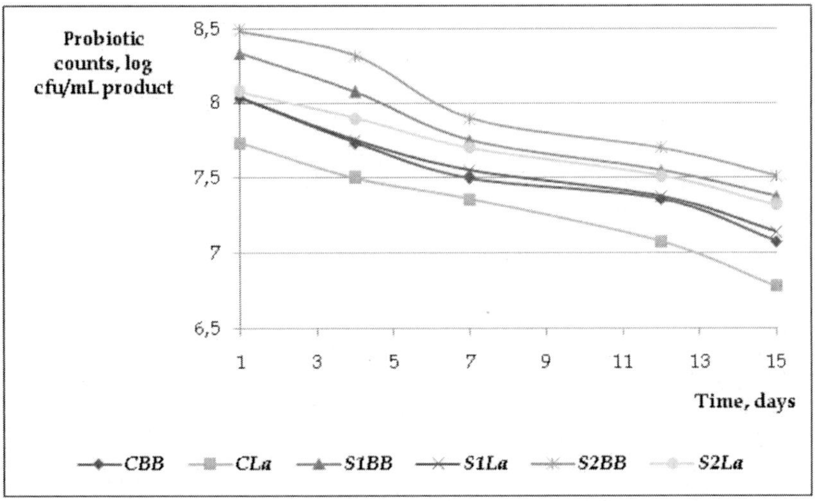

Figure 3. Viable counts variation during storage period

At the end of the storage period (after 15th days) the number of probiotic lactic bacteria for both, control samples and samples with *Spirulina platensis* biomass is still high, which shows that the product with *Spirulina platensis* biomass has been preserving its functional properties during storage period.

4.1.2. Cheeses

Perhaps no other fermented food starts with such a simple raw material and ends up with products having such an incredible diversity of color, flavor, texture, and appearance as does cheese. It is even more remarkable that milk, pale in color and bland in flavor, can be transformed into literally hundreds of different types of flavorful, colorful cheeses by manipulating just a few critical steps.

Just what happened to cause the milk to become transformed into a product with such a decidedly different appearance, texture, and flavor? To answer that question, it is first necessary to compare the composition of the starting material, milk, to that of the product, the finished cheese (figure 4).

In an attempt to diversify the range of probiotic dairy products, there has been made a series of research on the introduction of probiotic bacteria in cheese. According to [33], cheese is an interesting way of supplying probiotic bacteria due to the chemical composition of the raw milk that encourages their growth, metabolism and viability and also due to their relatively low acidity compared to other food products. The most of research has been focused on fresh cheese, but there are published some results on probiotic brined or ripened cheese, too.

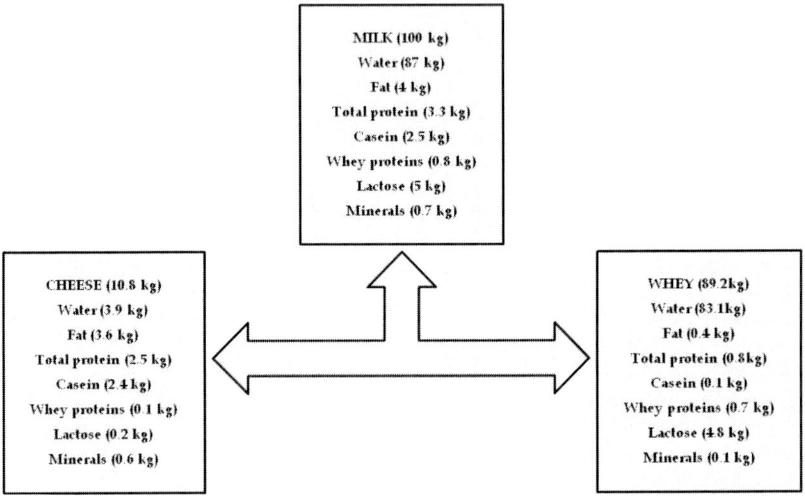

Figure 4. Partition of milk into cheese and whey (adapted from [**18**]).

Fresh cheese, mixt coagulated, is the most suitable cheese to carry probiotic bacteria, due to the high composition of nutrients, low acidity and low salt content. In 2009, [34] used probiotic fresh cheese and peach pulp in order to obtain a dessert, according to figure 5. Probiotic bacteria, *Lactobacillus acidophilus* La 5, was introduced in the fresh cheese as an agent of milk maturation, during coagulation stage. The product was rich in nutritive components (proteins: 10.9...11.3%; fat: 9.1...10.4% and minerals: 2...2.3%) and has a pseudoplastic rheological behaviour. This influenced the sensorial properties of the product, which achieved a creamy texture including in its structure the minced peach pulp and fat globules from the cream.

The research of the above mentioned authors continued, in the attempt to obtain a similar product using goat milk [35]. The amount of nutrients increased, comparing to the previous product (proteins: 12.4...12.5%; fat: 10.1...12.2% and minerals: 2.1...2.4%) but the rheological behaviour was not affected. Although there was expected a reserved attitude of the consumer because of the unpleasant flavour of goat milk, this was not observed.

In 2010 a new probiotic product based on fresh cheese was obtained, by mixing fresh cheese with caraway, cream and salt. The probiotic bacteria (*Bifidobacterium lactis* BB 12) were introduced in cheese at milk maturation stage. In figure 6 it can be observed that the caraway favourised the development of probiotic bacteria.

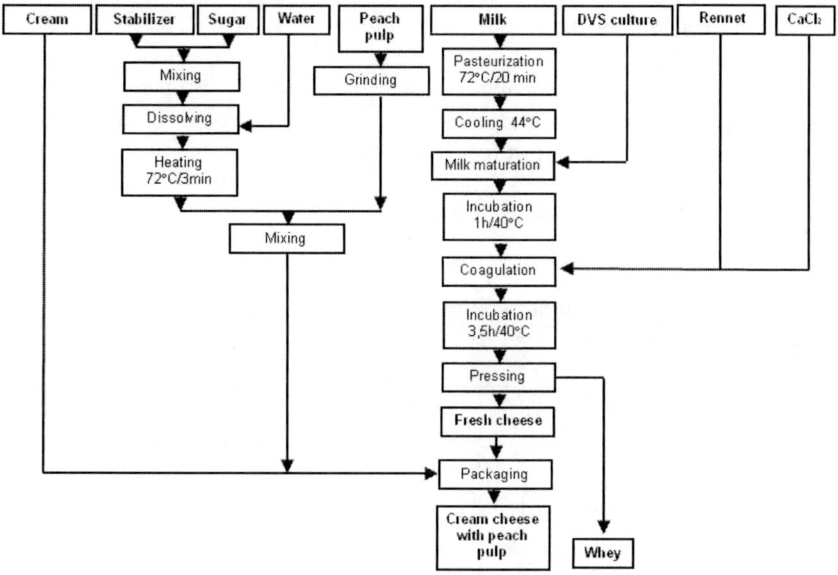

Figure 5. Technological flowchart for manufacturing the new product – Dessert based on fresh cheese and peach pulp

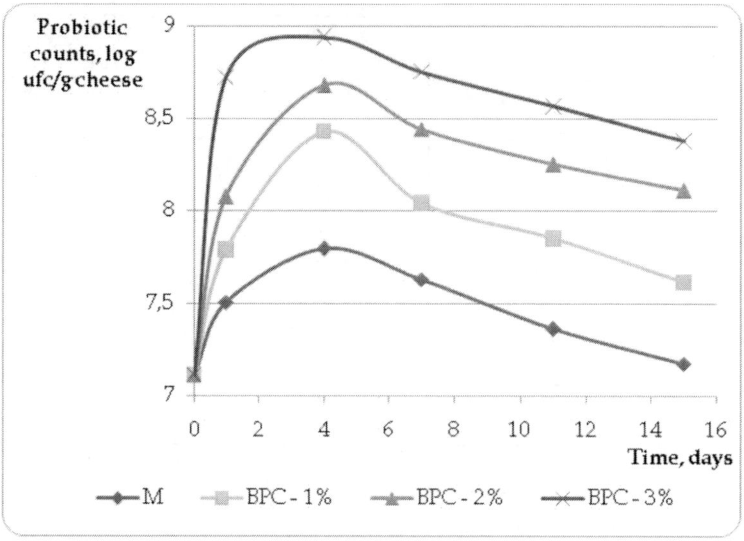

Figure 6. Evolution of bacteria during storage period

[36] and [37] studied the viability of probiotic bacteria *Bifidobacterium lactis, Lactobacillus acidophilus* and *Strepococcus thermophilus* in Telemea cheese during ripening and storage time. Telemea is a cheese variety originated in Romania, from where its manufacture spread to other Balkan countries and Turkey [38]. The specific of this variety of cheese is ripening in brine. Evolution of probiotic bacteria during different stages of manufacturing process is presented in table 4. Conclusion of the study is that Telemea cheese can be considered a probiotic product, even if the high salt concentration disadvantages probiotic bacteria growth, as long as the number of viable cells remains above 10^7 cfu g^{-1}.

Table 4. able 4.Evolution of probiotic bacteria during manufacturing of Telemea cheese (10^7 cfu g^{-1})

Stage	*Bifidobacterium lactis*	*Lactobacillus acidophilus*	*Lactobacillus bulgaricus*	*Streptococcus thermophilus*
Inoculated milk	2.71	2.50	2.90	8
Milk, 10 minutes after renneting	2.68	3.21	4	12
Coagulum, before pressing	2.60	7.49	9.40	24.9
Pressed coagulum	2.40	7.29	9.50	24.3
Salted coagulum	2.20	7.20	9.60	22.1
Cheese ripened for 5 days	1.84	6.90	8.90	15
Cheese ripened for 10 days	1.75	6.36	8.40	14.2
Cheese ripened for 15 days	1.65	4.67	7.30	13.4
Cheese ripened for 20 days	1.57	4.21	6.70	12.6
Cheese ripened for 25 days	1.48	3.48	5.30	11.3
Cheese ripened for 30 days	1.41	2.90	4.20	9.1
Cheese ripened for 35 days	0.99	2.31	3.90	7.2
Cheese ripened for 40 days	0.60	2.21	3.50	6

There are registered many other studies about probiotic cheese and methods of manufacturing probiotic cheese. Most of them introduce probiotic bacteria in the milk maturation stage, but there are reports about introducing them after pressing [39] or immobilized on fruit pieces [40].

5. IMPROVEMENT OF BENEFICAL EFFECT OF PROBIOTIC DAIRY PRODUCTS THROUGH THE USE OF BIOACTIVE COMPOUNDS FROM PLANTS

By sensorial analysis of several combinations milk-medicinal plants, as well as by physical and chemical analysis, there were selected the following medicinal plants: bilberry, sea-buckthorn, rosehip, liquorice, plants rich in active principles considered important for their pharmacological profile.

The research presented in this subchapter was realised on 14 variants of probiotic products (encoded according to table 5), manufactured from cow milk and medicinal plant extracts (bilberry, seabuckthorn, rosehip and liquorice), fermented by two types of probiotic cultures: ABY 3 (*Bifidobacterium lactis*, *Lactobacillus acidophilus*, *Lactobacillus delbrueckii* subsp. *bulgaricus* and *Streptococcus thermophilus*) and ABT 5 (*Bifidobacterium lactis*, *Lactobacillus acidophilus* and *Streptococcus thermophilus*).

The addition of aqueous medicinal plants extract positively influenced the number of viable probiotic microorganisms. At the end of storage period the number of probiotic lactic bacteria is high for both control samples and samples with medicinal plants, meaning that the products maintain their functional character [31, 41-46].

For the obtain products there was demonstrated the cytoprotective character, by studing the total antioxidant capacity, the total content of polyphenols, the superoxiddismutasic (SOD) activity, the minerals content, the ascorbic acid and anthocyaninis.

The results of the study reveal that probiotic dairy products with added medicinal plants contain a high level of total polyphenols and a high total antioxidant capacity. All these products are an excellent source of minerals with high biodisponibility in human diet. The addition of medicinal plants extract improved the SOD activity.

Table 5. able 5.Experimental variants

Crt. No.	Product	Code	Culture	Description
1.	Control	M – 3	ABY 3	Milk + 0.02% DVS culture
2.	Afinolact	A – 3		Milk + 0.02% DVS culture + 6% bilberry extract
3.		LD+A – 3		Milk + 0.02% DVS culture + 6% bilberry extract + 6% liquorice extract
4.	Cătinolact	C – 3		Milk + 0.02% DVS culture + 6% seabuckthorn extract
5.		LD+C – 3		Milk + 0.02% DVS culture + 6% seabuckthorn extract + 6% liquorice extract
6.	Rosalact	Mă– 3		Milk + 0.02% DVS culture + 6% rosehip extract
7.		LD+Mă – 3		Milk + 0.02% DVS culture + 6% rosehip extract + 6% liquorice extract
8.	Control	M – 5	ABT 5	Milk + 0.02% DVS culture
9.	Afinolact	A – 5		Milk + 0.02% DVS culture + 6% bilberry extract
10.		LD+A – 5		Milk + 0.02% DVS culture + 6% bilberry extract + 6% liquorice extract
11.	Cătinolact	C – 5		Milk + 0.02% DVS culture + 6% seabuckthorn extract
12.		LD+C – 5		Milk + 0.02% DVS culture + 6% seabuckthorn extract + 6% liquorice extract
13.	Rosalact	Mă – 5		Milk + 0.02% DVS culture + 6% rosehip extract
14.		LD+Mă – 5		Milk + 0.02% DVS culture + 6% rosehip extract + 6% liquorice extract

5.1. Identification of Bioactive Compounds From Plants

Medicinal plants are extremely valuable biological currants. The valorification of this potential represents a never-ending source of raw materials for pharmaceutics and food industry. World Health Organisation has recently announced that 75-80% of world's population is treated with natural remedies.

The plants do not cure all the diseases but they might be extremely helpful in rational treating of some diseases and are not to dangerous. The plants have favourable effect to human body and unfavourable effect to some pathogen agents due to certain substances from their composition. In every plant species there must be known that substance or substances which assure them the therapeutic effect (the active principles).

In order to test the chemical composition of studied plants (bilberry, sea-buckthorn, rosehip and liquorice) there were determined by chemical analysis: ascorbic acid (for seabuckthorn and rosehip), glycyrrhizic acid (for liquorice) and anthocyaninis (for bilberry). The concentrations of the active principles in medicinal plants samples are reported in table 6.

The values were expressed in mean ± standard errors of regression and values in parenthesis indicate minimum and maximum level recorded.

Table 6. able 6.Active principles in medicinal plants

Medicinal plants / Active principles	Bilberry (*Vaccinium myrtillus L.*)	Sea-buckthorn (*Hippophaë rhamnoides L.*)	Liquorice (*Glycyrrhiza glabra L.*)	Rosehip (*Rosa canina L.*)
Anthocyanins expressed as cyanidin-3-glucoside chloride	0.38 ± 0.06* (0.32÷0.47)	-	-	-
Ascorbic acid	-	0.8 ± 0.09* (0.66÷0.89)	-	1.24 ± 0.06* (1.18÷1.32)
Glycyrrhizic acid	-	-	5.03 ± 0.32* (4.6÷5.32)	-

The values were expressed in mean ± standard errors of regression and values in parenthesis indicate minimum and maximum level recorded.

Regarding the active principles content there was demonstrated that all the analysed medicinal plants respect the values presented in European Pharmacopee, V edition: minimum 0.3% cyanidin–3–glucozide chloride in bilberry, the analysed probes having a maximum content of 0.47%. The ascorbic acid in seabuckthorn must be minimum in 0.5% and in rosehip minimum 1%. The analysed samples registered values of 0.66÷0.98% for seabuckthorn and 1,18% for rosehip. The glycyrrhizic acid, the main active principle in liquorice, must be minimum 4% (according to European Pharmacopee) and the determined values varied between 4.6 and 5.38% [31, 41-46].

5.2. The Action of Bioactive Compounds From Plants On Probiotic Bacteria Metabolism

To have a probiotic effect, the strains of probiotic bacteria must be present in the product enough number. It is generally considered that the daily dose of probiotic strains must be between 1 10^8 and 1 10^9 cells. A portion of 100 g, the probiotic product should contain between 10^6 and 10^7 cfu mL^{-1}product. The addition of aqueous medicinal plant extracts has positively influenced the number of viable probiotic microorganisms due to the presence of fermentable sugars and some growth factors (mineral salts, non-protein nitrogen). At the end of incubation period (after 5 hours), for the samples with medicinal plant extracts the lowest number of microorganisms has established for the samples: Mă–3 ($7.8 \cdot 10^8$ cfu mL^{-1} probiotic bacteria) or A–5 and C–5 ($1.8 \cdot 10^9$ cfu mL^{-1} probiotic bacteria) instead the higher number of probiotic bacteria was registered for LD+Mă–3 ($4.5 \cdot 10^9$ cfu mL^{-1}probiotic bacteria) and LD+Mă–5 ($5.4 \cdot 10^9$ cfu mL^{-1} probiotic bacteria).

After 8th days of storage period (table 7) the higher number of viable microorganisms was found in the sample with ABT 5 culture (LD+A–5: $2.6 \cdot 10^8$ cfu mL^{-1} probiotic bacteria), and lowest number of probiotic bacteria cells was recorded for sample with ABY 3 culture (C–3: $0.9 \cdot 10^8$ cfu mL^{-1} probiotic bacteria). The storage in refrigerated conditions causes a reduction of the number of probiotic bacteria up to 4.37 times in the products obtained with ABY 3 culture and 3.62 times for those with ABT 5.

Table 7. able 7.Microbiological characteristics of fermented dairy products wit ABY 3 or ABT 5 after 8th days of storage

Microbiological characteristics			
Sample code	cfu·mL^{-1} product	Sample code	cfu·mL^{-1} product
M – 3	$3 \cdot 10^7$	M – 5	$8 \cdot 10^7$
A – 3	$1 \cdot 10^8$	A – 5	$1.4 \cdot 10^8$
LD+A – 3	$1.8 \cdot 10^8$	LD+A – 5	$2.6 \cdot 10^8$
C – 3	$9 \cdot 10^7$	C – 5	$1.1 \cdot 10^8$
LD+C – 3	$1.9 \cdot 10^8$	LD+C – 5	$1.8 \cdot 10^8$
Mă – 3	$1.2 \cdot 10^8$	Mă – 5	$1.3 \cdot 10^8$
LD+Mă – 3	$2.2 \cdot 10^8$	LD+Mă – 5	$2.1 \cdot 10^8$

At the end of storage period, the number of probiotic lactic acid bacteria for both control samples and for samples with medicinal plant extracts is still high ($1 \cdot 10^7 \div 1 \cdot 10^8$ cfu mL^{-1} probiotic bacteria), which shows that the products has been preserving its functional properties during storage period. Both cultures can be used in the production of probiotic products [31, 41-46].

Besides the cytoprotective effect conferred by the presence of probiotic bacteria, research has shown that products with added medicinal plants have a increased cytoprotective nature and because the content of biologically active compounds. Experimental results showed that the probiotic fermented dairy product with added medicinal plant extracts have a high content of total polyphenols with beneficial effects on human health, which help to prevent various diseases, such as cardiovascular disease, diabetes [47, 48] and consequently a higher total antioxidant capacity.

The higher amount of total polyphenols (table 8) was determined for

samples: LD+Mă–3 (280.78 μg mL^{-1}) or LD+Mă–5 (285.56 μg mL^{-1}).

Table 8. able 8.The total polyphenols content for samples with ABY 3 or ABT 5 culture

Total polyphenols expressed as catechin, μg·mL^{-1}			
Sample code	ABY 3	Sample code	ABT 5
M – 3	62.086	M – 5	82.086
A – 3	99.91	A – 5	106
LD+A – 3	152.95	LD+A – 5	158.6
C – 3	72.086	C – 5	87.73
LD+C – 3	135.56	LD+C – 5	147.3
Mă – 3	262.95	Mă – 5	239.47
LD+Mă – 3	280.78	LD+Mă – 5	285.56

Compared with control samples (not containing medicinal plant extracts) total antioxidant capacity (Table 9) increased by 3.25-9.94 times in products made with ABY 3 culture and 2.1-8.3 times the ABT 5 products.

TABLE 9. able 9.Total antioxidant capacity for sample with ABY 3 or ABT 5 culture

TEAC, mM·L^{-1}			
Sample code	ABY 3	Sample code	ABT 5
M – 3	0.16	M – 5	0.2
A – 3	0.57	A – 5	0.43
LD+A – 3	0.70	LD+A – 5	0.73
C – 3	0.52	C – 5	0.48
LD+C – 3	0.81	LD+C – 5	0.62
Mă – 3	1.19	Mă – 5	1.27
LD+Mă – 3	1.59	LD+Mă – 5	1.66

For products manufactured with ABY 3 culture and those with ABT 5 was observed that the total antioxidant capacity and total polyphenols content is higher for mixtures with liquorice extract, from the rest of the samples tested, except for samples Mă–3 şi Mă–5.

In addition to being an excellent source of protein, probiotic dairy products based on milk and medicinal plant extracts are a good source of minerals, calcium, potassium, phosphorus, magnesium, zinc. The minerals in these products are from raw milk, and medicinal plant extracts. Because the extracts of bilberry, sea-buckthorn, rosehip and liquorice have different mineral content, the products made have a different content in some microelements.

Distribution of broad in probiotic dairy products based on milk and medicinal plant extracts depends on the content of plant extracts and reactions/associations that occur during the technological process. The results of measurements are presented in Tables 10 and 11.

Table 10.able 10.Mineral concentration of fermented dairy products with ABY 3 culture

Crt. No.	Sample name	Microelements content, mg/100g product									
		Ca	Mg	Na	K	Mn	Fe	Zn	Cu	Pb	Cd
1.	M – 3	130	6	36	130	-	0.1	0.4	ND	ND	ND
2.	A – 3	135	7	39	115	0.1	0.15	0.45	ND	ND	ND
3.	C – 3	135	7.5	37.5	125	0.1	0.2	0.5	ND	ND	ND
4.	Mă – 3	137.5	9	41	135	0.1	0.27	0.5	ND	ND	ND
5.	LD+A – 3	140	7.8	47.5	135	0.1	0.21	0.5	ND	ND	ND
6.	LD+C – 3	141	7.8	40	140	0.1	0.25	0.5	ND	ND	ND
7.	LD+Mă - 3	139.5	9.3	46	150	0.1	0.3	0.5	ND	ND	ND

Table 11. Mineral concentration of fermented dairy products with ABY 5 culture

Crt. No.	Sample name	Microelements content, mg/100g product									
		Ca	Mg	Na	K	Mn	Fe	Zn	Cu	Pb	Cd
1.	M – 5	130	6	36	130	-	0.1	0.4	ND	ND	ND
2.	A – 5	135	7	34.5	110	0.1	0.14	0.45	ND	ND	ND
3.	C – 5	140	7.8	32.5	115	0.1	0.21	0.5	ND	ND	ND
4.	Mă – 5	140	8	35	120	0.1	0.25	0.5	ND	ND	ND
5.	LD+A – 5	150	7.5	39.5	140	0.1	0.17	0.5	ND	ND	ND
6.	LD+C – 5	145	8	38	135	0.1	0.23	0.5	ND	ND	ND
7.	LD+Mă–5	147	9	41	140	0.1	0.3	0.5	ND	ND	ND

Minerals in fermented dairy products based on milk and medicinal plant extract fulfill in human body the following functions:

- Are composed of hard tissue: Ca and Mg contribute in a major portion at the formation of the skeleton and teeth. Ca is also one of the most sensitive elements that regulate cellular functions. Is the regulator of enzymes involved in carbohydrate, lipid and protein metabolism, is also involved in important physiological processes such as muscle contraction, blood coagulation, apoptosis and necrosis;
- Are components of soft tissue: Fe and K in the form of organic compounds contribute to muscles, organs and blood. Fe are component of hemoglobin involved in oxygen transport, the of myoglobin, the body's oxygen tank. Fe is considered a major potential prooxidant metals from the human body;
- Are regulators of biological functions: as solubilized salts in body fluids contribute to sensitivity of nervous stimulus, maintain muscle elasticity, adjustment of pH digestive fluids and other secretions, maintaining of osmotic pressure.

Fermented dairy product based on milk and medicinal plant extracts had a higher superoxiddismutase activity. The relationship between the iron and SOD activity is presented in figures 7 and 8.

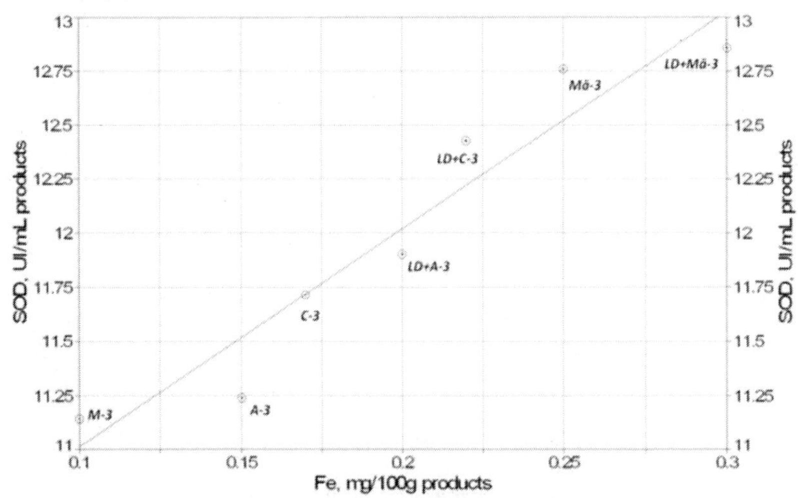

Figure 7. The relationship between SOD activity and iron content of products obtained with ABY 3 culture

For all samples of fermented dairy products with medicinal plant extracts is an increase in SOD activity compared with the control sample. Measured activity is total SOD-like activity (which contributes enzyme as such and superoxiddismutase-like activity of polyphenols and iron or zinc). SOD activity ranged from 11.142 to 12.857 IU mL^{-1} product; it was maximum for the sample LD+Mă–3. Samples obtained with ABY 3 culture had a higher SOD activity than samples with ABT 5.

5.3. Probiotic Dairy Products with Added Plant Extracts

To obtain the probiotic dairy products with medicinal plant extracts was used standardized cow milk to 1.5% fat. The technological process for production of fermented dairy products with medicinal plant extracts is presented in figure 8. The pasteurization of milk is achieved by maintaining standardized milk at 95 °C for 5 minutes. After pasteurization, milk is cooled to a temperature of 42 °C. Milk inoculation for these probiotic dairy products was made with two Probio-Tec probiotic cultures type: ABY 3 respectively ABT 5, at this time were added and aqueous extracts of medicinal plants (bilberries, sea-buckthorn, rosehip and liquorice) that have a number of characteristics presented in table 12.

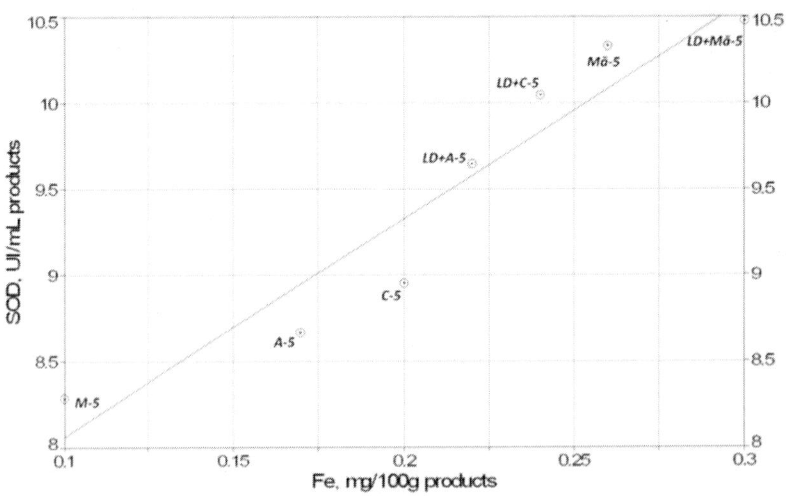

Figure 8. The Relationship Between SOD Activity And Iron Content Of Products Obtained With ABY 5 Culture

Table 12. Characteristics of concentrated medicinal plant extracts

Characteristics \ Sample	Bilberries extract	Sea-buckthorn extract	Rosehip extract	Liquorice extract
Dry matter, g/100g	4.84	4.84	5.29	4
Ash insoluble in hydrochloric acid, g/100g	0.74	0.83	0.67	0.81
Total carbohydrate, g/100g	4.69	0.2	7.19	7.29
Total proteins, g/100g	0.21	0.41	0.62	2.15
Calcium, mg/100g product	13.2	4.4	34.5	30
Magnesium, mg/100g product	15.4	4.4	13.8	60
Sodium, mg/100g product	6.6	4.4	8.05	9
Potassium, mg/100g product	132	110	300	230
Manganese, mg/100g product	2.2	2.2	1.38	0.4
Iron, mg/100g product	1.1	2.2	0.46	0.6
Zinc, mg/100g product	0.66	0.44	0.69	0.6
Copper, mg/100g product	0.22	0.22	0.23	0.6
Lead, mg/100g product	ND	ND	ND	ND
Cadmium, mg/100g product	ND	ND	ND	ND
Caffeic Acid, g/100g	1.47	1.69	1.63	0.54
Cyanidin-3-glucoside chloride, g/100g	0.55	-	-	-
Ascorbic acid, g/100g	-	0.26	0.18	-
Glycyrrhizic acid, g/100g	-	-	-	1.96

After inoculation follows the distribution and packaging and incubation was made at 42°C for 6 hours in the thermostats set at the optimal temperature for the development of these bacteria. Meanwhile yoghurt gel gets a specific consistency. Cooling and storage of obtained yoghurts is performed at 6 °C for 8 days. In this storage period, coagulum is more compact, the flavor and taste become more pleasant.

The characteristics of fermented dairy products studied, in terms of chemical properties are:

- Total dry matter have values between 12.05% and 12.5% (lowest for products that contain liquorice extract), exceeding the minimum specified in Romanian standard for fermented dairy products (12%);
- The fat content of the samples vary between 0.6% and 1.3% lowest in products with liquorice compared with other, because smaller proportions of milk of these products;

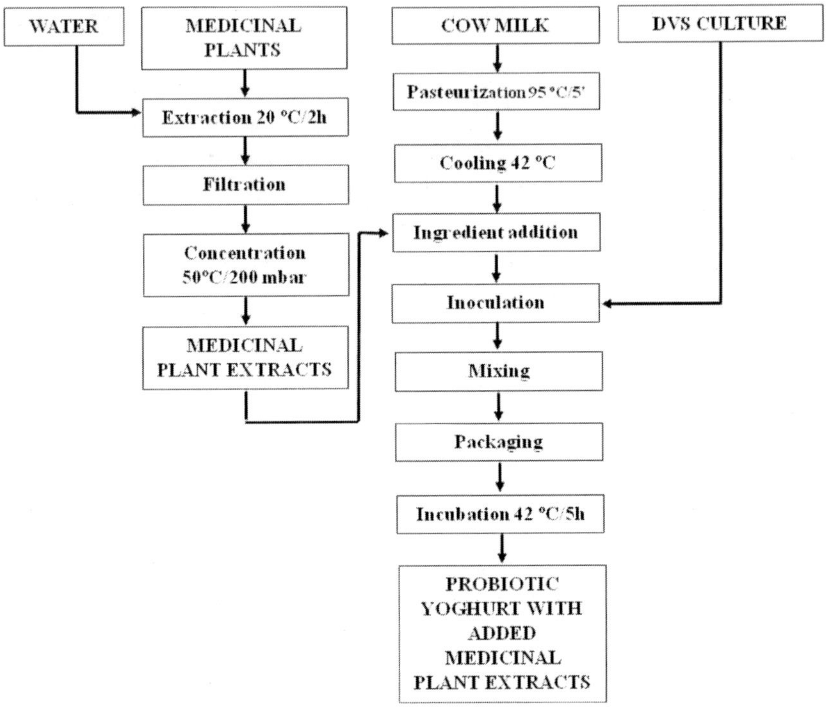

FIGURE 9. Technological flowchart for manufacturing the new product – Probiotic yoghurt with added medicinal plant extracts

- Lactic fermentation is faster for the samples with added plant extracts because of monosaccharides content (glucose, fructose, arabinose, xylose) and oligosaccharides (sucrose, raffinose, maltose, xiloglucan) from medicinal plants, which are fermented faster than lactose;
- Titratable acidity at the end of incubation period is between 67ºT and 78ºT, with higher values for products liquorice extracts. After 8th days of storage period, the titratable acidity is between 84ºT and 97ºT;
- The pH of products after incubation period varies between 5.035 and 5.287. After 8th days of storage it reaches values of 4.225-4.553, lowest value was obtained for the products with ABT 5 culture;
- ABT 5 probiotic culture which contains *Bifidobacterium lactis, Lactobacillus acidophilus* and *Strepococcus thermophilus* is more active than ABY 3 consists of *Bifidobacterium lactis, Lactobacillus acidophilus, Lactobacillus delbrueckii* subsp. *bulgaricus* and *Strepococcus thermophilus*;

- The established technological flowchart leads to obtaining some appropriate products in terms of physical-chemical characterization.

6. CONCLUSIONS

The researchers team of the Faculty of Food Science and Engineering, with many researchers in the scientific world, were concerned to investigate the possibility of obtaining probiotic products based on milk. Use milk as a vehicle for creating probiotic product was a constant concern of the staff of the Faculty of Food Science and Engineering in recent years. Probiotic character and functional role of probiotic products was obtained by adding fruit and vegetable juices, medicinal plant extracts, Spirulina platensis biomass, etc. We plan to continue research in this direction by investigating other products that may stimulate growth of probiotic bacteria.

ACKNOWLEDGEMENT

Authors are grateful to the S.C. Hofigal Export – Import S.A. Bucharest for the material support of this work (medicinal plants, biomass of Spirulina platensis).

REFERENCES

1. M. R. Gismondo, L. Drago, A. Lombardi, Review of probiotics available to modify gastrointestinal flora. International Journal of Antimicrobial Agents 199912287292
2. Lilly D.M. Stillwell R.H. Probiotics: Growth-promoting factors produced by microorganisms.Science 1965147747748
3. Parker R.B. Probiotics, the other half of the antibiotic story.Animal Nutrition & Health 19742948
4. J. Fioramonti, V. Theodorou, L. Bueno, What. Probiotics, they?. are, are. What, effects. their, gut. on, physiology?, Best Practice & Research Clinical Gastroenterology 200317711724
5. S. Salminen, Wright. A. von, L. Morelli, P. Marteau, D. Brassart, W. M. de Vos, R. Fondén, M. Saxelin, K. Collins, G. Mogensen, S. E. Birkeland, T. Mattila-Sandholm, Demonstration of safety of probiotics- A review. International Journal of Food Microbiology 19984493106
6. P. R. Marteau, M. de Vrese, C. J. Cellier, J. Schrezenmeir, Protection from gastrointestinal diseases with the use of probiotics. The American Journal of Clinical Nutrition (Suppl.) 200173430436
7. W. P. Charteris, P. M. Kelly, L. Morelli, J. K. Collins, detection. Selective, enumeration, of. identification, probiotic. potentially, Lactobacillus, species. Bifidobacterium,

mixed. in, populations. bacterial, International Journal of Food Microbiology 199735127
8. Food and Agriculture Organization/World Health Organization (FAO/WHO),2001Health and nutritional properties of probiotics in food including powder milk with live lactic acid bacteria, Report of a Joint FAO/WHO Expert Consultation on Evaluation of Health and Nutritional Properties of Probiotics in Food including Powder Milk with Live Lactic Acid Bacteria, 1289-25105-513-0Argentina, October 1-4, 2001
9. G. Reid, M. E. Sanders, H. R. Gaskins, G. R. Gibson, A. Mercenier, R. Rastall, New scientific paradigms for probiotics and prebiotics. Journal of Clinical Gastroenterology 200337105118
10. G. Reid, Safe and efficacious probiotics: What are they?. Trends in Microbiology 200614348352
11. R. Havenaar, Brink. B. Ten, Huis in't Veld J.H.J. Selection of Strains for Probiotic Use. In: R. Fuller (ed.) Probiotics: The Scientific Basis. Chapman & Hall; 1992151170
12. Soccol C.R.de Souza Vandenberghe L.P. Spier M.R. Pedroni Medeiros A.B. Yamaguishi C.T. De Dea Lindner J. Pandey A. Soccol V.T. The Potential of Probiotics. Food Technology and Biotechnology 2010484413434
13. Y. Young, Functional foods and the European consumer. In: J. Buttriss & M. Saltmarsh (ed.) Functional foods. II. Claims and evidence. The Royal Society of Chemistry; 2000
14. B. Mollet, I. Rowland, foods. Functional, the. At, between. frontier, food, pharma, Current Opinion in Biotechnology 200213483485
15. Roberfroid M.B.An European consensus of scientific concepts of functional foods. The American Journal of Clinical Nutrition 2000b16689691
16. K. Menrad, Market and marketing of functional food in Europe. Journal of Food Engineering 200356181188
17. L. Axelsson, acid. Lactic, classification. bacteria, In. physiology, S. Salminen, A. von, Write, A. Ouwehand, M. Dekker, (ed, Acid. Lactic, Microbiological. Bacteria, Aspects. Functional, Edition. Third, Inc, New York; 2004166
18. Hutkins R.W. Microorganisms and Metabolism. In: Microbiology and technology of fermented foods.Blackwell Publishing Professional; 20061566
19. Roberfroid M.B.Concepts and strategy of functional food science: The European perspective. The American Journal of Clinical Nutrition 2000aS1660S1664.
20. L. Kotilainen, R. Rajalahti, C. Ragasa, E. Pehu, enhancing. Health, Opportunities. foods, strengthening. for, sector. the, developing. in, countries, Agriculture and Rural Development Discussion 2006Paper 30.
21. N. S. Kwak, D. J. Jukes, foods. Functional, . Part, The development of a regulatory concept. Food Control 2001a1299107
22. C. Stanton, R. P. Ross, G. F. Fitzgerald, D. Van Sinderen, Fermented functional foods based on probiotics and their biogenic metabolites. Current Opinion in Biotechnology 200516198203
23. A. T. Diplock, P. J. Aggett, M. Ashwell, F. Bornet, E. B. Fern, M. B. Roberfroid, Scientific concepts of functional foods in Europe: Concensus document. British Journal of Nutrition 1999suppl. 1), S1S27.
24. P. Coppens, Fernandes Da Silva M. Pettman S. European regulations on nutraceuticals, dietary supplements and functional foods: A framework based on safety. Toxicology 20062215974
25. Spence J.T.Challenges related to the composition of functional foods. Journal of Food Composition and Analysis 2006S4S6.
26. I. Siró, E. Kápolna, B. Kápolna, A. Lugasi, food. Functional, development. Product, marketing, acceptance. consumer-A, review, Appetite 200851456467
27. M. Mäkinen-Aakula, Trends in functional foods dairy market. In Proceedings of the third functional food net meeting 2006
28. Gibson G.R. Roberfroid M.B.Dietary modulation of the human colonic microbiota: Introducing the concept of prebiotics. Journal of Nutrition 199512514011412

29. F. C. Prado, J. L. Parada, A. Pandey, C. R. Soccol, Trends in non-dairy probiotic beverages- review. Food Research International 200841111123
30. G. D. Mocanu, E. Botez, D. G. Andronoiu, L. Gîtin, Researches concerning the production of a fermented dairy drink with added carrot juice. Journal of Environmental Protection and Ecology 201112718727
31. G. D. Mocanu, E. Botez, O. V. Nistor, D. G. Andronoiu, Characterization of probiotic yoghurt obtained with medicinal plant extracts and modelling of bacteria cell growth during its production. Journal of Agroalimentary Processes and Technologies 2011176571
32. M. Guldas, R. Irkin, Influence of Spirulina platensis powder on the microflora of yoghurt and acidophilus milk. Mljekarstvo 201060237243
33. M. Gomes, F. X. Malcata, Development of Probiotic Cheese Manufactured from Goat Milk: Response Surface Analysis via Technological Manipulation. Journal of Dairy Science, 19988114921507
34. D. G. Andronoiu, E. Botez, O. V. Nistor, G. D. Mocanu, Technological variants for obtaining of a dessert based on fresh probiotic chees. Journal of Agroalimentary Processes and Technologies 2009XV, 102106
35. D. G. Andronoiu, L. Gîtin, E. Botez, G. D. Mocanu, Researches Concerning the Production and Characterisation of a Dessert Based on Fresh Cheese and Peach Pulp. Journal of Environmental Protection and Ecology 201112502508
36. M. Uliescu, G. Rotaru, D. Mocanu, V. Stanciu, Study of the probiotic telemea cheese maturation. The Annals of the University „Dunarea de Jos" Galaţi, Fascicle VI-Food Technology 2007XXX, 9299
37. G. Rotaru, D. Mocanu, M. Uliescu, D. Andronoiu, Research studies on cheese brine ripening. Innovative Romanian Food Biotechnology 200823039
38. M. H. Abd-Salam El, E. Alichanidis, Cheese Varieties Ripened in Brine. In: Fox, P.F.; P.L.H. McSweeney, T. M. Cogan, and T. P. Guinee (ed.) Cheese: chemistry, physics and microbiology, Third edition, 2Elsevier Academic Press, 2004
39. E. Songisepp, T. Kullisaar, P. Hutt, P. Elias, T. Brilene, M. Zilmer, M. A. Mikelsaar, Probiotic. New, with. Cheese, Antioxidative, Activity. Antimicrobial, Journal of Dairy Science 20048720172023
40. Y. Kourkoutas, L. Bosnea, S. Taboukos, C. Baras, D. Lambrou, M. Kanellaki, Probiotic Cheese Production Using Lactobacillus casei Cells Immobilized on Fruit Pieces. Journal of Dairy Science 20068914391451
41. D. G. Mocanu, G. Rotaru, A. Vasile, E. Botez, D. G. Andronoiu, O. Nistor, G. Vlăsceanu, A. Dune, Researches on maintaining functionality in the storage of probiotic dairy product- Afinolact. Journal of Agroalimentary Processes and Technologies 200915229233
42. D. G. Mocanu, G. Rotaru, A. Vasile, E. Botez, D. G. Andronoiu, O. Nistor, G. Vlăsceanu, A. Dune, Studies on the production of probiotic dairy products based on milk and medicinal plant extracts. Journal of Agroalimentary Processes and Technologies 200915234238
43. G. D. Mocanu, G. Rotaru, E. Botez, A. Vasile, D. Andronoiu, O. Nistor, L. Gîtin, G. Vlăsceanu, A. Dune, Research concerning the production of a probiotic dairy product with added medicinal plant extracts. The Annals of the University „Dunărea de Jos" of Galaţi. Fascicle VI- Food Technology 2009XXXII, 3744
44. G. D. Mocanu, G. Rotaru, E. Botez, A. Vasile, D. Andronoiu, O. Nistor, L. Gîtin, G. Vlăsceanu, A. Dune, Sensorial characteristics and rheological properties of probiotic product Cătinolact. The Annals of the University „Dunărea de Jos" of Galaţi. Fascicle VI- Food Technology 2009XXXII, 6469
45. G. D. Mocanu, G. Rotaru, E. Botez, L. Gîtin, D. Andronoiu, O. Nistor, G. Vlăsceanu, A. Dune, Sensory evaluation and rheological behavior of probiotic dairy products with Rosa canina L. and Glycyrriza glabra L. extracts. Innovative Romanian Food Biotechnology 200943239
46. D. Mocanu, G. Rotaru, E. Botez, D. Andronoiu, O. Nistor, Probiotic yogurt with medicinal plants extract: Physical-chemical, microbiological and rheological

characteristics. Journal of Agroalimentary Processes and Technologies 201016469476
47. C. Manach, A. Mazur, A. Scalbert, Polyphenols and prevention of cardiovascular diseases. Current Opinion in Lipidology. Nutrition and metabolism 20051617
48. M. Hărmănescu, A. Moisuc, F. Radu, S. Drăgan, I. Gergen, Total polyphenols content determination in complex matrix of medical plants from Romania by NIR spectroscopy. Bulletin UASVM, Agriculture 200865123128

CHAPTER 8

Microstructure and Composition of Full Fat Cheddar Cheese Made with Ultrafiltered Milk Retentate

Lydia Ong [1,2], Raymond R. Dagastine [1], Sandra E. Kentish [1] and Sally L. Gras [1,2,*]

[1]*Particulate Fluid Processing Centre, Department of Chemical and Biomolecular Engineering, The University of Melbourne, Parkville, VIC 3010, Australia*

[2]*The Bio21 Molecular Science and Biotechnology Institute, The University of Melbourne, Parkville, VIC 3010, Australia*

ABSTRACT

Milk protein is often standardised prior to cheese-making using low concentration factor ultrafiltration retentate (LCUFR) but the effect of LCUFR addition on the microstructure of full fat gel, curd and Cheddar cheese is not known. In this work, Cheddar cheeses were made from cheese-milk with or without LCUFR addition using a protein concentration of 3.7%–5.8% w/w. The fat lost to sweet whey was higher in cheese made from cheese-milk without LCUFR or from cheese-milk with 5.8% w/w protein. At 5.8% w/w protein concentration, the porosity of the gel increased significantly and the fat globules within the gel and curd tended to pool together, which possibly contributed to the higher fat loss in the sweet whey. The microstructure of cheese from cheese-milk with a higher protein concentration was more compact, consistent with the increased hardness, although the cohesiveness was lower. These results highlight the potential use of LCUFR for the standardization of protein concentration in cheese-milk to 4%–5% w/w (equivalent to a casein to total protein ratio of 77%–79% w/w) to increase yield. Beyond this concentration, significant changes in the gel microstructure, cheese texture and fat loss were observed.

Keywords: Cheddar cheese; fat retention; microstructure; ultrafiltration

1. INTRODUCTION

The protein concentration of milk varies according to the season, weather, feed type, stage of lactation and breed of lactating cows. This variation has a major influence on the coagulation of milk and production of cheese. Milk standardization minimizes the effect of the variations in milk on the composition and quality of cheese. Several studies have reported the use of milk ultrafiltration (UF) retentate for milk standardization prior to cheese-making, highlighting the potential benefits of standardization with UF, including an increase in cheese yield and plant productivity [1,2,3,4,5,6,7]. Higher milk protein concentrations also result in less whey being expelled during cheese making, reducing the volume of whey that requires processing [3].

The UF preparation used most widely in the standardization of cheese-milk for Cheddar cheese manufacture is low concentration UF retentate (LCUFR), with a concentration factor of approximately 1.6–1.7 fold [8]. The process of ultrafiltration changes the milk composition, increasing the total solids content and decreasing the non-protein nitrogen (NPN), soluble calcium and lactose, as these latter components permeate through the UF membrane. Ultrafiltration can also cause a decrease in micelle size and number of hydrophobic sites on the surface of milk proteins, as assessed by 1-anilinonaphthalene-8-sulfonate (ANS) binding [9]. These surface changes in turn further decrease the surface potential of casein micelles.

The changes in the milk properties after ultrafiltration are known to affect rennet induced coagulation [10,11,12,13] and the rheological properties of rennet gels have been comprehensively examined. Studies on the effect of LCUFR on the microstructure of the gel, curd and cheese are, however, limited.

In a previous study where skim milk retentate was used to give 3%–19% w/w protein, the gels formed from UF concentrated milk were much firmer, as the increased milk protein concentration decreased the mean distance between casein micelles [10]. This proximity increased the rate of aggregation of para-casein micelles, as aggregation depends on the number of effective collisions. A study by Green [7] looked at the microstructure of the gel, curd and cheese progressively during Cheddar cheese making from skim milk that was concentrated 1.7–4 fold and then blended with cream. The samples were examined by a conventional scanning electron microscopy (SEM) and transmission electron microscopy (TEM), however, only the microstructure of the protein was shown as fat is removed during sample preparation for conventional SEM or TEM [7]. Green found that the 2–4 fold UF concentration lead to a coarser protein network, a reduced volume of whey and more fat in the whey compared to cheese made using unconcentrated milk. Curd formation was faster when a higher concentration of milk protein was used, despite the reduced ratio of rennet to protein and the final texture of the cheese tended to be granular and hard. The authors used light microscopy to examine the distribution of the fat globules but only details of the surface microstructure could be obtained using this method [7].

Other microstructure studies have only focused on the very early stages of cheese making. Confocal laser scanning microscopy (CLSM), for example, was used to look at the microstructure of the gel from UF concentrated skim milk with a casein content enriched from 2.75% to 19.8% w/w [14]. The UF gel contained smaller aggregates and fewer hydrolysed kappa casein molecules as compared to in unconcentrated milk (20% c.f. 50%). After 24 h at 30 °C, syneresis was observed and the protein strands within the concentrated samples were longer and thicker. Although this study provides useful insight into the structure of gels made using UF concentrated skim milk, no cheese making was performed to examine the effect of the altered gel structure on the microstructure of curd and final cheese.

The present study looks beyond the changes in the microstructure of the gel to later processing steps. Curd and cheese are prepared using pilot scale manufacturing conditions with cheese-milk and UF retentate from an industrial scale process; so that results are highly applicable to commercial production. The present study also applies advanced microscopy techniques including cryo SEM and CLSM to investigate the microstructure of samples made using cheese-milk standardized with LCUFR. The ability to preserve the fat within the curd microstructure under cryogenic conditions provides new insights into the distribution and changes to the fat microstructure during manufacturing. The influence of microstructure on the composition, texture and yield of the final cheese product is also examined.

2. EXPERIMENTAL SECTION

2.1. Acidification of Cheese-Milk by Starter Bacteria

The effect of protein concentrations of 4%, 5% or 6% w/w on the acidification of milk by starter bacteria was investigated prior to cheese making experiments. Target protein concentrations were achieved by blending raw milk with raw low concentration factor ultrafiltration retentate (LCUFR, protein = 7.46% w/w and fat = 8.62% w/w) obtained from an ultrafiltration process at 14 °C. Cream was then added to the milk to achieve a final fat to protein ratio of 0.84 before pasteurization at 72 °C for 15 s. All milk preparations were obtained from a local cheese factory (Murray Goulburn, Melbourne, VIC, Australia). Freeze dried mixed strain direct vat set (DVS) mesophilic starter culture (0.13 U/kg; Chr. Hansen, Melbourne, VIC, Australia) was then added to the pasteurized and standardized cheese-milk at 33 °C and incubated for 330 min. The pH of the milk was recorded at 15 min intervals in the first 150 min then at 60 min intervals until the end of the fermentation. The experiment was repeated 3 times (n = 3).

2.2. Manufacture of Cheddar Cheese

Four batches of Cheddar cheeses were made in triplicate at the Murray Goulburn pilot plant, Cobram, VIC, Australia. The experimental plan is shown in Table 1. Batch 1 was made using pasteurized raw milk with no protein standardization (the fat was standardized as described in Section 2.1). Batches 2–4 were made using milk prepared with a target protein concentration of 4%, 5% and 6% w/w respectively. The protein and fat concentration of the milk was also achieved using the standardization method described in Section 2.1. These milk preparations were then used for cheese making and are referred to as cheese-milk.

The total protein in each cheese vat was normalised to approximately 8 kg of protein by decreasing the volume of milk used for preparations with an increased protein content (Table 1). This ensured that the curds prepared using preparations with a higher protein content could fit within the 12 kg limitation of the cheese press. Four cheese-presses were available and two batches of cheese were made each day giving 4 blocks of cheese each 10–12 kg in weight. After pasteurization, the cheese-milk was cooled to 33 °C before inoculation with 25 U/vat starter culture (Chr. Hansen, Melbourne, VIC, Australia) to give final starter culture concentration as shown in Table 1. The milk was then ripened for 20 to 40 min until the pH of the milk reached approximately pH 6.5 as shown in Table 1. Rennet (Hannilase, 690 IMCU/mL; Chr. Hansen, Melbourne, VIC, Australia) was added to give a final concentration of 0.06 mL/kg of milk for all treatments. The milk was allowed to coagulate at 33 °C for 20–45 min as determined in Section 2.3. The consistency of the gel made using the different protein concentrations before cutting was classified as "medium set" by an experienced cheese maker. The gel was cut over a period of 20 min and cooked from 33 °C to 38 °C with stirring over a period of 45 min. The sample was then cooked at 38 °C with stirring until the pH of the samples reached approximately pH 6.2. The sweet whey was released and collected for compositional analysis.

Table 1. Experimental plan for pilot scale trial investigating the effect of protein concentration on the microstructure and composition of Cheddar cheese.

Batch	Target milk protein (% w/w)	Weight of milk in cheese vat (kg)	Milk fat (% w/w) #	Milk protein (% w/w) #	Starter culture concentration (U/kg milk) *	Starter culture concentration in relation to milk protein content (U/kg protein) *	Ripening time ^ (min)	Rennet concentration (mL/kg milk)	Rennet concentration in relation to milk protein content (mL/kg protein) #
1	3.5	235	4.37 ± 0.06	3.69 ± 0.05	0.11	3.0 ± 0.2	20	0.06	1.7 ± 0.1
2	4.0	200	4.83 ± 0.02	4.02 ± 0.01	0.13	3.1 ± 0.0	20	0.06	1.5 ± 0.0
3	5.0	160	5.67 ± 0.61	4.79 ± 0.34	0.16	3.3 ± 0.2	30	0.06	1.2 ± 0.1
4	6.0	135	6.97 ± 0.10	5.76 ± 0.06	0.19	3.2 ± 0.0	40	0.06	1.0 ± 0.0

* 10 U of freeze dried culture corresponds to 1 L of active bulk culture, defined here as active starter culture ready to be used for cheese-making; ^ Ripening time is the time from starter addition to rennet addition; # Results are expressed as mean ± standard deviation of mean ($n = 3$).

Cooked curd samples were piled into blocks ~15 cm in height and turned every 15 min during cheddaring until the pH of the curd reached pH 5.3. Curd was milled and salted with 770 g salt per vat before pressing at 200 psi. The salty whey that was released during pressing was collected and the composition analysed.

2.3. Viscoelastic Properties of Cheese-Milk with Different Protein Concentrations

The storage modulus (G') of cheese-milk with different protein concentrations were analysed before the main cheese-making experiments using a rheometer (ARES-TA Instruments, New Castle, DE, USA), as described previously [15]. A dynamic time sweep analysis at angular frequency of 0.8 Hz and 0.1% strain was used to analyze the changes in the storage modulus (G') as the milk gelled. The time taken for the cheese-milk to first reach a G' of 60 Pa was recorded and used as a guide for the cutting time in the pilot scale cheese-making experiments. The experiment was performed twice (n = 2).

2.4. CLSM, Cryo SEM and Image Analysis

The microstructure of the gel, curd and cheese samples were analysed using a Confocal Laser Scanning Microscopy (Leica Microsystems, Heidelberg, Germany) and a cryo Scanning Electron Microscopy (Quanta; Fei Company, Hillsboro, OR, USA) using the method reported in a previous study [16].

The cooked curd samples were collected from the cheese vat immediately after the whey draining process. Milled curd was collected prior to salting. The cheese samples were collected after 16 h of pressing. Samples were sealed in air tight containers to prevent drying and kept at 4 °C for analysis within 3 days of sample collection. Gel samples were prepared separately in the laboratory using the same cheese-milk used for the pilot scale cheese-making trial and observed immediately within 15 min after gel formation. The details of sample preparation for CLSM and cryo SEM observations have been reported in a previous study [16].

The 3D CLSM image reconstruction and analysis were performed using a commercial software package (Imaris, Bitplane, South Windsor, CT, USA), as described previously [15]. Quantitative outputs such as the number of fat globules per unit volume, their sphericity, mean volume and diameter, the total volume of fat and the fraction of pore volume with respect to the total sample volume (porosity) were obtained from the image analysis. Two 3D images were taken for each batch of the cheese produced. Three batches of cheeses were produced for each milk protein concentration, giving a total of 6 analyses for each data point presented from image analysis.

2.5. Compositional Analysis, Fat and Protein Retention and Cheese Yield

A Milko Scan FT120 (FOSS, North Ryde, NSW, Australia) was used to analyse the protein and fat content of the milk and sweet whey. Non-casein nitrogen (NCN) and non protein nitrogen (NPN) within the milk samples were determined using Kjeldahl analysis from the filtrate obtained after selective precipitation at pH 4.6 or the filtrate from a 12% trichloroacetic acid solution, as described in AOAC Methods 991.20 and 991.21 [17]. The casein content was calculated by difference from the NCN concentration and total protein concentration.

The fat, protein, total solids, salt, pH and moisture content of the cheese were analysed as described previously [15]. The fat and protein lost in the whey or retained in the cheese, the cheese yield and dry matter cheese yield were also calculated as described in Ong *et al.* [15]. The total amount of protein and fat lost in the whey plus that retained in the cheese was in all cases between 93% and 99% of the initial amount of each ingredient added.

2.6. Texture Analysis

Texture Profile Analysis was performed on cheese samples using a texture analyser TAXT (Stable Micro System, Surrey, UK) as described previously [15]. Texture analyses were performed six times for six independent samples from each batch of cheese giving six pseudo replicates which were then averaged. Three batches of cheeses were made for each milk protein concentration and the results were expressed as the average of three means for each protein concentration.

2.7. Determination of Total Calcium Concentration in Milk and Cheese

Inductively Coupled Plasma Optical Emission Spectroscopy (ICP-OES; Varian Inc., PaloAlto, CA, USA) was used to analyse the total calcium of milk and cheese. The details of these methods have been reported elsewhere [15]. Two samples were collected from each batch and three batches of cheeses were made for each milk protein concentration, giving a total of 6 analyses for each data point presented.

2.8. Enumeration of Starter Bacteria

To assess the viability of the starter bacteria and the non-starter lactic acid bacteria, the cheese samples (5 g) were diluted in 45 mL of sterile 2% (w/v) trisodium citrate (Oxoid Ltd., West Heidelberg, VIC, Australia). The sample was homogenized at 10,000 g using a high speed homogeniser (Polytron, Kinematica, Lucerne, Switzerland) for 1 min to obtain a cheese slurry for the first dilution. The homogenizer was equipped with a shaft that could be detached from the rotor. This shaft was rinsed thoroughly with warm sterile distilled water to remove any residual cheese. The shaft was then sterilized with 80% (v/v) ethanol and dried between use for samples from different cheese batches.

Subsequent serial dilutions were performed in sterile 0.15% (w/v) peptone solution (Oxoid Ltd.). The starter culture was enumerated on

LM17 agar (Merck, South Granville, NSW, Australia) and incubated at 30 °C under aerobic conditions for 72 h [18]. The NSLAB was enumerated on Rogosa agar (Merck) and incubated at 30 °C in an anaerobic jar (Oxoid Ltd.) with a Gas Generating Kit® (Oxoid Ltd.) for 72 h [18]. The result is presented as the mean of three results obtained from three batches of cheeses from the same milk protein concentration.

2.9. Statistical Analysis

Data was analysed using a statistical package from Minitab (Minitab Inc., State College, PA, USA). The differences between means were determined by one way analysis of variance (ANOVA) and Tukey's paired comparison, with a significance level of $\alpha = 0.05$.

3. RESULTS AND DISCUSSION

3.1. The Effect of Milk Protein Concentration on the Acidification by Starter Bacteria

A set of laboratory experiments was conducted prior to the pilot scale cheese-making experiments, to determine the effect of milk protein concentration on the acidification and setting time of the gel. As shown in Figure 1a, the milk with a higher protein concentration (5.8% w/w protein) had a higher buffering capacity with a higher pH after incubation at 33 °C for an extended period of 330 min.

Rennet is normally added within the first 30 min of starter culture addition during Cheddar manufacture. While large variations in pH at this point may affect coagulation and the resulting microstructure, the minor differences in milk pH at the time when rennet is usually added (Figure 1a arrow) are not likely to affect the coagulation process. The starter culture concentration was increased to compensate for the increased buffering capacity so that ~3.1 U of starter was added for every kilogram of milk protein (Table 1). This adjustment prevents an increase in cooking time at the whey-off period and also ensures a similar concentration of starter culture in each cheese.

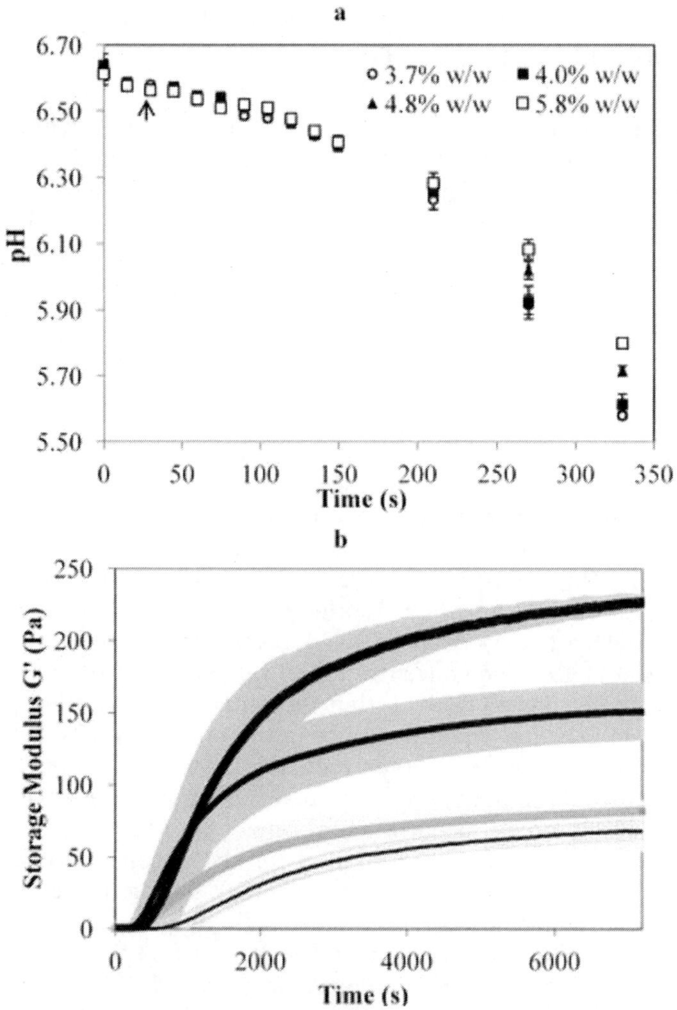

Figure 1. (**a**) The pH of unconcentrated cheese-milk with 3.7 % w/w milk protein and cheese-milk standardized with low concentration factor ultrafiltration retentate (LCUFR) to 4% w/w, 4.8% w/w or 5.8% w/w milk protein after starter culture addition. The arrow indicates the point where rennet is normally added. (**b**) Storage modulus (G') measured from the time of rennet addition for unconcentrated cheese-milk with 3.7% w/w milk protein (thin line) and cheese-milk standardized with LCUFR to 4% w/w (grey line), 4.8% w/w (medium thick line) or 5.8% w/w (thick line) milk protein incubated at 33 °C. The setting time is the total time required for the sample to reach a G' of 60 Pa. Results are expressed as mean ± standard deviation of mean from three (*n* = 3) or two (*n* = 2) independent trials for (**a**) and (**b**) respectively.

3.2. The Effect of Milk Protein Concentration on the Elastic Modulus and Processing Time

The storage modulus (G') of the gel formed from milk with a higher protein concentration was significantly higher than for milk with a lower protein concentration (Figure 1b). The rate of gel firming also increased rapidly in milk with a higher protein concentration, as reflected by the gradient of the slope of the storage modulus and the decrease in the time required to obtain a given value of G'. The G' increased rapidly between 200 and 400 s for all gels (Figure 1b), indicating a similar onset of gelation.

In our previous experiments, the time from rennet addition to cutting, known as the setting time of the gel, was set at 2700 s corresponding to a G' of 60 Pa, as this gave a Cheddar cheese with acceptable moisture content using standardised milk with a protein concentration of 4% w/w [15]. To maintain a similar level of stiffness of 60 Pa, the setting time of the gel made with 4.8% or 5.8% w/w milk protein was set to 1200 s (Figure 1b). This adjustment prevents the curd becoming too firm and avoids curd fines in the whey. Another way to reduce the firmness of a gel is to reduce the coagulation temperature or reduce the amount of rennet used [1,19]. The concentration of rennet added per unit mass of milk was similar here (Table 1), but the rennet concentration per kg of milk protein was consequently lower in the milk with a higher protein concentration (Table 1). Further reducing the rennet for high protein treatments would impact on the residual rennet in the final cheese. For this reason, the same concentration of rennet per unit mass of milk was used in this study. The time required for the unconcentrated milk to reach G' = 60 Pa was about 5000 s (Figure 1b). During the actual cheese-making experiment, however, the milk in the cheese vat reached a gel consistency of medium firmness and was ready for cutting at 2700 s. The longer time indicated by the rheology analysis could possibly be due to a small degree of shearing applied to the sample during analysis that disturbed the gel formation. This was not apparent in the concentrated sample, but the setting time indicated by the rheology analysis in this study was less accurate for gel made with the unconcentrated milk.

The processing time from the addition of starter culture to whey draining or milling in the pilot scale cheese-making was not significantly different ($p > 0.05$) (data not shown), suggesting that adjustments made to the starter concentration and the setting time described above were effective.

3.3. Composition of Cheese-Milk, Sweet Whey and Salty Whey

Ultrafiltration is known to alter the composition of milk. Approximately 81% of the protein within the LCUFR used in this study was casein as compared to 77% in unconcentrated milk. The LCUFR also contains less non protein nitrogen (NPN) arising from low molecular weight nitrogen containing compounds, such as creatine and creatinine and urea [20] as a portion of the NPN permeates through the UF membrane during ultrafiltration [21]. Consequently the concentration of NPN as a percentage of total nitrogen (TN) differed significantly in the LCUFR and unconcentrated milk from 2.5% w/w to 4.6% w/w, respectively.

The impact of standardization of cheese-milk on casein and NPN content is shown in Table 2. As the casein concentration increased with LCUFR addition, the ratio of casein nitrogen over TN increased and the ratio of NPN over TN decreased. These different CN/TN or NPN/TN concentrations in the cheese-milk are expected to impact the final cheese yield [22].

Table 2. Protein composition, total solids and total calcium content of cheese-milk used for pilot scale cheese-making #.

Batch	Casein (% w/w)	CN/TN (% w/w)	NPN/TN (% w/w)	Total solids (% w/w)	Total calcium (mg/kg)	P/F ratio
1	2.83 ± 0.04^{d}	76.7 ± 0.00^{d}	4.61 ± 0.00^{a}	13.6 ± 0.14^{d}	985 ± 62^{d}	0.84 ± 0.00
2	3.12 ± 0.01^{c}	77.5 ± 0.15^{c}	4.19 ± 0.09^{b}	14.4 ± 0.07^{c}	1108 ± 31^{c}	0.83 ± 0.01
3	3.78 ± 0.28^{b}	78.9 ± 0.19^{b}	3.42 ± 0.10^{c}	15.9 ± 1.20^{b}	1223 ± 52^{b}	0.85 ± 0.04
4	4.60 ± 0.04^{a}	79.9 ± 0.00^{a}	2.83 ± 0.02^{d}	18.1 ± 0.53^{a}	1500 ± 34^{a}	0.83 ± 0.00

Results are expressed as mean ± standard deviation of mean ($n = 3$); CN = casein, NPN = non protein nitrogen and TN = total nitrogen; A–d Means across a single column with different superscripts are significantly different ($p < 0.05$); Means across a single column without superscripts are not significantly different ($p > 0.05$).

The weight and composition of the whey generated during cheese-making at pilot scale is shown in Figure 2a,b. The weight of sweet whey and salty whey was significantly reduced ($p < 0.05$) for samples made with higher protein milk. This was expected and is consistent with previous reports that less processing of whey is required when milk with a higher protein concentration is used for cheese-making [2]. The percentage of fat lost to the sweet whey was significantly higher ($p < 0.05$) in cheese made from milk without the addition of LCUFR (Figure 2c). Greater fat loss also occurred when the concentration of milk protein was the highest (5.8% w/w) despite the increase in CN/TN, with minimum levels of fat loss occurring for samples made using cheese-milk with 4% or 4.8% (w/w) protein (Figure 2c). Conversely, while the percentage of protein in the sweet whey increased continuously as the protein in the cheese milk increased (Figure 2a), the total protein lost to this whey stream reduced continuously as the protein concentration in the cheese milk increased (Figure 2c).

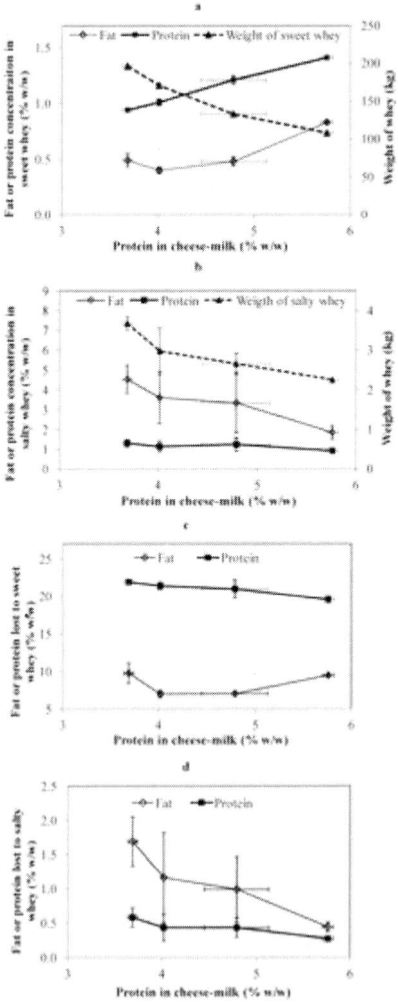

Figure 2. The composition and weight of sweet whey (**a**) and salty whey (**b**) collected during cheese-making at pilot scale using cheese-milk with different protein concentrations. The percentage of fat lost or protein lost to the sweet whey (**c**) and salty whey (**d**) were calculated on the basis of total fat or protein in the cheese-milk. Results are expressed as mean ± standard deviation of mean (n = 3).

The variability in fat composition measured for the salty whey was quite high. This could be due to the sampling technique used to obtain the salty whey at a pilot scale, as it was noted that some fat adhered to the cheese-press instead of flowing out with the salty whey, leading to greater variability. There is still clear evidence, however, of both a decline in the concentration of fat in the salty whey and the total fat loss (Figure 2b,d). The trends in the protein data are less clear but there is still some evidence of a decrease in protein loss into the salty whey as the protein content of the original cheese milk increases (Figure 2d).

3.4. Fat and Protein Retention, the Composition and Yield of Cheese

Consistent with the changes in the sweet and salty whey, the percentage of protein retained in the cheese increased with the protein in the cheese milk, although the trend is within experimental error (Figure 3a). The percentage of fat retained in the cheese also increases up to a level of 4.8% w/w total protein in the cheese milk. No further increase is observed at 5.8 w/w protein. In a similar study, Govindasamy-Lucey *et al.* [3] reported no significant difference in fat retention when increasing the milk protein concentration from 3.2% to 5.0% w/w using whole milk UF retentate [3]. Overall, the mean values of fat retention for all cheeses (Figure 3a) were within the range of values (~83%–92% w/w) reported elsewhere for full fat Cheddar cheese [23].

The composition of all cheeses was also within the acceptable range for Cheddar cheese. The fat content of the cheese made with LCUFR standardized cheese-milk with 4.8% or 5.8% w/w protein was significantly higher ($p < 0.05$) than cheese made using unconcentrated cheese-milk (Figure 3b). This could be due to the higher fat retention but also the lower moisture content of these cheeses (Figure 3b). The effect on cheese moisture suggests that the addition of LCUFR could be used as a tool to reduce the moisture content of Cheddar cheese when seasonal differences make it difficult to achieve a low moisture Cheddar [24]. The protein and salt content were not significantly different between different treatments ($p > 0.05$, Figure 3b). Our results are in agreement with a previous study [1], where milk protein level was increased in the range of 30–70 g/L with little effect on Cheddar composition.

The weight of the cheeses made using cheese-milk with ~3.7%, 4%, 4.8% and 5.8% w/w protein were 24.4 ± 1.7, 24.1 ± 0.56, 22.5 ± 1.3 and 23.0 ± 0.41 kg respectively. The yield calculation on the basis of milk volume and the dry matter yield increased linearly with protein content (Figure 3c), due to the higher protein and total solids content of the LCUFR standardized cheese-milk (Table 2).

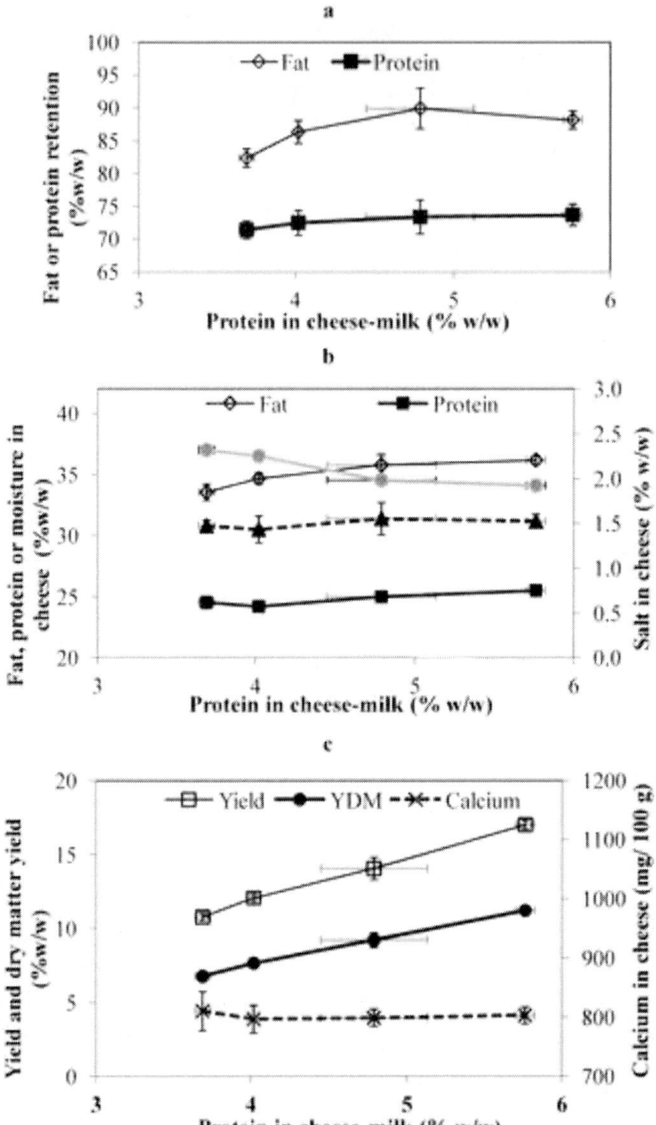

Figure 3. The fat and protein retention (**a**), composition (**b**) and yield (**c**) of cheese prepared at pilot scale using cheese-milk with different protein concentrations. Fat retention or protein retention in the cheese were calculated on the basis of total fat or protein in the cheese-milk. Results are expressed as mean ± standard deviation of mean (n = 3).

The concentration of total calcium in the cheese-milk increased with the increased concentration of protein (Table 2), although the ratio of total calcium (TC) over total protein (TP) in the cheese-milk was similar at 24 mg/g to 28 mg/g ($p > 0.05$). The concentration of TC in the whey was not significantly different ($p > 0.05$). The TC in the whey was 380 ± 32, 392 ± 40, 383 ± 26 and 416 ± 29 mg/kg from cheese-making with 3.7%, 4%, 4.8% and 5.8 % w/w milk protein, respectively. Increasing the concentration of milk protein did not affect the total calcium in the final cheese (Figure 3c). Less milk was added to the cheese vat for the treatments with increasing concentrations of protein. This adjustment led to similar amount of TC at a range between 230 g to 200 g in all vats regardless of protein concentration.

3.5. Microstructure of Gel, Curd and Cheese

The microstructure of the gel observed using cryo SEM is shown in Figure 4a–d. After the addition of rennet, casein micelles aggregate to form a protein network in which the fat globules are entrapped. Qualitatively, there was no clear distinction between the microstructure of the gels made using cheese-milk with ~3.7%–4.8% w/w milk protein (Figure 4a–c respectively). The microstructure of the gel with 5.8% w/w milk protein, however, was denser with smaller pores (indicated by the black areas in Figure 4d). This dense network might arise due to the increased aggregation of casein micelles. As discussed previously, the high concentration of protein in the samples decreased the mean free distance between casein micelles, resulting in a densely aggregated CM network.

It is interesting to note that the gel stiffness was similar for all samples regardless of the protein concentration during cutting but the structure of the gel made from milk with 5.8% protein was very different. The stiffness of the gel could be attributed to two factors; the number of aggregating CM and the strength of the bonds between the CM and the relative importance of these factors may differ between treatments.

Representative CLSM images showing the microstructure of gel made from ~3.7% and 5.8% w/w milk protein are shown in Figure 5 and a complete set of 2D CLSM images is given in supplementary Figure S1.

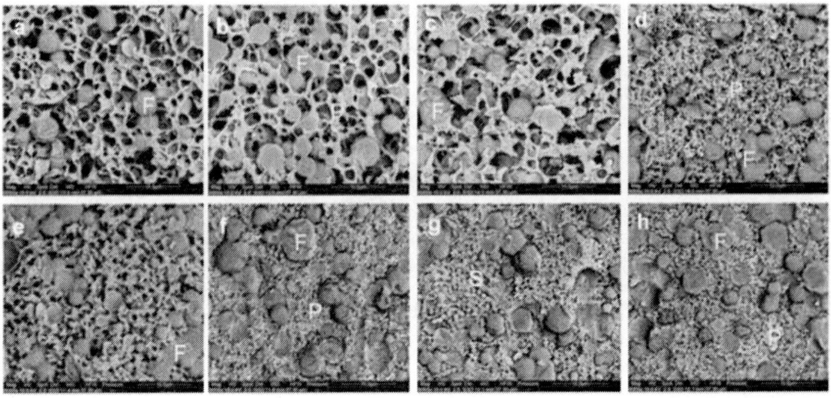

Figure 4. Cryo SEM micrographs of gels (**a–d**) and cooked curd (**e–h**) made using (**a**,**e**) unconcentrated cheese-milk with ~3.7% w/w protein or cheese-milk standardized with LCUFR to (**b**,**f**) 4% w/w, (**c**,**g**) 4.8% w/w or (**d**,**h**) 5.8% w/w protein, where F = fat, P = protein network and S= starter culture bacteria. The gel samples were fixed in slushed liquid nitrogen when it reached the cutting time at 60 Pa. Scale bars within the images are 10 μm in length.

Despite a difference of ~2% in fat content between cheese-milk with ~3.7% w/w and 5.8% w/w protein samples, the number of fat globules and the total volume of the fat globules were not significantly different within the gel ($p > 0.05$, supplementary Figure S2a,e). The image analysis output shows a variation greater than 2% which made it hard to detect the small variation in fat content. The sphericity, fat droplet volume and fat droplet diameter were also similar ($p > 0.05$, supplementary Figure S2b–d). A lower porosity was observed for the gel with 5.8% w/w milk protein (supplementary Figure S2f), consistent with the denser structure observed by cryo SEM.

CLSM provides a poorer resolution of the protein network compared to cryo SEM but a better overview of the distribution of fat and an even distribution of fat was seen within the gel made using the unconcentrated milk (Figure 5b). The porous nature of the protein network in this sample (Figure 5c), however, suggests some of this fat may be lost to the whey after cutting. In contrast, the fat globules tended to pool together at higher protein concentrations and the fat distribution was more heterogeneous (Figure 5e).

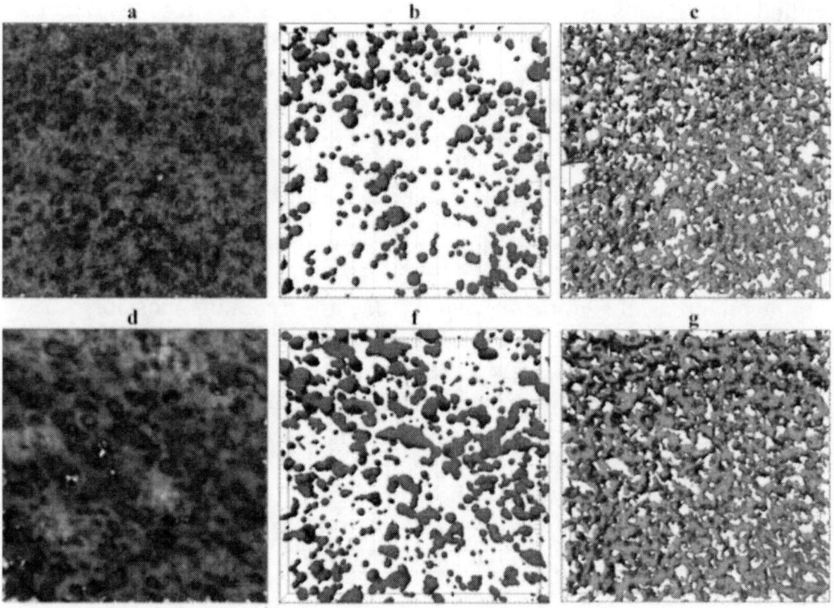

Figure 5. Confocal laser scanning microscopy (CLSM) microstructure of gels made using unconcentrated cheese-milk with ~3.7% w/w milk protein (**a**–**c**) or cheese-milk standardized with LCUFR to 5.8% w/w milk protein (**d**–**f**). All images are 3D reconstructions consisting of 40 layers where the separation between layers is 0.25 μm giving a total observation depth of 10 μm. Nile red stained fat appears red and fast green stained protein appears green. Images **b**–**f** are the rendered volume of the CLSM images, where **b** and **e** show the rendered volume of the fat; **c** and **f** show the rendered volume of the protein. All scale bars are 20 μm in length.

Due to the nature of the pilot study occurring at a manufacturing site, samples of cooked curd, milled curd and cheese were stored at 4 °C for 2 days prior to observation by CLSM. A separate laboratory study was conducted to assess the impact of sample storage and ensure that any small changes in structure did not cause differences between treatments. A small degree of compaction was observed for all these samples regardless of protein content (data not shown) and a small degree of syneresis observed for the cooked curd. The trend of samples made from 5.8% w/w protein displaying a denser structure was not found to be obscured by storage.

The microstructure of the cooked curd samples produced at a pilot scale as observed by cryo SEM, is shown in Figure 4e–h. The heat treatment applied during cooking at 38 °C resulted in further fusion of CM particles within each sample compared to the gel (Figure 4a–d). Some bacterial cells could also be observed within the cooked curd sample (Figure 4g). The only difference in the observed microstructure was the bigger pores remaining within the cooked curd prepared using the unconcentrated cheese-milk (Figure 4e), as compared to cooked curd made using milk with a higher protein concentration (Figure 4f–h).

Representative CLSM images showing the microstructure of the cooked curd produced at a pilot scale are shown in Figure 6. The number of fat globules decreased significantly ($p < 0.05$) after cooking, possibly due to the coalescence of fat globules (supplementary Figure S2a). The sphericity of the fat globules was reduced in some cases (supplementary Figure S2b) and a concurrent increase in the size of the fat globules was observed (supplementary Figure S2c,d). The porosity of the cooked curd samples was not significantly different ($p > 0.05$) for different treatments but the protein strands within the cooked curd made with ~3.7% w/w milk protein (Figure 6c) were qualitatively thinner and less dense than strands observed within cooked curd made with 5.8% w/w milk protein (Figure 6f).

The microstructure of the curd after cheddaring and milling at a pilot scale observed by cryo SEM is shown in Figure 7a–d. Micro-pores of about 1 μm diameter could readily be observed within the milled curd made with a lower concentration of milk protein (Figure 7a,b). This was consistent with the higher moisture content of the lower protein cheeses (Figure 3b). Conversely smaller micro-pores of diameter <1 μm were observed within the milled curd made from a higher concentration of milk protein (Figure 7c,d). The starter culture bacteria could also be observed randomly within the curds formed in all treatments.

The micro-pores seen by cryo SEM (Figure 7) could not readily be observed within the CLSM images and fewer differences between the protein treatments in the milled curd microstructure could be observed by CLSM, as also reflected in the data obtained by image analysis for these samples (supplementary Figures S1 and S2).

The microstructure of the cheese after pressing is shown in Figure 7e–h. Pressing completely closes the micro-pores observed earlier within the milled curd (Figure 7a–d). Qualitatively, the microstructure observed by cryo SEM or CLSM for cheese made using the unconcentrated cheese-milk and cheese-milk with 5.8 % w/w protein were not different (Figure 7e–h and supplementary Figure S1). The size of the fat globules within cheeses made from cheese-milk with 5.8% w/w protein, however, was significantly bigger than in cheese made using unconcentrated milk (supplementary Figure S2c,d). This difference may influence the flavour perception and texture of the cheese.

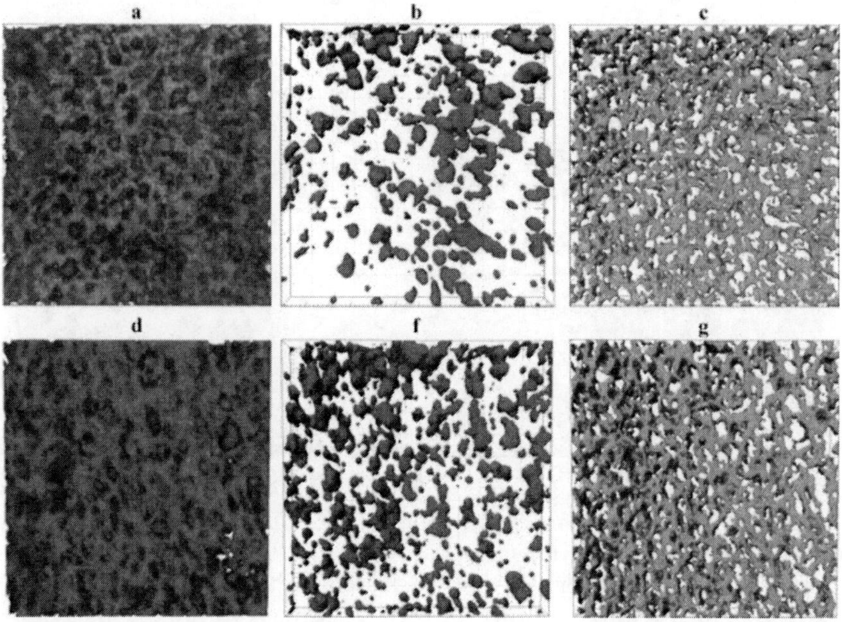

Figure 6. CLSM microstructure of cooked curd made using unconcentrated cheese-milk with ~3.7% w/w milk protein (**a**–**c**) or cheese-milk standardized with LCUFR to 5.8% w/w milk protein (**d**–**f**). All images are 3D reconstructions consisting of 40 layers where the separation between layers is 0.25 μm giving a total observation depth of 10 μm. Nile red stained fat appears red and fast green stained protein appears green. Images **b**–**f** are the rendered volume of the CLSM images, where b and e show the rendered volume of the fat; **c** and **f** show the rendered volume of the protein. All scale bars are 20 μm in length.

3.6. Texture Profile Analysis of the Cheese

The textural profile of the cheese measured within a week of cheese manufacture is shown in Figure 8. The resistance to compression also known as the hardness, increased as the milk protein level increased ($p < 0.05$) (Figure 8). This result indicates that the cheese structure was more rigid, possibly due to the higher concentration of protein in the cheese-milk during gelation, the difference in microstructure or the lower moisture content of the cheese made using milk with greater protein. No significant measured differences in gumminess were observed (Figure 8). Interestingly, the cohesiveness for cheese made from 5.8% w/w milk protein was significantly lower ($p < 0.05$) than for cheese made from unconcentrated milk (Figure 8). Further, we observed that cheeses made with 5.8% w/w milk protein crumbled more easily during the compression test.

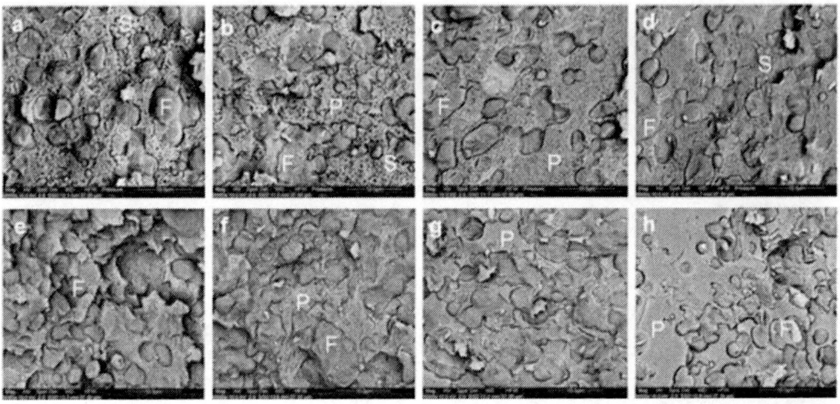

Figure 7. Cryo SEM micrographs of the milled curd (**a**–**d**) and Cheddar cheese (**e**–**h**) made using (**a**,**e**) unconcentrated cheese-milk with ~3.7% w/w protein or cheese-milk standardized with LCUFR to (**b**,**f**) 4% w/w, (**c**,**g**) 4.8% w/w or (**d**,**h**) 5.8% w/w protein, where F = fat, P = protein network and S = starter culture bacteria. Scale bars within the images are 10 μm in length.

3.7. The Starter Bacteria and Total Lactobacilli Count in the Cheese

Increasing the concentration of starter culture bacteria as the concentration of milk protein increased (Table 1) resulted in a similar population of bacteria in all cheeses (supplementary Table S1). Any variations observed in the microstructural or textural properties of the samples were therefore not due to a different concentration of starter bacteria. The number of the total lactobacilli present was variable between treatments, possibly reflecting variation that can be encountered at a pilot scale in an industrial setting.

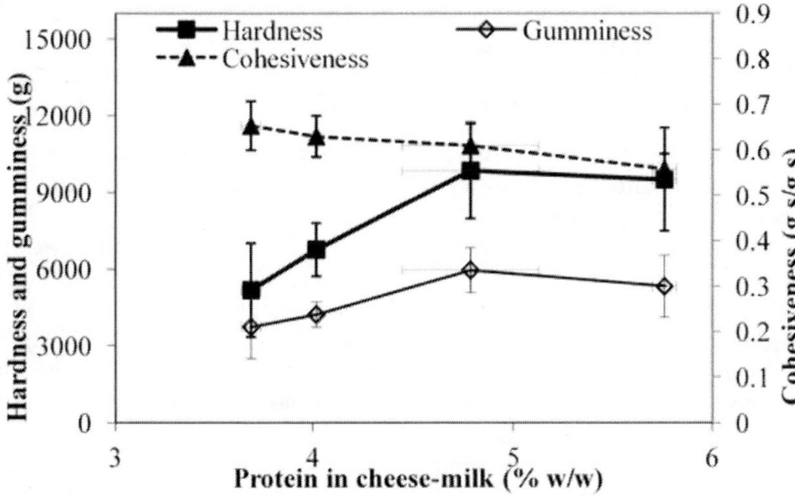

Figure 8. Texture profile analysis showing the hardness, gumminess, and cohesiveness of Cheddar cheeses made using unconcentrated cheese-milk with ~3.7% w/w milk protein or cheese-milk standardized with LCUFR to 4% w/w, 4.8% w/w or 5.8% w/w milk protein. Data are presented as mean ± standard deviation of mean (*n* = 3).

4. CONCLUSIONS

This pilot scale study highlights the potential of using cheese-milk with increasing concentrations of protein to increase the total cheese yield of Cheddar cheese by up to 6% w/w and yield in dry matter by up to 4% w/w for every kg of cheese-milk. These modifications result in minimal changes to the cheese composition, provided the setting time and starter concentration are adjusted during cheese making.

During manufacture, the increased protein concentration decreased the volume of the sweet and salty whey generated, potentially reducing the cost associated with processing this by-product. Less fat was also lost in the salty whey for 5.8% w/w protein samples. Protein addition could also be used as a tool when cheese with a lower moisture content is required and led to only subtle changes in the microstructure, with denser gels, denser milled curds and larger fat globules in the pressed cheese made with milk with 5.8% w/w protein. The cheeses made with higher milk protein (4.8% w/w or 5.8% w/w) were harder than cheeses made with unstandardised milk but not significantly harder than cheeses made with 4% w/w milk protein, which is often used for cheese manufacture. Cheeses made with milk with 5.8% w/w protein were less cohesive, however, suggesting that the optimal concentration may be 4.8% w/w milk protein, where the cheese fat retention was improved by ~7.5% w/w on the basis of the fat levels in the cheese-milk and the cheese yield increased without associated changes in cohesiveness.

Our results show that structural differences were most significant at the early stage of cheese-making. The bigger pores observed within the cooked curd from the unconcentrated milk may contribute to the higher fat loss observed for these samples and confirms that the practice of standardising milk protein has a positive effect on cheese production. At 5.8% w/w protein concentration, the fat globules within the gel and curd tended to pool together, which possibly contributed to the higher fat loss in the sweet whey. The ability to observe the microstructure of the fat within the curd and cheese using CLSM and cryo SEM gives insights into how fat may be lost during processing. This knowledge may help manufacturers to further optimize the cheese-making process. Our findings also highlight how LCUFR can be added to cheese milk as a tool to improve cheese production and to potentially generate new cheese varieties with different textural characteristics.

ACKNOWLEDGMENTS

This work was funded by Dairy Innovation Australia Limited and the Australia Research Council (LP0883300). The authors would like to thank the Bio21 Molecular Science & Biotechnology Institute and the Particulate Fluids Processing Centre, which is a Special Research Centre of the Australian Research Council at The University of Melbourne, for access to equipment. We also thank Murray Goulburn Co Op. for providing access to the cheese-making facility in Cobram (Victoria, Australia) and for the supply of cheese-milk. The assistance of the staff and the cheese-makers at Murray Goulburn Cobram during the cheese-making trials is also gratefully acknowledged. Warrnambool Cheese and Butter and Lion were also involved in this project.

CONFLICTS OF INTEREST

The authors declare no conflict of interest.

REFERENCES

1. Guinee, T.P.; Predrag, D.P.; Mulholland, E.O. Effect of milk protein standardisation, by ultrafiltration, on the manufacture, composition and maturation of Cheddar cheese. J. Dairy Res. 1994, 61, 117–131.
2. Govindasamy-Lucey, S.; Jaeggi, J.J.; Johnson, M.E.; Wang, T.; Lucey, J.A. Use of cold ultrafiltered retentates for standardization of milks for pizza cheese: Impact on yield and functionality. Int. Dairy J. 2005, 15, 941–955.
3. Govindasamy-Lucey, S.; Jaeggi, J.J.; Bostley, A.L.; Johnson, M.E.; Lucey, J.A. Standardization of milk using cold ultrafiltration retentates for the manufacture of parmesan cheese. J. Dairy Sci. 2004, 87, 2789–2799.
4. Govindasamy-Lucey, S.; Jaeggi, J.J.; Martinelli, C.; Johnson, M.E.; Lucey, J.A. Standardization of milk using cold ultrafiltration retentates for the manufacture of Swiss cheese: Effect of altering coagulation conditions on yield and cheese quality. J. Dairy Sci. 2011, 94, 2719–2730.
5. Guinee, T.P.; O'Kennedy, B.T.; Kelly, P.M. Effect of milk protein standardization using different methods on the composition and yields of Cheddar cheese. J. Dairy Sci. 2006, 89, 468–482.
6. Green, M.L.; Glover, F.A.; Scurlock, E.M.W.; Marshall, R.J.; Hatfield, D.S. Efefct of use of milk cocentrated by ultrafiltration on the manufacture and ripening of Cheddar cheese. J. Dairy Res. 1981, 48, 333–341.
7. Green, M.L.; Turvey, A.; Hobbs, D.G. Development of structure and texture in Cheddar cheese. J. Dairy Res. 1981, 48, 343–355.
8. Banks, J.M. Ultrafiltration of Cheesemilk. In Cheese Problems Solved; McSweeney, P.L.H., Ed.; Woodhead Publishing Limited: Cambridge, England, 2007; pp. 30–33.
9. Erdem, Y.K. Influence of ultrafiltration on modification of surface hydrophobic sites of the milk protein system in the course of renneting. J. Food Eng. 2000, 44, 63–70.
10. Guinee, T.P.; O'Callaghan, D.J.; Pudja, P.D.; O'Brien, N. Rennet coagulation properties of retentates obtained by ultrafiltration of skim milks heated to different temperatures. Int. Dairy J. 1996, 6, 581–596.
11. Garnot, P.; Rank, T.C.; Olson, N.F. Influence of protein and fat contents of ultrafiltered milk on rheological properties of gels formed by chymosin. J. Dairy Sci. 1982, 65, 2267–2273.
12. Upreti, P.; Mistry, V.V.; Acharya, M.R. Characterization of renent coagulation of milk concentrated by vacuum condensing and ultrafiltration. Dairy Sci. Technol. 2011, 91, 383–395.
13. Dalgleish, D.G. Effect of milk concentration on the nature of curd formed during renneting: A theoretical discussion. J. Dairy Res. 1981, 48, 65–69.
14. Karlsson, A.O.; Ipsen, R.; Ardö, Y. Rheological properties and microstructure during rennet induced coagulation of UF concentrated skim milk. Int. Dairy J. 2007, 17, 674–682.
15. Ong, L.; Dagastine, R.R.; Kentish, S.E.; Gras, S.L. The effect of pH at renneting on the microstructure, composition and texture of Cheddar cheese. Food Res. Int. 2012, 48, 119–130.
16. Ong, L.; Dagastine, R.R.; Kentish, S.E.; Gras, S.L. Microstructure of gel and cheese curd observed using cryo scanning electron microscopy and confocal microscopy. LWT Food Sci. Technol. 2011, 44, 1291–1302.
17. AOAC Association of Official Analytical Chemists: Official Methods of Analysis 991.20 and 991.21, 18th ed.; AOAC International: Gaithersburg, MD, USA, 2005.
18. Ong, L.; Henriksson, A.; Shah, N.P. Development of probiotic Cheddar cheese containing Lactobacillus acidophilus, Lb. casei, Lb. paracasei and Bifidobacterium spp. and the influence of these bacteria on proteolytic patterns and production of organic acid. Int. Dairy J. 2006, 16, 446–456.

19. Ong, L.; Dagastine, R.R.; Kentish, S.E.; Gras, S.L. Coagulation temperature affects the microstructure and composition of full fat Cheddar cheese. Dairy Sci. Technol. 2011, 91, 739–758.
20. Fox, P.F.; McSweeney, P.L.H. Milk Protein. In Dairy Chemistry and Biochemistry; Blackie Academic & Professional: London, UK, 1998.
21. Green, M.L.; Scott, K.J.; Anderson, M.; Griffin, M.C.A.; Glover, F.A. Chemical characterization of milk concentrated by ultrafiltration. J. Dairy Res. 1984, 51, 267–278.
22. Neocleous, M.; Barbano, D.M.; Rudan, M.A. Impact of low concentration factor microfiltration on milk component recovery and Cheddar cheese yield. J. Dairy Sci. 2002, 85, 2415–2424.
23. Guinee, T.P.; Mulholland, E.O.; Kelly, J.; Callaghan, D.J.O. Effect of protein-to-fat ratio of milk on the composition, manufacturing efficiency, and yield of Cheddar cheese. J. Dairy Sci. 2007, 90, 110–123.
24. Broome, M.C.; Tan, S.E.; Alexander, M.A.; Manserm, B. Low-concentration ratio ultrafiltration for Cheddar cheese manufacture. Part 1. Effect on seasonal cheese composition. Austr. J. Dairy Technol. 1998, 53, 5–10.

CHAPTER 9

Functional Dairy Probiotic Food Development: Trends, Concepts, and Products

Aziz Homayouni[1], Maedeh Alizadeh[2], Hossein Alikhah[3] and Vahid Zijah[4]

[1] *Department of Food Science and Technology, Faculty of Health and Nutrition, Tabriz University of Medical Sciences, Tabriz, I.R. Iran*

[2] *Pediatric Nursing, Faculty of Nursing and Midwifery of Maragheh, Tabriz University of Medical Sciences, Tabriz, I.R. Iran*

[3] *Faculty of Medicine, Tabriz University of Medical Sciences, Tabriz, I.R. Iran*

[4] *Department of Dentistry, Behbood Hospital, Tabriz University of Medical Sciences, Tabriz, I.R. Iran*

1. INTRODUCTION

In recent years, scientific investigators have moved from primary role of food as the source of energy and nutrients to action of biologically active food components on human health. On the other hand, consumer interest about the active role of food in well-being and life prolongation has been increased. In this way, a novel term -functional food- was introduced which refers to preventional and/or curing effects of food beyond its nutritional value. There is a wide rage of functional foods that were developed recently and many of them are being produced in all over the world including probiotic, prebiotic and symbiotic foods as well as foods enriched with antioxidants, isoflavones, phytosterols, anthocyanins and fat-reduced, sugar-reduced or salt-reduced foods. Among these foods, probiotic functional food has exerted positive effects on the overall health. We can divide it in both probiotic dairy foods and probiotic non-dairy foods. The market of probiotic dairy foods is increasing annually. An increased demand for dairy probiotic products comes from health promotion effects of probiotic bacteria which are originally initiated from milk products, bioactive compounds of fermented dairy products and prevention of lactose intolerance. Therefore, development of these products is a key research priority for food design and a challenge for both industry and science sectors.

Literatures about probiotic application in pediatrics have some characteristics including numerous, randomized, controlled clinical trials or meta-analyses but the substantial heterogeneity of these works greatly complicates the interpretation of the results and thus makes it difficult to draw univocal and general conclusions. Despite these complications, it is possible to draw some conclusions about the clinical effectiveness of probiotics by examining the most significant literature on each pathology. In particular, there is strong evidence indicating that probiotics have preventive and therapeutic effect on pathologies such as acute diarrhea, antibiotic-associated diarrhea, NEC, and allergic pathology. It was reported that administration of L.GG to 50 infants, for a period of 6 weeks, did not improve abdominal pain but did reduce the incidence of abdominal tension compared to the placebo (Bausserman and Michail, 2005) But in other works it was clearly demonstrated that L. acidophilus did improve the symptoms in about half of the patients with IBS, that the blend of VLS#3 probiotics decreased abdominal swelling, while the combined use of L. plantarum and B. breve reduced pain intensity (Halpern, et al., 1996; Kim, et al., 2003; Saggioro, 2004). L. acidophilus and B. infantis for 4 weeks were administered alone or in combination with antibiotics ciprofloxacin for the first week to three different groups with IBS: diarrhea, constipation, and alternating diarrhea and constipation. Both therapeutic approaches have improved the quality of life and reduced symptoms in all three groups (Faber, 2000). In conclusion, although the use of some types of probiotics on IBS appears promising, additional studies are needed. Food supplementation with pre- and probiotics may reduce the prevalence for the infant in high-risk families developing an atopic eczema during the first 2 years of life. Those pregnant women should be advised to take probiotics (L. GG) in late pregnancy and the first 6 months postnatally during nursing. If breast-feeding is not possible, pro- or prebiotics can be supplemented to the infant. There are no known adverse reactions and it might prevent atopic eczema, especially in neonates after cesarean delivery. Therapeutic use of probiotics to improve atopic eczema is only supportive in infants 18 months and with IgE sensitization.

Recent experimental studies have shown that certain gut bacteria, in particular species of Lactobacillus and Bifidobacterium, may exert beneficial effects in the oral cavity by inhibiting Streptococci and Candida sp. Probiotic lactic acid bacteria can produce different antimicrobial components such as organic acids, hydrogen peroxide, carbon peroxide, diacetyl, low molecular weight antimicrobial substances, bacteriocins, and adhesion inhibitors, which also affect oral microflora. However, data is still sparse on the probiotic action in the oral cavity. More information is needed on the colonization of probiotics in the mouth and their possible effect on and within oral biofilms. There is every reason to believe that the putative probiotic mechanisms of action are the same in the mouth as they are in other parts of the gastrointestinal tract. Because of the increasing global problem with antimicrobial drug resistance, the concept of probiotic therapy is interesting and pertinent, and merits further research in the fields of oral medicine and dentistry (Meurman, 2005).

The number of microbial cells in the human gut is 10 times more than the number of cells in the adult body (Mountzouris and Gibson, 2003). So, the change of microbial balance in human intestine can impress the host health. The ratio between the beneficial microbes (probiotics) and harmful microbes would have an important effect on host health. One way to keeping up the probiotic cells in the gut, is to entering probiotics into the intestine through the regular consumption of food containing these bacteria. Among the functional foods, the dairy probiotic products, especially ice cream and cheese are good vehicle to transfer probiotics to the human intestinal tract (Homayouni, 2008a; Homayouni et al., 2012). Dairy products have an important role in human health and form the main part of the food pyramid. The therapeutical and health care characteristic of fermented dairy products has been used over long years. Another way to keeping up the probiotic cells in the gut is to entering prebiotics into the intestine through the regular consumption of foods containing these components. It is clear that versus probiotics the amounts of prebiotics do not changes during the passage from upper intestinal tract (Homayouni, 2008a).

The main role of food is providing enough nutrients to meet metabolic requirements in human body, while giving the consumer a satisfaction feeling and well-being (Homayouni, 2008a). Beyond meeting nutrition needs, food may have different physiological functions and may play detrimental or beneficial roles in some diseases (Koletzko et al., 1998). Functional foods were developed in order to promote a well-being state, improving health, and reducing the diseases risk. "Functional food" means; special foods which have preventional and/or curing effects beyond its nutritional (Homayouni, 2008a). There is a wide rage of functional foods that were developed recently and many of them are being produced in all over the world including probiotic, prebiotic and symbiotic foods as well as foods enriched with antioxidants, isoflavons, phytosterols, anthosyanins and fat-reduced, sugar-reduced or salt-reduced foods. Among these foods, probiotic functional foods are the first choice to exert positive effects on the human health. Probiotic functional foods were divided into dairy probiotic foods and non-dairy probiotic foods. Some of dairy probiotic foods including probiotic ice cream, frozen fermented dairy deserts, probiotic cheese, bio-yoghurt, drinking yoghurt, kefir, Freeze-dried yoghurt and spray dried milk powder have been employed as possible delivery vehicles for probiotic bacteria (Haynes and Playne, 2002; Homayouni et al., 2008b; Homayouni et al., 2012; Ejtahed et al., 2011; Ejtahed et al., 2012; Mirzaei et al., 2012 Kailasapathy and Rybka, 1997; Ravula and Shah, 1998; Stanton et al., 2001). Probiotics are distinct as live micro-organisms which, when administered in sufficient amounts present a health benefit on the host (Food and Agriculture Organization of United Nations; World Health Organization - FAO/WHO, 2002; Homayouni, 2009). In recent years probiotic bacteria have increasingly been incorporated into dairy foods as dietary adjuncts. *Lactobacillus* and *Bifidobacterium* are the most common probiotic bacterial cells that were used in the production of fermented and non-fermented dairy products.

Consumption of probiotic bacteria via dairy food products is an ideal way to re-establish the intestinal micro-floral balance. It must conform to certain requirements for a dairy food product to be considered as a valuable alternative for delivery of probiotic bacteria in one hand and for variety of probiotic cultures to use as a dietary adjunct and to exert a positive influence in the other hand. The culture must be native of the human gastrointestinal tract, having the ability to ferment prebiotics, survives passage through the stomach and small bowel in adequate numbers, be capable of colonizing in site of action, and have beneficial effects on human health. In order to survive, the strain must be resistant to acidic conditions (gastric pH 1-4), alkaline conditions (bile salts present in the small bowel), enzymes present in the intestine (lysozyme) and toxic metabolites produced during digestion (Homayouni et al., 2008d). For example in traditional yoghurt production, *Lactobacillus bulgaricus* and *Streptococcus thermophilus* were used as starter culture. These bacteria do not belong to the indigenous intestinal flora, are not bile-acid resistant and do not survive passage through the gut. So, the traditional yoghurt culture is not to be considering as probiotic. In the case of dairy food product to be considered as a valuable alternative for delivery of probiotics, it must to match definite necessities such as neutral pH, high enough total solids level, absence of oxygen and near to ambient temperatures (Homayouni et al., 2008b; Homayouni et al., 2008d; Homayouni et al., 2012). A number of dairy food bio-products have been employed and developed as delivery vehicles of probiotic bacteria. Around 80 bifido containing products are estimated to be on the world markets. Most of these products are from dairy origin including fresh milk, fermented milk, dairy beverages, ice cream, dairy desserts, cheese, cottage cheese and powdered milk (Tamime et al., 1995). Since the more interest in probiotics, different types of functional products were proposed as carrier foods for probiotic micro-organisms by which consumers can take in large amounts of probiotic bacteria for the therapeutic effects. Therefore, development of these products is a key research priority for food design and a challenge for both industry and science sectors. This chapter presents an overview of functional foods development with emphasizing probiotic dairy foods.

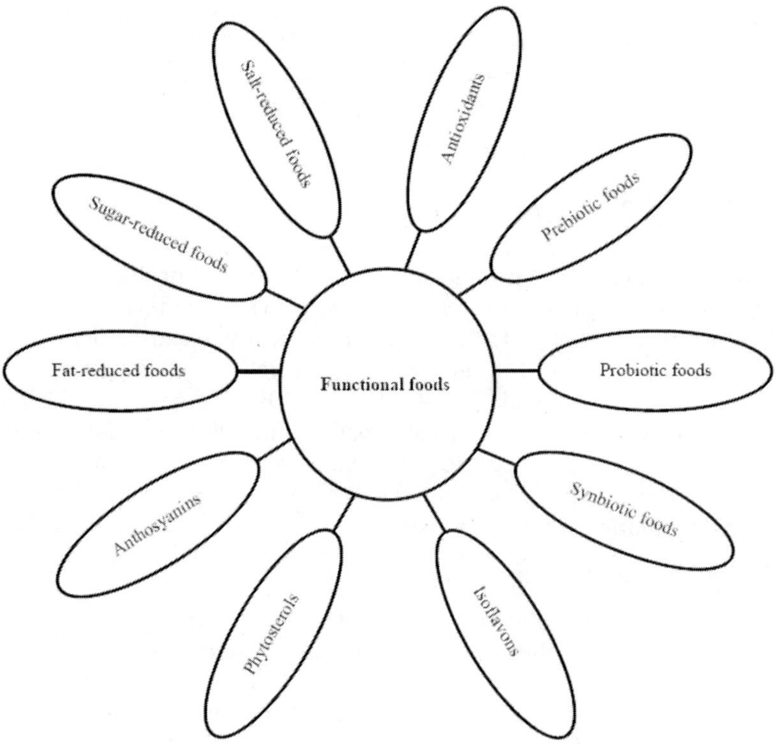

Figure 1. Classification of functional foods

2. DAIRY PROBIOTIC FOODS

As mentioned before, dairy functional foods beyond its basic nutritional value has physiological benefits. Milk has an outstanding position in the development of functional foods because it has Omega-3, phytosterols, isoflavins, conjugated linoleic acid, minerals, and vitamins. Dairy products such as ice cream, cheese, yogurt, Acidophilus-Bifidus-milk, Ayran, Kefir, Kumis, Doogh containing probiotics and dairy beverages (both fermented and non-fermented) have long been considered as important vehicles for the delivery of probiotics. In fermentation process, acids such as lactic acid, acetic acid and citric acid are naturally produced. These acids are commonly used as organic acids to enhance organoleptic qualities as well as safety of food products. Lactic acid bacteria are found to be more tolerant to acidity and organic acids than most of the pathogens and spoilage micro-organisms.

2.1. Probiotic Ice Cream

Probiotic ice cream can be produced by incorporation of probiotic bacteria in both of fermented and unfermented mix (Homayouni et al., 2008b; Homayouni et al., 2012). Ice cream is ideal vehicle for delivery of these micro-organisms in the human diet (Akin et al., 2007; Kailasapathy and Sultana, 2003; Ravula and Shah, 1998; Homayouni et al., 2008d; Homayouni et al., 2012). *Lactobacillus* and *Bifidobacterium* are the most common species of lactic acid bacteria used as probiotics for fermented dairy products. Among the frozen dairy products with live probiotics, probiotic ice cream is also gaining popularity for its neutral pH. The pH of non-fermented ice cream is near to seven which is providing to survive probiotic bacteria (Akin et al., 2007; Christiansen et al., 1996; Homayouni et al., 2008b; Homayouni et al., 2008c; Homayouni et al., 2012). The high total solids level in ice cream including the fat and milk solids provides protection for the probiotic bacteria (Homayouni et al., 2012). Because the efficiency of added probiotic bacteria depends on dose level, type of dairy foods, presence of air and low temperature (Homayouni et al., 2008b), their viability must be maintained throughout the product's shelf-life and they must survive the gut environment (Kailasapathy and Chin, 2000). The therapeutic value of live probiotic bacteria is more than unviable cells; therefore, International Dairy Federation (IDF) recommends that a minimum of 10^7 probiotic bacterial cells should be alive at consumption time per gram/mililiter of product. Studies indicate, however, the bacteria may not survive in high enough numbers when incorporated into frozen dairy products unless a suitable method is used against freeze injury and oxygen toxicity (Dave and Shah, 1998; Kailasapathy and Sultana,

2003; Ravula and Shah, 1998; Homayouni et al., 2008d). The methods of increasing probiotic survival depend on type of food products. Selection of resistant probiotic strains to tolerate production, storage and gastrointestinal tract conditions, is one of the important methods (Homayouni et al., 2008d). Another way is to adjust the conditions of production and storage for more survival rates. The physical protection of probiotics by microencapsulation is a new method for increasing the survival of probiotics (Homayouni et al., 2007; Homayouni et al., 2008b). Encapsulation helps to isolate the bacterial cells from the adverse environment of the product and gastrointestinal tract, thus potentially reducing cell loss. Encapsulation thus may enhance the shelf-life of probiotic cultures in frozen dairy products (Kebary et al., 1998; Shah and Ravula, 2000; Homayouni et al., 2008b). Selecting of suitable probiotic strains depends to ability survive simulated conditions of ice cream (high sucrose concentrations, high oxygen, refrigeration and freezing temperatures), acidic (to simulate gastric) and alkaline conditions (to simulate intestinal). Microencapsulation of probiotics can further protect these bacteria from the mentioned conditions (Homayouni et al., 2008d).

Homayouni et al. (2008d) studied the survival of probiotics in simulated ice cream and gastrointestinal conditions in order to select appropriate probiotic strains for use in probiotic ice cream. The growth and survival rate of *Lactobacillus acidophilus, Lactobacillus casei, Bifidobacterium lactis* and *Bifidobacterium longum* in varying amount of sucrose concentrations (10, 15, 20 and 25%), oxygen scavengering components (0.05% L-cysteine and 0.05% L-ascorbate) and low temperatures (4°C and 20°C) during different periods of time (30, 60 and 90 days) in MRS-broth medium was studied. All of above stress factors have been able to influence the growth and survival of four probiotic strains. Results have demonstrated that it is possible to select the appropriate probiotic strains for use in probiotic ice cream. *Lactobacillus casei* (Lc01) and *Bifidobacterium lactis* (Bb12) had the highest resistance to simulated acidic, alkaline and ice cream conditions in comparison with other probiotic strains, making them suitable probiotic strains for use in probiotic ice cream (Homayouni et al., 2008b; Homayouni et al., 2008d).

2.2. Probiotic Cheese

Survival in processing conditions, presence of oxygen, degree of acidity, ability to grow well in milk-based products and to rapidly acidify milk, thus reducing the fermentation time and, consequently, contamination risk during preparation of inoculums are important factors for probiotic bacteria such as *Lactobacillus spp.* and *Bifidobacterium spp* in order to apply these bacteria in probiotic dairy products. Probiotic bacterial cells have to fulfill the basic technological necessities when used in commercial probiotic dairy products. Since probiotic bacteria have to be presented in sufficient numbers in product at consumption time, their survival have to be maintained up to shelf-life date. In addition, no adverse effects on taste and aroma of the product should be exerted by the probiotic organisms. Various types of cheese have a good potential to maintain the probiotic survival. So, it is a good vehicle to transfer probiotics to the human intestinal tract. There are two ways for development of probiotic cheese: in the first step, the manufacture processes of cheese products may have to be modified and adapted to the requirements of probiotics and in second step, appropriate probiotic strains may be applied or new cheese products may have to be developed. Dairy products containing living bacteria have to be cooled during storage. Cooling is necessary to guarantee high survival rates of probiotics and to bring sufficient stability of the product (Roy et al., 1997). In addition, oxygen content and water activity of the probiotic cheese have to be considered in prepackaged cheese (Dave and Shah, 1997a). Interaction of the live probiotic microorganisms with the components of the cheese have to be inhibited by cooling of product. The degree of interaction depends on the kind and amount of carbohydrates available, degree of hydrolysis of milk proteins and thus availability of essential amino acids, and composition and degree of hydrolysis of milk lipids, determining the availability of short chain fatty acids (Fox et al., 1996). However, the proteolytic and lipolytic properties of the probiotic bacterial cells may have important effects on taste and flavor of the probiotic cheese (Kunji et al., 1996). The strength of interactions between probiotics and starter organisms in probiotic cheese depends on when the probiotics are added to the product. If they are added after fermentation, interactions may be kept to a minimum, since addition is possible immediately before or even after cooling below 8°C and metabolic activities of starters and probiotics are considerably reduced at refrigerated temperatures.

Antagonism between bacteria is often based on the production of metabolites that inhibit or inactivate more or less specifically other related starter organisms or even unrelated bacteria. While antagonism caused by bacteriocins, peptides, or proteins exhibiting antibiotic properties has been described as a limiting factor for combinations of starters and probiotics (Joseph et al., 1998), antagonism caused by hydrogen peroxide, benzoic acid, biogenic amines, and lactic acid may have considerable effects on probiotics in probiotic cheese. The physiological state of the probiotics may be of considerable importance for survival during ripening and/or storage if probiotics are added to the probiotic cheese after fermentation (Desmazeaud, 1996; Lankaputhra et al., 1996; Leuschner et al., 1998; Weber, 1996).

In probiotic cheese, probiotic cells must be able to grow and/or multiply in the human intestine and therefore should be able to survive during the passage through the gastrointestinal tract (GIT), which involves exposure to hydrochloric acid in stomach and bile in small intestine (Stanton et al., 2003). In fact, cheese provides a valuable vehicle for probiotic delivery, due to creation of a buffer against the high acidic environment in the gastrointestinal tract, and thus creates a more favorable environment for probiotic survival throughout the gastric transit, ought to higher pH. Moreover, the dense matrix and relatively high total solids as well as fat content of cheese may offer additional protection to probiotic bacteria in stomach (Bergamini et al., 2005; Ross et al., 2002). The presence of the prebiotics inulin and oligofructose can promote growth rates of *bifidobacteria* and *lactobacilli*, besides increased lactate and short chain fatty acids production in petit-suisse cheese (Cardarelli et al., 2007).

2.3. Probiotic Yoghurt

Yoghurt has been historically recognized to be 'a healthy food' with therapeutically effects. There has been a considerable increase in the popularity of yoghurt especially probiotic yoghurt in recent years. The conventional yoghurt starter bacteria, *L. bulgaricus* and *Streptococcus thermophilus*, do not have ability to survive passage through intestinal tract and consequently so, they are not considered as probiotics. But the addition of *L. acidophilus* and *B. bifidum* into yoghurt can add extra nutritional and physiological values.

Similar processing to traditional yoghurt is applied for production of bio-yoghurt with incorporation of live probiotic starter cultures. Heat treated homogenized milk with an increased protein content (3.6–3.8%) is inoculated with the conventional starter culture at 45°C or 37°C and incubated for 3.5 and 9 h, respectively. The probiotic culture can be added prior to fermentation simultaneously with the conventional yoghurt cultures or after fermentation to cooled (4°C) product before packaging. Bio-yoghurt, containing *L. acidophilus* and *B. bifidum* is a potential vehicle for delivery of these probiotic cells to consumers. *L. acidophilus* and *B. bifidum* have to retain viability and activity in yoghurt as a probiotic at consumption time. Viability of probiotic bacteria in yoghurt products at refrigeration temperature is reported to be unsatisfactory over a long shelf life (Dave and Shah, 1997a). The survival of probiotic bacteria in fermented dairy products depends on the chemical composition of the fermentation medium (e.g. carbohydrate source), final acidity, milk solids content, availability of nutrients, growth promoters and inhibitors, strains used, interaction between species present, culture conditions, concentration of sugars (osmotic pressure), dissolved oxygen (especially for *Bifidobacterium* spp.), level of inoculation, incubation temperature, fermentation time and storage temperature. The lack of acid tolerance of some probiotic species and strains in fermented products based on milk is an important factor. During fermentation, pH levels decreases when the lactic acid content increases. 'Over-acidification' or 'post acidification' is due to decrease in pH after fermentation and during storage at refrigerated temperature. Excessive acidification is mainly due to the uncontrollable growth of strains of *L. bulgaricus* at low pH values and refrigerated temperatures. The 'overacidification' can be prevented to a limited extent by applying 'good manufacturing practice' and by using cultures with reduced 'over-acidification' behavior.

Viability of both *Lactobacillus* and *Bifidobacterium* species reduces at low pH levels during refrigerated storage. So, strain selection and survival monitoring are necessary to produce high quality bio-yoghurt. Probiotic yoghurt contains metabolic products secreted by starter microorganisms, which influence the viability of *L. acidophilus* and *B. bifidum*. The inhibition of *bifidobacteria* in probiotic yoghurt is due to antagonism effects among starter bacteria rather than hydrogen peroxide or organic acids (Dave and Shah, 1997a). The ideal procedure for probiotic yoghurt manufacturing is growing the *Bifidobacterium* spp. separately, followed by washing out of free metabolites and the transfer of the cells to the probiotic yoghurt. Oxygen toxicity is a critical problem for *Bifidobacterium* spp. because they are strictly anaerobic. Low initial oxygen content in milk may obtain the low redox potential required in the early phase of incubation to guarantee *Bifidobacteria* growth. Oxygen easily dissolves in milk during yoghurt production and also permeates through packages during storage. It has been suggested to inoculate *S. thermophilus* and *Bifidobacterium* simultaneously during fermentation to avoid the oxygen toxicity problem. *S. thermophilus* has a high oxygen utilization ability, which results in reduction of dissolved oxygen in probiotic yoghurt and an enhancement in viability of *bifidobacteria*. Higher survival rates of lactic acid bacteria were obtained at lower storage temperatures (Foschino et al., 1996). Low storage temperature restricts the growth of *L. bulgaricus* and consequently also over-acidification. *Bifidobacteria* are substantially less tolerant to low storage temperature when compared to *L. acidophilus*.

2.4. Probiotic Milk

Lactobacillus acidophilus does not rapidly grow in milk because it is an acid-loving bacterium. Therefore, it is essential to maintain the inoculum active by daily transfers of mother culture in acidophilus milk production. The probiotic milk is to market in liquid form. During fermentation, milk pH often goes beyond the narrow range of optimal pH of *Lactobacillus acidophilus* (5.5-6.0). This eventually leads to decrease these bacterial counts. In traditional acidophilus milk production, the milk is heated at 95°C for 1 h or at 125°C for 15 min (Vedamuthu, 2006). Such a high heat treatment stimulates the growth of *Lactobacillus acidophilus* by providing denatured proteins and released peptides. High-heat-treated milk is cooled to 37°C and kept at this temperature for a period of 3-4 h to allow any spores present to germinate. Then, milk is re-sterilized to destroy almost all vegetative cells. Unless skim milk is used, the heat-treated milk is homogenized and cooled down to inoculation temperature (37°C). *Lactobacillus acidophilus* is added as active bulk culture. The level of inoculation is usually 2-5% and the inoculated milk is left to ferment until pH 5.5-6.0 or ~1.0% lactic acid is obtained, with no alcohol (Surono and Hosono, 2002). The fermentation takes about 18-24 h under inactive conditions. After the fermentation, the number of viable *Lactobacillus acidophilus* colonies is about $2\text{-}3\times10^{9}$ cfu mL^{-1}, but this number decreases up to consumption time. In extended incubation period reduction in counts of *Lactobacillus acidophilus* may occur. To overcome this problem, replacement of 25% of *Lactobacillus acidophilus* culture by a mixture of *Streptococcus thermophilus* and *Lactobacillus delbrueckii* subsp. *Bulgaricus* can be used. Following fermentation, the warm product is rapidly cooled to <7°C before agitation and pumped to a filler where it is filled into bottles or cartons (Kosikowski and Mistry, 1997; Vedamuthu, 2006). Protein quality and total amino acid content are similar in both fermented and non-fermented milk. Acidophilus milk has higher free amino acids than milk. As the milk lactose is hydrolyzed by β-galactosidase of *Lactobacillus acidophilus*, acidophilus milk is more suitable for individuals suffering from lactose intolerance. It is also possible to enrich acidophilus milk with calcium, iron and vitamins. Undesirable sour milk flavor caused acidophilus milk is gained limited popularity by consumers. So, sweet acidophilus milk has been developed. When *Lactobacillus acidophilus* is incorporated into pasteurized milk at about 5°C and bottled aseptically, these bacteria are able to keep their viability up to 14 days without reducing the pH of milk due to it does not grow at low temperatures (<10°C). Freeze-dried cultures may keep their viability up to 58% after 23 days at 4°C in sweet acidophilus milk. *Lactobacillus acidophilus* remained viable in sweet acidophilus milk over 28 days at 7°C. Addition of 200 g of frozen culture concentrate to 2000 L of pasteurized milk is satisfactory to reach the target level of *Lactobacillus acidophilus* in the probiotic milk (Vedamuthu, 2006).

Technology of bifidus milk and acidophilus-bifidus milk manufacturing is similar to acidophilus milk. Milk is standardized to desire protein and fat levels in both products. Then, for manufacture of bifidus milk, milk is heat-treated at 80-120°C for 5-30 min and rapidly cooled to 37°C. Heat-treated milk is inoculated with frozen culture of *Bifidobacterium bifidum* and *Bifidobacterium longum* at a level of 10% and left to ferment until pH 4.5. After fermentation, the product is cooled to <10°C and packaged. Final product has a slightly acidic flavor and the ratio of lactic acid to acetic acid is 2:3. Milk used for acidophilus-bifidus milk production is usually enriched with protein prior to fat standardization and homogenization. The standardized milk is heat-treated at 75°C for 15 s or 85°C for 30 min. After cooling the milk to 37°C, frozen cultures of *Lactobacillus acidophilus* and *Bifidobacterium bifidum* are inoculated and fermentation is allowed until pH 4.5–4.6 is reached (~16 h). Following fermentation, the fermented milk is cooled to <10°C. The shelf life of the product is about 20 days. Acidophilus-bifidus milk has a characteristic aroma and slightly acidic flavor. High viscosity of product cause to producing it in set form. It is also possible to produce probiotic milks by simply adding mix culture of *Lactobacillus acidophilus* and *Bifidobacterium bifidum* to cold pasteurized milk.

3. DEVELOPMENT OF FUNCTIONAL DAIRY FOODS

Innovation is today's business demand and development of a new functional food is an expensive process and is very important for both food companies and consumers. Regulations should encourage food companies to follow functional food development. Development of dairy probiotic products requires detailed knowledge of both products and customers. It needs to manage customer knowledge effectively (Walzem, 2004; Jousse, 2008). Fundamental risks can affect the development of new functional food products and may leads to fail the development process. Development of new functional food products is very challenging and it has to complete the consumer's expectations for palatable and healthy products (Fogliano and Vitaglione, 2005; Granato et al., 2010; Shah, 2007). So, the development and commerce of functional food products is rather complex, expensive, and uncertain. Key points regarding for a successful functional food product development are consumer demands, technological conditions, and legislative regulatory background. However, consumer's knowledge of the health effects of specific ingredients can affect the acceptance of specific functional food. Therefore, functional ingredients that are in consumers mind for a long period of time, such as minerals, fiber, and vitamins, achieve considerably higher rates of consumer acceptance than new products, such as foods enriched with probiotics, prebiotics, flavonoids, carotenoids, and conjugated linilenic acid (CLA). Several ways to make a functional food product is to eliminating an allergenic protein, lactose, phenylanine and etc from the natural food product; by fortification with a micronutrient; by adding antioxidants, probiotics or prebiotics); by replacing a component, or by increasing bioavailability or stability of a component known to produce a functional effect or to reduce the disease-risk potential of the food (Roberfroid, 2000; Siro, et al., 2008; Granato, et al., 2010). Field of functional probiotic foods requires the cooperation of food technologists, nutritionists, medical doctors, and food chemists in order to obtain innovative products. In this way, these foods may be able to adjust physiological parameters related to health status or disease prevention in human. So, the design and development of functional probiotic foods is a scientific work (Hasler, 1998; Walzem, 2004; Fogliano and Vitaglione, 2005) which is an expensive and multistage process that takes into account many factors, such as sensory acceptance, physical and microbial stability, price, and chemical and other intrinsic functional properties to be successful in the marketplace. Moreover, consumer attitude toward the functional probiotic product also needs to be understood and taken into consideration.

4. CONSUMER ATTITUDE TOWARD FUNCTIONAL DAIRY FOODS

The development of functional probiotic foods is increasing, as their market increases day by day, although the consumer's information about these foods is increasing without relation to gender, age, and educational or economic levels of the consumers. The therapeutical effect of a functional probiotic food may depend on the consumer's characteristics and the type of carrier and enrichment considered. For instance, yoghurt is most preferred by its enrichment with calcium and fiber. Ingredients such as vitamins and minerals applied in fortification of functional foods are widely recognized and accepted by consumers, but new functional ingredients such as probiotics and prebiotics are not common to them. So, there is a need for increasing the consumer knowledge with respect to these new special ingredients (Hillian, 2000; Luckow and Delahunty, 2004; Ares and Gambaro, 2007; Vianna et al., 2008).

The sensory properties of prebiotic functional foods in comparison with conventional products can lead to different acceptance level. Oligofructose provides some suitable sensory properties such as rounder mouth feel, reduced aftertaste, and slight sweetness to the products. These properties are responsible for high score values for taste, creaminess, and overall acceptability of functional food products. The first important marker in choosing a functional food is flavor, and health consideration is in the second order. If the ingredients added give unpleasant flavors to the product, consumers are not interested in consume such functional probiotic food even if this results in health advantages. This means that flavor is correlated to intrinsic sensory properties of the product such as overall acceptability. In general, as functional products consumption increases, the acceptance of such products may increase, even if the sensory profiles are different from conventional products. When functional ingredients such as probiotics are added to dairy foods, consumers must be aware of probiotics health benefits in order to recognize the functional probiotic foods as being more beneficial than the conventional ones. Functional probiotic food industry should communicate with consumer in a clear way and this is one of the most important aspects for success (Tepper and Trail, 1998; Matilla-Sandholm et al., 1999; Roberfroid, 2000; Tuorila and Cardello, 2002; Nicolay, 2003; Vieira, 2003; Homayouni, 2008a).

5. CONCLUSION

The future success of functional probiotic dairy foods in marketplace depends on consumer acceptance of such products. The consumers must be convinced by its health claims through clear, honest, and definite messages to agree to pay the cost associated with functional probiotic dairy foods. Development of probiotic dairy products is a key research priority for food design and a challenge for both industry and science sectors. Among the functional foods, the dairy probiotic products, especially ice cream and cheese are good vehicle to transfer probiotics to the human intestinal tract. Additional way to keeping up the probiotic cells in the gut is to entering prebiotics into the intestine through the regular consumption of food containing these components. It is clear that versus probiotics the amounts of prebiotics do not changes during the passage from upper intestinal tract.

REFERENCES

1. M. B. Akin, M. S. Akin, Z. Kirmaci, Effects of inulin and sugar levels on the viability of yogurt and probiotic bacteria and the physical and sensory characteristics in probiotic ice creamFood Chemistry20079399
2. Ares, G., and G´ambaro, A. Influence of gender, age and motives underlying food choice on perceived healthiness and willingness to try functional foods, Appetite 49 (2007), pp. 148-158.
3. M. Bausserman, S. Michail, 2005The use of Lactobacillus GG in irritable bowel syndrome in children: a double-blind randomized control trial.The Journal of Pediatrics1472197200
4. C. V. Bergamini, E. R. Hynes, A. Quiberoni, V. B. Sua´rez, C. A. Zalazar, Probiotic bacteria as adjunct starters: influence of the addition methodology on their survival in a semi-hard Argentinean cheeseFood Research International2005597604
5. Cardarelli, H. R., Saad, S. M. I., Gibson, G. R., and Vulevic, J. Functional petitsuisse cheese: Measure of the prebiotic effect, Anaerobe 13 (2007), pp. 200-207.

6. P. S. Christiansen, D. Edelsten, J. R. Kristiansen, E. W. Nielsen, Some properties of ice cream containing Bifidobacterium bifidum and Lactobacillus acidophilusMilschwissenschaft 51 (1996502504
7. J. H. Cummings, G. R. Macfarlane, H. N. Englyst, Prebiotic digestion and fermentation,American Journal of Clinical Nutrition 73 (2001415420
8. R. Dave, N. P. Shah, Viability of probiotic bacteria in yoghurt made from commercial starter cultures,International Dairy Journal 71997a3141
9. R. I. Dave, N. P. Shah, Effect of cysteine on the viability of yoghurt and probiotic bacteria in yoghurts made with commercial starter culturesInternational Dairy Journal1997b537545
10. R. I. Dave, N. P. Shah, Ingredient supplementation effects on viability of probiotic bacteria in yogurt,Journal of Dairy Science199828042816
11. M. Desmazeaud, 1996Growth inhibitors of lactic acid bacteria, in Dairy Starter Cultures, Cogan, T.M. and Accolas. New York: VCH Publishers, (131155
12. H. S. Ejtahed, J. Mohtadi-Nia, A. Homayouni-Rad, M. Niafar, M. Asghari-Jafarabadi, V. Mofid, A. Akbarian-Moghari, 2011Effect of probiotic yogurt containing Lactobacillus acidophilus and Bifidobacterium lactis on lipid profile in individuals with type 2 diabetes mellitusJournal of Dairy Science9432883294
13. H. S. Ejtahed, J. Mohtadi-Nia, A. Homayouni-Rad, M. Niafar, M. Asghari-Jafarabadi, V. Mofid, 2012Probiotic yogurt improves antioxidant status in type 2 diabetic patientsNutrition28539543
14. S. M. Faber, 2000Comparison of probiotics and antibiotics to probiotics alone in treatment of diarrhea predominant IBS (D-IBS), alternating (A-IBS) and constipation (C-IBS) patients.Gastroenterology118687688
15. Food and Agriculture Organization of United Nations; World Health Organization. FAO/WHO (2002). Guidelines for the evaluation of probiotics in food, Food and Agriculture Organization of the United Nations and World Health Organization Expert Consultation Report. Available from. http://www.who.int/foodsafety/publications/fs_management/ probiotics2/en.
16. Fogliano, V., and Vitaglione, P. Functional foods: planning and development, Molecular Nutrition and Food Research 49 (2005), pp. 256-262.
17. R. Foschino, E. Fiori, A. Galli, Survival and residual activity of Lactobacillus acidophilus frozen cultures under different conditionsJournal of Dairy Research 63 (1996295303
18. P. F. Fox, J. M. Wallace, S. Morgan, C. M. Lynch, E. J. Niland, J. Tobin, Acceleration of cheese ripening,Antonie van Leeuwenhoek1996175201
19. D. Granato, I. A. Castro, L. S. N. Ellendersen, M. L. Masson, Physical stability assessment and sensory optimization of a dairy-free emulsion using response surface methodologyJournal of Food Science2010149155
20. D. Granato, G. F. Branco, A. G. Cruz, J. A. F. Faria, N. P. Shah, Probiotic dairy products as functional foodsComperhensive reviews in Food Science and food safety 9 (2010455470
21. G. M. Halpern, T. Prindville, M. Blankenburg, 1996Treatment of irritable bowel syndrome with Lacteol forte: A randomized, double-blind, crossover trial. The American Journal of Gastroenterology, 9115791585
22. L. T. Hansen, P. M. A. Wojtas, Y. L. Jin, A. T. Paulson, Survival of Ca-alginate microencapsulated Bifidobacterium spp. In milk and simulated gastrointestinal conditionsFood Microbiology20023545
23. Hasler, C. M. Functional foods: their role in disease in developing new food products for a changing prevention and health promotion, Food Technology 52 (1998), pp. 57-62.
24. I. N. Haynes, M. J. Playne, Survival of probiotic cultures in low-fat ice-creamAustralian Journal of Dairy Technology20021014
25. Hillian, M. Functional food: how big is the market? World Food Ingredients 12 (2000), pp. 50-53.

26. A. Homayouni, M. R. Ehsani, A. Azizi, M. S. Yarmand, S. H. Razavi, Effect of lecithin and calcium chloride solution on the microencapsulation process yield of calcium alginate beadsIranian Polymer Journal2007597606
27. A. Homayouni, 2008aTherapeutical effects of functional probiotic, prebiotic and symbiotic foods. (1st ed.). Tabriz: Tabriz University of Medical Sciences.
28. A. Homayouni, A. Azizi, M. R. Ehsani, S. H. Razavi, M. S. Yarmand, Effect of microencapsulation and resistant starch on the probiotic survival and sensory properties of synbiotic ice creamFood Chemistry2008b5055
29. A. Homayouni, M. R. Ehsani, A. Azizi, S. H. Razavi, M. S. Yarmand, Spectrophotometrically evaluation of probiotic growth in liquid media, Asian Journal of Chemistry 20(3) (2008c24142420
30. A. Homayouni, M. R. Ehsani, A. Azizi, S. H. Razavi, M. S. Yarmand, Growth and survival of some probiotic strains in simulated ice cream conditionsJournal of Applied Sciences 8(2) (2008d379382
31. A. Homayouni, Letter to the editorFood Chemistry20091073
32. A. Homayouni, A. Azizi, M. Javadi, S. Mahdipour, H. Ejtahed, (1 EOF10 EOF .3923/ijds.2012.
33. P. J. Joseph, R. I. Dave, N. P. Shah, Antagonism between yogurt bacteria and probiotic bacteria isolated from commercial starter cultures, commercial yogurts, and a probiotic capsule, Food Australia 50 (19982023
34. F. Jousse, Modeling to improve the efficiency of product and process developmentComprihensive Review of Food Science and Food Safety 7 (2008175181
35. K. Kailasapathy, S. L. Rybka, acidophilus and Bifidobacterium spp. Their therapeutic potential and survival in yoghurt, Australian Journal of Dairy Technology 52 (1997), 2835 .
36. K. Kailasapathy, J. Chin, Survival and therapeutic potential of probiotic organisms with reference to Lactobacillus acidophilus and Bifidobacterium spp.Immunology and Cell Biology20008088
37. K. Kailasapathy, K. Sultana, Survival and β-D-galactosidase activity of encapsulated and free Lactobacillus acidophilus and Bifidobacterium lactis in ice cream, Australian Journal of Dairy Technology 58 (2003223227
38. K. M. K. Kebary, S. A. Hussein, R. M. Badawi, Improving viability of bifidobacterium and their effect on frozen ice milk, Egyptian Journal of Dairy Science 26 (1998319337
39. H. J. Kim, M. Camilleri, S. Mc Kinzie, 2003A randomized controlled trial of a probiotic, VSL 3, on gut transit and symptoms in diarrhoea-predominant irritable bowel syndromeAlimentary Pharmacology and Therapeutics, 17895904
40. Koletzko, B., Aggett, P.J., and Bindels, J.G. Growth, development and differentiation: a functional food science approach, Brazilian Journal of Nutrition 8019983545
41. F. V. Kosikowski, V. V. Mistry, 1997Cheese and fermented milk foodsin Origin and Principles: Fermented Milks. Westport: F.V. Kosikowski, (5774
42. E. R. S. Kunji, I. Mierau, A. Hagting, B. Poolman, W. N. Konings, The proteolytic systems of lactic acid bacteriaAntonie van Leeuwenhoek 70 (1996187221
43. W. E. V. Lankaputhra, N. P. Shah, M. L. Britz, Survival of bifidobacteria during refrigerated storage in the presence of acid and hydrogen peroxideMilchwissenschaft19966570
44. R. G. Leuschner, M. Heidel, W. P. Hammes, Histamine and tyramine degradation by food fermenting microorganisms,International Journal of Food Microbiology1998110
45. T. Luckow, C. Delahunty, Consumer acceptance of orange juice containing functional ingredientsFood Research International2004805814
46. Matilla-Sandholm, T., Blum, S., Collins, J.K., Crittenden, R., DeVos, W., Dunne, C., et al. Probiotics: towards demonstrating efficacy, Trends in Food Science and Technology 10 (1999), pp. 393-399.

47. J. H. Meurman, 2005Probiotics: do they have a role in oral medicine and dentistry? Europian Journal of Oral Sciences, 1133188196
48. H. Mirzaei, H. Pourjafar, A. Homayouni, 2012Effect of calcium alginate and resistant starch microencapsulation on the survival rate of Lactobacillus acidophilus La5 and sensory properties in Iranian white brined cheeseFood Chemistry13219661970
49. K. C. Mountzouris, G. R. Gibson, Colonization of the gastrointestinal tractAnnales Nestle20034354
50. C. Nicolay, Language is a key to marketing digestive health products, Functional Foods and Nutraceuticals 6 (20032022
51. R. R. Ravula, N. P. Shah, Viability of probiotic bacteria in fermented frozen dairy dessertsFood Australia1998136139
52. M. Roberfroid, (2000). Inulin-type fructans. Boca Raton: CRC Press.
53. M. B. Roberfroid, Concepts and strategy of functional food science: the European perspective,American Journal of Clinical Nutrition 71 (200016601664
54. Ross, R. P., Fitzgerald, G., Collins, K., and Stanton, C. Cheese delivering biocultures: probiotic cheese, Australian Journal of Dairy Technology 57(2) (2002), pp. 71-78.
55. D. Roy, I. Mainville, F. Mondou, Bifidobacteria and their role in yogurt-related products, Microecology Therapy 26 (1997167180
56. A. Saggioro, 2004Probiotics in the treatment of irritable bowel syndrome.Journal of Clinical Gastroenterology38104106
57. N. P. Shah, R. R. Ravula, Microencapsulation of probiotic bacteria and their survival in frozen fermented dairy dessertsAustralian Journal of Dairy Technology2000139144
58. N. P. Shah, Functional cultures and health benefits, International Dairy Journal 17 (200712621277
59. Siro, I., Kapolna, E., Kapolna, B., and Lugasi, A. Functional food: product development, marketing and consumer acceptance-A review, Appetite 51 (2008), pp. 456-467.
60. C. Stanton, G. Gardiner, H. Meehan, K. Collins, G. Fitzgerald, P. B. Lynch, et al.Market potential for probiotics,American Journal of Clinical Nutrition 73 (200147654835
61. C. Stanton, C. Desmond, M. Coakley, J. K. Collins, G. Fitzgerald, R. P. Ross, 2003Challenges facing development of probioticcontaining functional foods. In E. R. Farnworth (Eds.). Handbook of fermented functional foods (2758Boca Ranton: CRC Press.
62. I. S. Surono, A. Hosono, 2002Fermented milks: Types and standards of identity. In H. Roginski, J. Fuquay, and P.F. Fox (Eds.). Encyclopedia of Dairy Microbiology. (10181023
63. A. Y. Tamime, V. M. E. Marshall, R. K. Robinson, Microbiological and technological aspects of milks fermented by bifidobacteria,Journal of Dairy Research 62 (1995151187
64. Tepper, B., and Trail, A. Taste or health: a study on consumer acceptance of corn chips, Food Quality and Preference 9 (1998), pp. 267-272.
65. H. Tuorila, A. V. Cardello, Consumer responses to an off-flavour in juice in the presence of specific health claims, Food Quality and Preference 13 (2002561569
66. E. R. Vedamuthu, 2006Other fermented and culture-containing milks. In R. Chandan, C.H. White, A. Kilara, and Y.H. Hui (Eds.), Manufacturing Yogurt and Fermented Milks (295308Blackwell Publishing.
67. C. L. Vernazza, B. A. Rabiu, G. R. Gibson, 2006Human colonic microbiology and the role of dietary intervention: Introduction to prebiotics. In G. R. Gibson and R. A. Rastall (Eds.), Prebiotics: Development and application (112England: John Wiley and Sons Ltd.
68. Vianna, J. V., Cruz, A. G., Zoellner, S. S., Silva, R., and Batista, A. L. D. Probiotic foods: consumer perception and attitudes, International Journal of Food Science and Technology 43 (2008), pp. 1577-1580.

69. Z. Vickers, L. Mullan, E. Holton, Impact of differences in taste ratings on the consumption of milk in both a laboratory and a foodservice setting,Journal of Sensory Studies1999249262
70. P. Vieira, How to create brand awareness for new products, Functional Foods and Nutraceuticals 6 (20033840
71. R. L. Walzem, foods. Functional, In. Trends, Science. Food, . Technology, 2004518
72. H. Weber, 1996Starter cultures in dairy industry. In H. Weber (Eds.), Mikrobiologie der Lebensmittel: Milch und Milchprodukte (in German) (105152Hamburg: Behr's Verlag.

CHAPTER 10

Reduction of Aerobic and Lactic Acid Bacteria in Dairy Desludge using an Integrated Compressed CO_2 and Ultrasonic Process

Tim W. Overton[1,2] & Tiejun Lu[1] & Narinder Bains [3] & Gary A. Leeke [1]

[1].*School of Chemical EngineeringUniversity of BirminghamBirminghamUK*

[2].*Institute of Microbiology & InfectionUniversity of BirminghamBirminghamUK*

[3].*Sere-Tech Innovation Ltd.BirminghamUK*

ABSTRACT

Current treatment routes are not suitable to reduce and stabilise bacterial content in some dairy process streams such as separator and bactofuge desludges which currently present a major emission problem faced by dairy producers. In this study, a novel method for the processing of desludge was developed. The new method, elevated pressure sonication (EPS), uses a combination of low frequency ultrasound (20 kHz) and elevated CO_2 pressure (50 to 100 bar). Process conditions (pressure, sonicator power, processing time) were optimised for batch and continuous EPS processes to reduce viable numbers of aerobic and lactic acid bacteria in bactofuge desludge by ≥3-log fold. Coagulation of proteins present in the desludge also occurred, causing separation of solid (curd) and liquid (whey) fractions. The proposed process offers a 10-fold reduction in energy compared to high temperature short time (HTST) treatment of milk.

Keywords: Desludge Milk byproduct processing Sonication Pressure Supercritical

This paper is part of the Special issue dedicated to the 9th International Cheese Symposium held in Cork, Ireland and organized by Teagasc in collaboration with University College Cork and INRA, 12th & 13th November 2014.

1 INTRODUCTION

There are a number of treatments commonly applied to reduce the viability of bacteria in raw milk to generate products fit for human consumption. Fresh milk is commonly pasteurised using a continuous high temperature short time (HTST) pasteurisation process, typically 72 °C for ≥15 s, or sterilised using an ultra-high temperature treatment (UHT; 150 °C for 1 s). Alternative routes of bacterial deactivation or removal from dairy streams, either industrially or in academic research settings, include centrifugation/bactofugation, reduction in pH (in casein manufacture), sonication, pulsed electric fields and high pressure treatments, each of which have different impacts on the microbiological safety and chemical and sensory properties of the product (Sampedro et al. 2005; Ortega-Rivas and Salmerón-Ochoa 2014). These techniques are typically used in combination to reduce viability of spoilage and pathogenic bacteria to acceptable limits in the final product. Many of the above processes typically require high energy inputs (Pereira and Vicente 2010) or require additional downstream separation processes (for example removal of organic or mineral acids in casein coagulation). Rapid decompression treatment has also been reported in the literature (Foster et al. 1962; Fraser 1951; Hemmingsen and Hemmingsen 1980), but this is only suitable for bacteria that contain gas vacuoles and is therefore not broadly applicable for food processing.

However, many of these treatment routes are not suitable to reduce the bacterial content in some process streams found in the dairy industries. Desludge streams generated by the bactofuge (containing large numbers of bacteria) and separator (containing particles even heavier than the skim, such as sediment, somatic cells and some bacteria) cannot be treated by thermal routes and are currently disposed of at high cost to the processor. A large milk processing site can produce up to

15,000 L · day^{-1} of desludge which is disposed of through water dilution and discharged to sewer at costs of between US$ 150 k and US$ 1.5 M per annum in effluent discharge. In light of this, UK dairy producers have tried local initiatives to treat the desludge; however, several challenges have arisen: bacterially and thermally induced solidification of the desludge, classification of the desludge as waste as it is not effectively pasteurised or heat treated, high off-site waste disposal costs, high transport costs and high tanker cleaning costs.

In this paper, we describe the development of elevated pressure sonication (EPS), a novel technology for the treatment of desludges (Bains and Leeke 2015). The technique uses a combination of elevated CO_2 pressure (up to 100 bar; lower than high pressure and rapid decompression treatments) and relatively low frequency sonication (20 kHz as opposed to 18 to 500 MHz commonly used in other antibacterial processes). We optimised process parameters (sonicator power, pressure, process time, temperature) to maximise reduction of viable aerobic and lactic acid bacterial counts. The process was optimised in both batch and continuous modes.

2 MATERIALS AND METHODS

2.1 Materials

Bactofuge and separator desludges were supplied by a leading dairy company. Samples were collected, stored at 4 °C and used within 24 h of collection; after this time, the bactofuge desludge had phase separated and resultant solids led to blockage of the pump inlet. CO_2 and N_2 were obtained from BOC (Wolverhampton, UK) and had given purities of 99.8% *v*/*v* and 99.998%, respectively.

2.2 Batch tests

A schematic of the experimental setup used for the high pressure sonication is shown in Fig. 1a. The vessel was a 1-L stainless steel reactor modified for ultrasound (Parr, USA; maximum working conditions 350 °C, 200 bar), with heating jacket, temperature control and thermowell. An ultrasound probe (Sonics and Materials, USA) was integrated into the vessel head and sealed to maintain the pressure at the operating conditions. Although not mechanically agitated, the ultrasound probe allowed sufficient dispersion of the gas with the desludge, as validated by Cenci et al. (2014). Then, 700 mL of desludge was charged into the vessel, which resulted in the probe tip being immersed in the desludge to a depth of 2 mm, and the headspace and ancillaries were evacuated using the gas of choice (CO_2 or N_2). The pressure and temperature were then raised to the desired conditions. Sonication was introduced using pulsed mode (1 s on, 1 s off) for the desired time at 20 kHz at different percentages of power (1500 W = 100% power). The internal pressure was observed using a transducer connected to a display (Druck, Leicester, UK) to within ±0.1 bar, while the internal temperature was measured using a type-K thermocouple to within ±0.1 °C. Following the experiment, the solid (curd) and liquid (whey) fractions were collected and analysed for cell viability. For decontamination, the vessel was heated to 125 °C for 30 min and then washed with Virkon (DuPont, Stevenage, UK) to minimise cell contamination between tests.

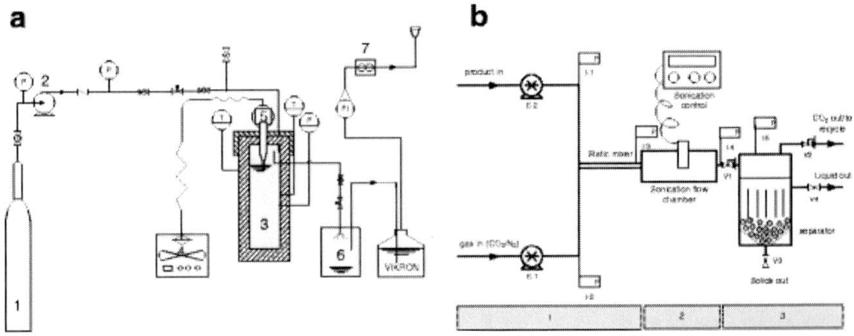

Fig. 1. Schematic of experimental setup. **A** Elevated pressure sonication batch tests. *1* gas cylinder, *2* pump, *3* high pressure vessel, *4* ultrasonic controller, *5* ultrasonic probe, *6* collection vessel, *7* mass flow meter. **B** Continuous flow experimental setup. Section 1 = product/gas contact; Section 2 = high pressure sonication chamber; Section 3 = separation

2.3 **Continuous flow apparatus**

A similar setup to batch was used for the continuous flow tests. A schematic of the process is shown in Fig. 1b and consists of three sections, (1) product/gas contact, (2) sonication chamber and (3) separator. In (1), the product and gas (CO_2 or N_2) were pumped into a stainless steel static mixer where they were contacted. A diaphragm pump (LEWA LDB M9, GmbH) was used for delivering desludge samples into the sonication chamber. Inside the sonication chamber (2) the bacteria present in the product stream were subjected to relatively low frequency sonication (20 kHz). Sonication was provided by a Sonics VCX130 Ultrasonic Processor connected to a 1/4-inch solid Titanium Alloy Probe to give a maximum power output of 130 W. Typical residence times were 30 s to

1 min. The product left the chamber and entered the separation vessel (3) where the solid fraction was separated by gravity from the liquid fraction. The liquid fraction was removed over the weir plates for analysis. The gas disengaged from the liquid and could be vented or recycled. There will be a small amount of gas in the liquid phase which could either be vented (and recycled) or, depending on its use, kept within the liquid. The solid fraction can be periodically removed from the separation vessel using a rotating valve. Desludge is typically produced at the dairy at 50 °C and this temperature was largely used; however, lower temperatures were also evaluated.

2.4 Bacteria viability tests

Bacterial counts were performed on desludge samples before and after processing to generate bacterial deactivation data. Samples were serially diluted in sterile maximum recovery diluent (MRD; 8.5 $g \cdot L^{-1}$ NaCl and 1 $g \cdot L^{-1}$ tryptone in distilled water) and plated onto nutrient agar (Oxoid, Basingstoke, UK; for enumerating total aerobic viable count) or MRS (de Man, Rogosa, Sharpe) agar (Oxoid; for enumerating lactic acid bacteria). Nutrient agar plates were incubated aerobically for 24–48 h at 37 °C and MRS agar plates were incubated in candle jars (microaerobic environment with decreased O_2 and elevated CO_2 level) for 24–72 h at 30 °C prior to counting. For analysis of bacterial content of curd samples, fresh wet curd obtained after batch processing was weighed, added to 100 mL of water and stomached using a Seward 400 Circulator Stomacher at 200 rpm for 5 min before serial dilution and plating as above.

2.5 Chemical analysis methods

The Rose Gottlieb method (the standard method of ISO, FAO and WHO for milk fat content measurement) was used to determine the fat content ($mg \cdot 100\ mL^{-1}$) in desludge before and after treatment (AOAC 2000; Official method 905.02). Hexane was used to replace petroleum ether.

A MaxSignal® No-Mel Milk Test Kit (Bioo Scientific, Austin, TX) was used to measure protein content in liquid samples before and after treatment. The standard curve was obtained using UV on-line detection using standards provided in the test kit. This kit was not suitable for solid samples, which were sent for analysis at an independent accredited laboratory (Butterworths, London, UK). The classic Kjeldahl method was used in accordance with USP 35,461 Method 1 to determine the nitrogen content in a sample which was then converted into protein content. Liquid samples were also sent to them for comparison with our results.

COD reagent (Merck Millipore Spectroquant) with concentration range of 500–10,000 mg · $L^{-1}$was used to measure the COD (chemical oxygen demand) of liquid samples before and after treatment.

3 RESULTS AND DISCUSSION

3.1 Batch processing of bactofuge desludge

Power sonication is a technique used to pasteurise many dairy products (Bermúdez-Aguirre and Barbosa-Cánovas 2011); however, it is used at high frequency (typically 18 to 500 MHz) to kill bacteria to acceptable limits. Sonication causes cycles of compression and expansion as well as cavitation in microbial cells. The implosion of bubbles generates points with very high temperatures and pressures which are able to inactivate cells. Sonication is most effective when combined with other microbiological control measures as smaller microbial cells are often more resistant to this treatment (Geciova et al. 2002). One such combination is thermosonication which is undertaken at temperatures between 80 and 95 °C. However, since bactofuge desludge is periodically discharged during dairy processing at 50 °C, this technique would be economically unviable due to the energy input needed to raise the temperature. In addition, processing desludge at high temperatures also results in solidification to a latex-like product which cannot flow and would therefore lead to blockages in an industrial process.

In light of the above, lower frequency sonication (20 kHz) was tested in combination with gases at elevated pressures to investigate their effect on the viable cell count of aerobic and lactic acid bacteria in desludge. Lower frequency sonication has lower energy input than the power sonication techniques described above. A series of batch tests were undertaken on bactofuge desludge using the apparatus shown in Fig. 1a and different process conditions (Figs. 2 and 3).

3.2 Effect of sonication power, pressure and processing time on viable cell counts

The sonication power was varied from 1500 to 150 W at a CO_2 pressure of 100 bar and a processing temperature of 50 °C (Fig. 2a, b). The sonication power could be reduced to 150 W and a 3 to 3.5 log reduction in viable cell count for both aerobic and lactic acid bacteria could still be obtained. Viable cell counts for samples at 300, 450 and 750 W sonication power were undertaken up to 2 and 3 days post-processing. The results demonstrate the effectiveness of the combined treatment to stabilise cell count levels. It is evident that the kill rate of aerobic bacteria is better than that for lactic acid bacteria.

The aerobic and lactic acid bacteria counts were repeated for some tests and were found to have a ST_{DEV} of 11% for aerobic bacteria and a ST_{DEV} of 21% for lactic acid bacteria. The reproducibility of aerobic bacteria is better than for lactic acid bacteria although both are within an acceptable experimental error.

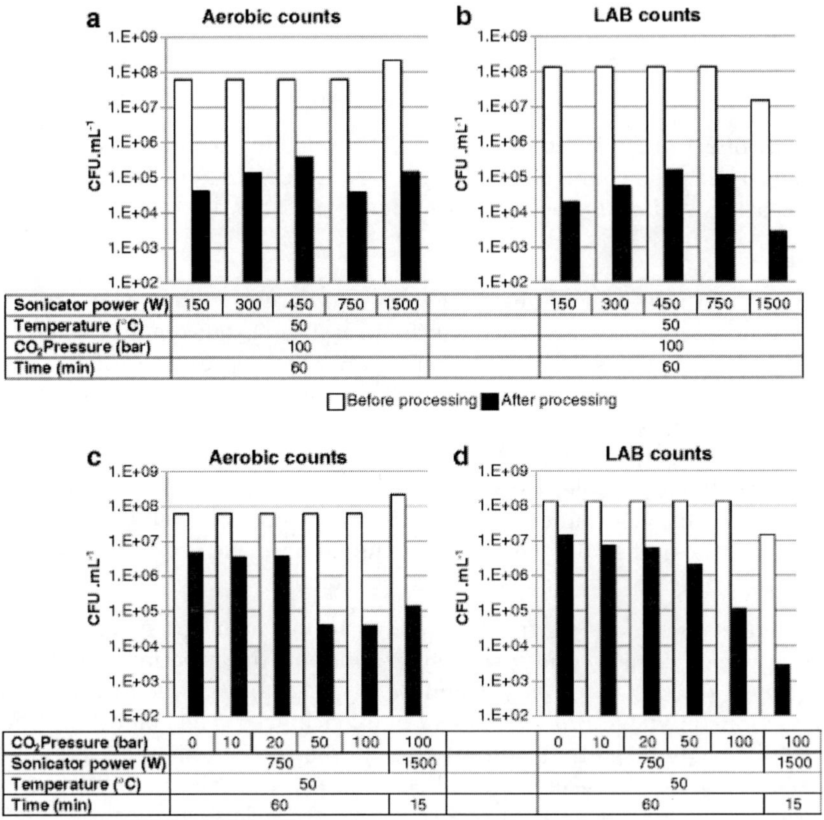

Fig. 2. Effect of sonication power (**A, B**) and pressure (**C, D**) on viable aerobic (**A, C**) and lactic acid (**B, D**) bacterial counts in bactofuge desludge before and after batch EPS processing. *White bars*, before processing; *black bars*, after processing

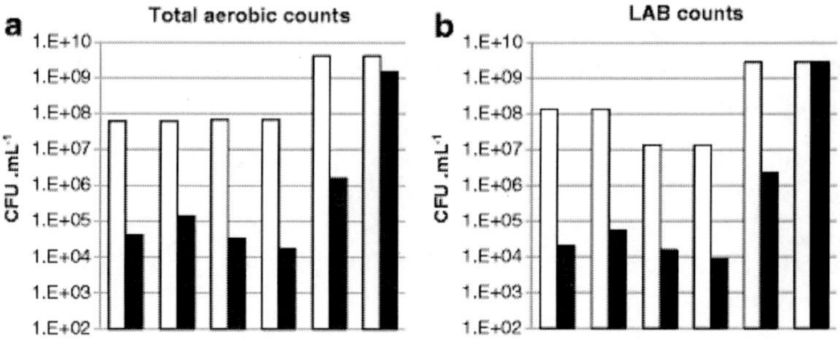

(a)

Time (min)	30	15	2.5	1	0.5	0.5
Pressure (bar)	100					
Gas	CO_2					N_2
Temperature (°C)	50					
Sonicator power (W)	300					

(b)

Time (min)	30	15	2.5	1	0.5	0.5
Pressure (bar)	100					
Gas	CO_2					N_2
Temperature (°C)	50					
Sonicator power (W)	300					

□ Before processing ■ After processing

Fig. 3. Effect of sonication time and gas on viable aerobic (**A**) and lactic acid (**B**) bacterial counts in bactofuge desludge before and after batch EPS processing. *White bars*, before processing; *black bars*, after processing

In light of the results shown in Fig. 2a and b, tests were undertaken at lower pressure in order to reduce the pressure design requirements for the treatment process. Figure 2c and d shows that a pressure ≥50 bar was required to achieve a 3.5-log-fold reduction in aerobic viable cell count levels, whereas a pressure ≥100 bar was required to achieve a 3.5-log-fold reduction in lactic acid viable cell count levels. In the absence of any gas pressure (0 bar, ambient pressure conditions), sonication had a small effect on viable cell count with approximately 1-log-fold reduction in both total aerobic and LAB counts. The combination of CO_2 and sonication is therefore necessary for the process to be effective.

The effect of sonication time was investigated in order to reduce the energy requirement and processing time of the treatment process. Figure 3 shows that the time could be reduced to 1 min with 3-log-fold reductions in viable cell counts still being achieved. The sonication time was further reduced to 0.5 min (with all other conditions the same) and showed favourable results toward cell death (~3.4-log-fold reduction in aerobic counts, ~3.1-log-fold reduction in LABs). This short process time makes the EPS treatment process potentially suitable for continuous processing of desludge. Tests were also undertaken in the presence of N_2 at 100 bar instead of CO_2. The results show that N_2 had an insignificant effect on cell viability, demonstrating that CO_2 was required for process effectiveness.

3.3 Characteristics of processed desludge

Photographs of untreated and treated bactofuge desludge are shown in Fig. 4. Due to the relatively low temperature, frequency of the sonication and power input, the solid fraction did not form a latex-like material following processing and exhibited the ability to flow due to the presence of the liquid phase. During exposure to high pressure CO_2, the pH in the desludge would have been reduced to approximately 3 at the process conditions (Leeke et al. 2005). The literature reports that when sonication is combined with a low pH environment, organisms are not able to grow (Bermúdez-Aguirre and Barbosa-Cánovas 2011). It is thought that the combination of sonication, pressure and lowering of pH results in bacterial death in this process, in a similar manner to the hurdle technology approach used in food processing. The lowered pH also caused the desludge to separate into liquid (whey) and solid (curd) fractions, without the addition of solid chemicals dissolved in the process liquor. The lack of bacterial killing in the presence of N_2 (Fig. 3) may be explained by its lower solubility in water (assumed to be the main liquid component); 1.011 cm^3 STP · g^{-1} water as opposed to 25.6 cm^3 STP · g^{-1} water for CO_2 at 100 bar and 50 °C (Wiebe and Gaddy 1939; Wiebe and Gaddy 1940; Wiebe et al. 1933). In addition, the chemical properties of N_2 would not lead to a decrease in pH during processing.

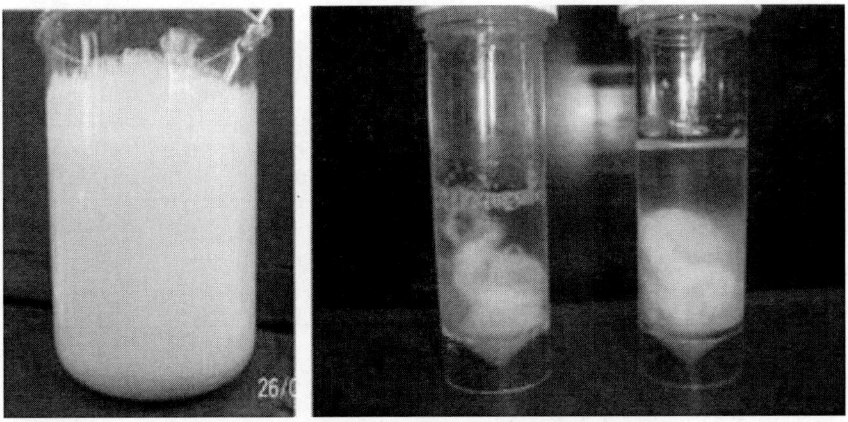

Fig. 4. Images of bactofuge desludge pre- (*left*) and post- (*right*) EPS treatment. Process conditions: batch; 100 bar CO_2, 300 W pulsed sonication power, 3 min

Prior to treatment the COD of the desludge was 1.2 million mg · L^{-1} making it costly to dispose of. After high pressure sonication treatment, the COD of the whey fraction was 33,000–42,000 mg · L^{-1} with the remainder in the solid fraction. Thus, the treatment does not lower the COD but partitions it into the solid fraction. This offers an added advantageous separation process and retains the majority of the COD in a dewatered solid fraction making it suitable for use as a value-added product such as animal feed or an energy source. The whey fraction is also likely to contain valuable components, similar to those found in cheese whey (de Wit 2001).

3.4 **Energy requirements for elevated pressure sonication treatment**

Table 1 shows the energy requirements of the sonicator (kJ · L^{-1}) to treat 700 mL of bactofuge desludge using the batch EPS process. Selecting test 6 as the optimal case, the additional energy requirements for CO_2 compression and desludge compression were calculated as 6.1 and 8.6 kJ · L^{-1}, respectively, giving the total energy required to process 1 L of desludge as 24.7 kJ. A comparison is made with the energy required to pasteurise milk using HTST treatment. The EPS process requires less energy than that required to pasteurise milk and is therefore highly attractive as a commercial process.

Table 1. Energy requirement to treat bactofuge desludge at different conditions

Method/test number		Pressure (bar)	Temperature (°C)	Power (W)	Time (min)	Energy needed (kJ·L^{-1})
EPS	Test 1	100	50	1500	15	854
	Test 2	50	50	750	30	701
	Test 3	100	50	750	30	1241
	Test 4	100	50	300	15	171
	Test 5	100	50	300	2.5	34
	Test 6	100	50	300	1	10
Pasteurisation of milk by HTST		Continuous	72		0.25	196
		Batch	63		15	158
		High fat	75		0.25	209

All tests gave a 3- to 3.5-log-fold reduction in total aerobic counts. All tests, except Test 2, gave a 3- to 3.5-log-fold reduction in lactic acid bacteria. Energy needed for EPS process based on sonicator energy input only. HTST based on process without heat integration

HTST high temperature short time

3.5 **Continuous tests on desludge samples**

Using the optimum batch conditions identified above, the process was modified to treat desludge under continuous flow conditions (Fig. 1b) to test its suitability for large volume processing. Tests were carried out on both bactofuge and separator desludge. Since dairy plants produce similar quantities of separator and bactofuge desludge, a mixture containing 50:50 vol.% of both desludges was tested using the continuous EPS process (Fig. 5).

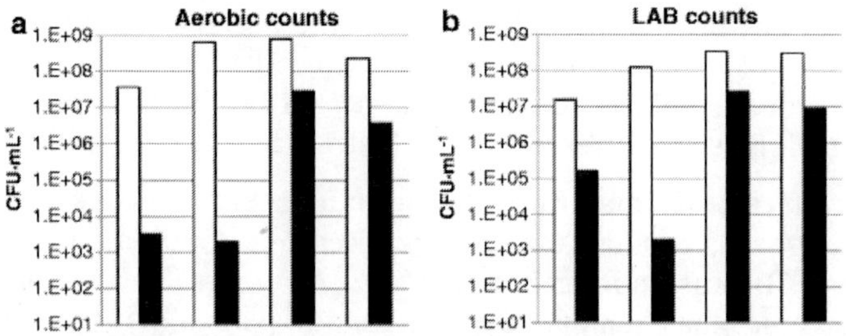

Process	Batch	Continuous				Batch	Continuous		
Time (min)	1	1				1	1		
Sonicator power (W)	300	32.5	32.5*	32.5		300	32.5	32.5*	32.5
Temperature (°C)	50					50			
Pressure (bar)	100			50		100			50

□ Before processing ■ After processing

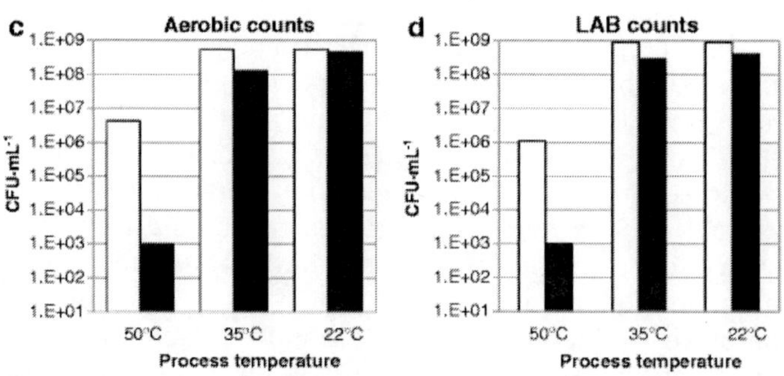

Fig. 5. A, B The effect of different batch and continuous EPS processing conditions on aerobic (**A**) and lactic acid (**B**) bacterial counts. *White bars*, before processing; *black bars*, after processing. Batch process was performed on bactofuge desludge (Test 6 in Table 1), continuous process was performed on a 1:1 mixture of bactofuge and separator desludges. Continuous process used flow rates of 10 mL · min^{-1} desludge and ~4 g · min^{-1} CO_2, giving a residence time of 1 min. *Pulsed sonication, 1 s on and 1 s off. **C, D** Effect of processing temperature on aerobic (**C**) and lactic acid (**D**) bacterial counts. Process conditions were as follows: continuous process; pressure, 100 bar; temperature, as shown; sonicator power = 32.5 W; sample flow, 10 mL · min^{-1}, CO_2 flow, 4 g · min^{-1}

Similar to the batch tests (Figs. 2 and 3), lowering the pressure or sonicator energy input resulted in poorer reductions in viable cell numbers. Cell viability was significantly reduced for the separator sample with the flow process giving an even better reduction in viable counts than batch processing, possibly from the enhanced mixing induced in the flow process. In addition, pulsed sonication was also tested. The flow rate of the samples was kept constant at 10 $mL \cdot min^{-1}$ while the sonication was either constant (18 $kJ \cdot L^{-1}$ energy input) or pulsed modality (11.9 $kJ \cdot L^{-1}$ energy input). Temperature (50 °C), pressure (100 bar) and CO_2 flow rate (~4 $g \cdot min^{-1}$) were the same for both tests. For either continuous or pulsed sonication, aerobic bacterial counts reduced from 4.3×10^6 $CFU \cdot mL^{-1}$ before treatment to $<10^3$ $CFU \cdot mL^{-1}$ after treatment; LAB counts similarly reduced from 1.1×10^6 $CFU \cdot mL^{-1}$ before treatment to $<10^3$ $CFU \cdot mL^{-1}$ after treatment (see Fig 5c and d).

The effect of temperature on continuous EPS processing of bactofuge desludge was tested. Tests were carried out at 50 and 35 °C (conditions where CO_2 was supercritical) and 22 °C (where CO_2 was liquid). It was discovered that temperature and state play a key role in desludge treatment (Fig. 5c, d). Both liquid CO_2 and supercritical CO_2 close to the critical point (31 °C) did not effectively kill bacteria.

3.6 Shelf-life of treated and untreated bactofuge desludge

The shelf-life of untreated bactofuge desludge was compared to treated desludge (100 bar CO_2, 50 °C, 32.5 W power, pulsed, 1 min residence time and energy input of 12 $kJ \cdot L^{-1}$) in terms of lactic acid and aerobic counts. Treated and untreated samples were incubated at either room temperature or 4 °C and viable counts determined at daily intervals (Fig. 6a, b). EPS-treated bactofuge desludge could be stored at 4 °C for 5 days without an obvious increase in LAB counts and with less than 1 log increase in total aerobic counts. The treated sample stored at room temperature attained similar viable counts to the untreated sample after 1-day storage at room temperature. After 3 days of storage at room temperature, both untreated and treated bacterial counts were high; the untreated desludge solidified, after which it could not be further assessed.

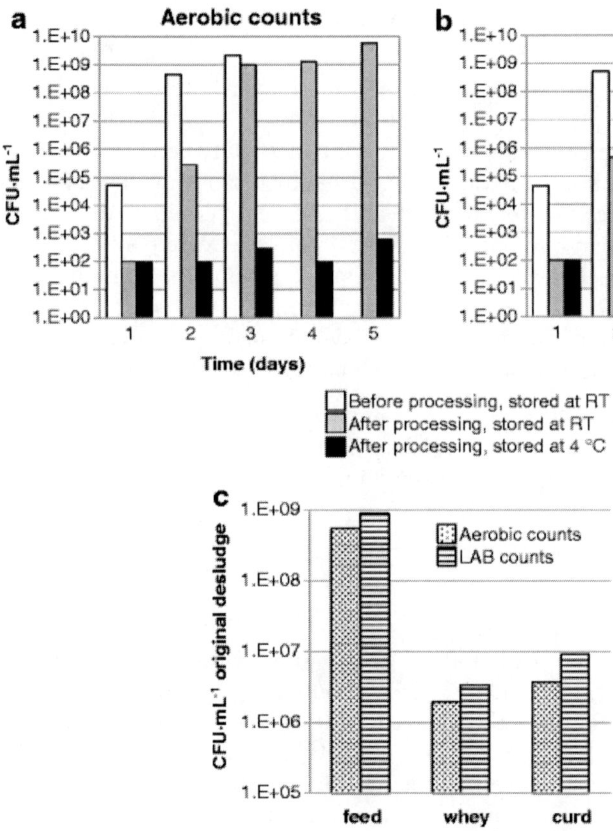

Fig. 6. **A**, **B** Shelf-life of bactofuge desludge before (feed) and after processing. Total aerobic bacteria (**A**) and lactic acid bacteria (**B**) counts were determined for bactofuge desludge without EPS treatment (*white bars*) and with EPS treatment, after storage at room temperature (*grey bars*) or 4 °C (*black bars*) for up to 5 days. Processing conditions: 100 bar CO_2; 50 °C; 32.5 W sonicator power; (pulsed) 1 min residence time; energy input of 12 kJ · L^{-1}. **C** Partitioning of bacteria into curd and whey fractions following EPS processing. Bactofuge desludge was processed (batch EPS process; 100 bar CO_2; 50 °C; 300 W sonicator power pulsed 1 s on, 1 s off; 2 min; energy input of 19 kJ · L^{-1}) and viable aerobic (*spotted bars*) and LAB (*striped bars*) determined before processing and after processing in the whey and curd (following stomaching) fractions. Numbers of bacteria are expressed as CFU per millilitre of original desludge sample, taking into consideration the quantity of curd and whey fractions generated during EPS processing

3.7 **Analysis of curd and whey fractions**

The volume of curd and whey fractions formed after batch processing of bactofuge desludge (100 bar, 50 °C, 300 W power, 1:1 pulse for 1 min) were measured. Typically the liquid fraction was between 71 and 77% with a 22–29% wet solid fraction. The percentage mass of the dry solid was found to be approximately 5 to 8% and was determined by drying the wet solid sample in an oven at 25 °C for 15 consecutive days until there was no mass change. The drying process caused gradual hardening of the sample and resulted in a colour change from white cream to semi-transparent. The solid merged together when drying to form a hard, strong and hollow material of roughly the shape of its container.

The bacterial content of the curd and whey fractions was determined; the curd fraction was stomached to release bacteria (Fig. 6c). Both aerobes and LABs were preferentially partitioned into the curd fraction.

3.8 **Fat and protein content of desludge and processed fractions**

The fat content of the curd and whey fractions obtained after batch processing was determined using the Rose Gottlieb method (Table 2) and compared to values found in the literature (de Wit 2001). Dry solids were obtained after drying at 25 or 70 °C. Protein concentrations of EPS-derived whey fractions determined using the Rose Gottlieb method were comparable to the protein content of casein, lactic acid and cheese whey (de Wit 2001). Protein concentrations of whey samples were also determined by an external laboratory; data were comparable between the two analysis methods. The protein content of cheese is typically between 17 and 42% with the majority of cheeses being between 20 and 32%. The fresh wet curd obtained from the EPS process contained 35.3% protein and so has potential for use as animal feed or other purposes.

The protein content of the bactofuge curds were corrected on a dry basis and are shown in Table 3. The calculation was based on a 60% moisture content in the fresh curd, 12% moisture content in the 25 °C dried curd and 8% moisture content in the 70 °C dried curd. The data suggest that drying the processed curd at 70 °C might damage the protein in the curd. Yellow-brownish areas were seen on the 70 °C dried curd surface and suggest that the Maillard reaction had occurred in the presence of the water leading to protein damage; its lower protein content agreed with this assumption.

Table 2. Fat and protein content of desludge samples before and after batch EPS treatment

Sample	State	Fat content mg·100 mL^{-1} (liquid) or mg·100 g^{-1} (solid)		Protein content %		
		This work	Literature value	This work	External lab	Literature value
Unprocessed bactofuge desludge	Liquid	60	–	6.89	10.55	–
Whey (sample 1)	Liquid	19.6	30–50	0.67	1.1	0.6–0.62[a]
Whey (sample 2)	Liquid	16.6	30–50	0.53	0.7	0.6–0.62[a]
Fresh curd of sample 2	Wet solid	455.2	ND	33.7	35.3	ND
Curd of sample 1 dried at 25 °C	Solid	ND	ND	ND	77.5	ND
Curd dried at 70 °C	Solid	ND	ND	ND	70.95	ND
Unprocessed separator desludge	Liquid	143.6	ND	ND	ND	ND
Milk[b]	Liquid	ND	4110	ND	ND	3.3

All tests processed bactofuge desludge in batch EPS process at 100 bar, 50 °C, 300 W power for 1 min; samples 1 and 2 are repeat analyses to show reproducibility. Fat content units depend on solid or liquid samples. Fat content compare our data (Rose Gottlieb method) and literature data (de Wit 2001); protein content compare our data (MaxSignal® No-Mel Milk Test Kit), external laboratory data (Kjeldahl method) and literature data (de Wit 2001)

ND not determined

[a] Range of values for casein, lactic acid and cheese whey

[b] Data for milk from literature (de Wit 2001)

Table 3. Protein determination of bactofuge curd on dry weight basis

Sample	Protein %	Protein % (dry base)
Fresh (wet) curd	35.3	88
25 °C dried curd	77.5	88
70 °C dried curd	70.95	77

4 CONCLUSION

Elevated pressure sonication (EPS) offers a way to effectively process bactofuge desludge. Pressure alone or sonication alone do not kill bacteria; however, in combination, a 3-log-fold decrease in aerobic bacteria and 3.5-log-fold decrease in lactic acid bacteria can be obtained. Low CO_2 pressure (50 bar) can be used to reduce viable aerobic bacteria whereas high CO_2pressure (100 bar) must be used to reduce numbers of lactic acid bacteria. High pressure sonication in the presence of N_2 resulted in insignificant cell death. Sonication power as low as 150 or 300 W can be used to kill aerobic and lactic acid bacteria in a very short treatment time (<0.5 min). In comparison to HTST of fresh milk, the technology leads to a 10-fold reduction in energy requirement. Comparison with other emerging dairy processing techniques such as high pressure or pulsed electric field treatment is difficult as they have not been applied to desludge treatment; this could form the basis of future work.

The good flow properties of the desludge post-processing meant that this treatment technique has real potential for industries that rely on high volume processing. The treatment of desludge by conventional thermal treatments leads to a latex-like product that cannot flow. The curd fraction could potentially be used as a food or feed product (e.g. cheese or animal feed), or could be used to raise energy by digestion. The whey fraction has value in its own right as a potential source of bioactive peptides. The induced separation can therefore also be seen as a dewatering process of the solid curd fraction. The EPS process thereby generates a value product and the technique has potential to cross over to other dairy products, brewery waste and beverage processing. If the product is considered as a food product, further testing (e.g. for endotoxins in the lysate) and approval is required.

ACKNOWLEDGEMENTS

The authors greatly acknowledge the funding from Innovate UK, Project number 101382, grant number TS/K004018/1

Data access

Data sets from the above work may be requested from the authors by email (T.W.Overton@bham.ac.uk or G.A.Leeke@bham.ac.uk).

REFERENCES

1. AOAC (2000) In: Horwitz W (ed) Official methods of analysis of AOAC International, 17th edn. Association of Official Analytical Chemists, USA

2. Bains NS, Leeke GA (2015) Inhibition of microbial and cellular growth in substances. Patent number WO2015052506
3. Bermúdez-Aguirre D, Barbosa-Cánovas GV (2011) Ultrasound technologies for food and bioprocessing. Food Eng Ser. doi:10.1007/978-1-4419-7472-3_16. Springer
4. Cenci S, Cox LR, Leeke GA (2014) Ultrasound-induced emulsification of subcritical carbon dioxide and water with and without surfactant as a strategy for enhanced mass transport. Ultrason Sonochem 21:401–408
5. de Wit JN (2001) Lecturer's handbook on whey and whey products. European Whey Products Association, Brussels
6. Foster JW, Cowan RM, Maag TA (1962) Rupture of bacteria by explosive decompression. J Bacteriol 83:330–334
7. Fraser D (1951) Bursting bacteria by release of gas pressure. Nature 167:33–34
8. Geciova J, Bury D, Jelen P (2002) Methods for disruption of microbial cells for potential use in the dairy industry—a review. Int Dairy J 12:541–553
9. Hemmingsen BB, Hemmingsen EA (1980) Rupture of the cell envelope by induced intracellular gas phase expansion in gas vacuolate bacteria. J Bacteriol 143:841–846
10. Leeke GA, Santos RC, Seville J, Al-Duri B, Smith CJ, Holmes AB (2005) Solubilities of iodobenzene, phenylboric acid, biphenyl and 4-phenyltoluene in carbon dioxide from the measurements of the relative permittivity. J Chem Eng Data 50:1370–1374
11. Ortega-Rivas E, Salmerón-Ochoa I (2014) Nonthermal food processing alternatives and their effects on taste and flavor compounds of beverages. Crit Rev Food Sci Nutr. doi:10.1080/10408398.2011.579362
12. Pereira RN, Vicente AA (2010) Environmental impact of novel thermal and non-thermal technologies in food processing. Food Res Int 43:1935–1943
13. Sampedro F, Rodrigo M, Martínez A, Rodrigo D, Barbosa-Cánovas GV (2005) Quality and safety aspects of PEF application in milk and milk products. Crit Rev Food Sci Nutr 45:25–47
14. Wiebe R, Gaddy VL (1939) The solubility in water of carbon dioxide at 50, 75 and 100 °C from 12 to 700 atmospheres. J Am Chem Soc 61:315–318
15. Wiebe R, Gaddy VL (1940) The solubility of carbon dioxide in water at various temperatures from 12 to 40 °C and at pressures to 500 atmospheres. Crit Phenom J Am Chem Soc 62:815–817
16. Wiebe R, Gaddy VL, Heins C (1933) The solubility of nitrogen in water at 50, 75 and 100 °C from 25 to 1000 atmospheres. J Am Chem Soc 55:947–953

CHAPTER 11

Change in Color and Volatile Composition of Skim Milk Processed with Pulsed Electric Field and Microfiltration Treatments or Heat Pasteurization †

Anupam Chugh [1], Dipendra Khanal [1], Markus Walkling-Ribeiro [1,2,*], Milena Corredig [1], Lisa Duizer [1] and Mansel W. Griffiths [1]

[1]*Department of Food Science, University of Guelph, Guelph, ON N1G 2W1, Canada*

[2]*Department of Food Science, Cornell University, Ithaca, NY 14853, USA*

ABSTRACT

Non-thermal processing methods, such as pulsed electric field (PEF) and tangential-flow microfiltration (TFMF), are emerging processing technologies that can minimize the deleterious effects of high temperature short time (HTST) pasteurization on quality attributes of skim milk. The present study investigates the impact of PEF and TFMF, alone or in combination, on color and volatile compounds in skim milk. PEF was applied at 28 or 40 kV/cm for 1122 to 2805 μs, while microfiltration (MF) was conducted using membranes with three pore sizes (lab-scale 0.65 and 1.2 μm TFMF, and pilot-scale 1.4 μm MF). HTST control treatments were applied at 75 or 95 °C for 20 and 45 s, respectively. Noticeable color changes were observed with the 0.65 μm TFMF treatment. No significant color changes were observed in PEF-treated, 1.2 μm TFMF-treated, HTST-treated, and 1.4 μm MF-treated skim milk ($p \geq 0.05$) but the total color difference indicated better color retention with non-thermal preservation. The latter did not affect raw skim milk volatiles significantly after single or combined processing ($p \geq 0.05$), but HTST caused considerable changes in their composition, including ketones, free fatty acids, hydrocarbons, and sulfur compounds ($p < 0.05$). The findings indicate that for the particular thermal and non-thermal treatments selected for this study, better retention of skim milk color and flavor components were obtained for the non-thermal treatments.

Keywords: pulsed electric field (PEF); microfiltration (MF); thermal pasteurization; skim milk; volatile compounds; color degradation; non-thermal processing; hurdle technology

1. INTRODUCTION

Milk is a widely consumed beverage due to its nutritional importance, a pleasant aroma and mouth-feel, and a slightly sweet taste. Color and flavor are important sensorial attributes of milk that are influenced by several factors, such as milk composition, cow's feed and metabolism, environmental factors and processing conditions [1,2]. Few studies are available on the impact of conventional thermal treatments and emerging processing technologies like pulsed electric field (PEF), high hydrostatic pressure (HHP) and ultra high pressure (UHP) on volatile compounds in milk or milk based beverages [3,4,5,6]. Pulsed electric field (PEF) and microfiltration (MF) are emerging innovative technologies that could meet the increasing consumer demand for "fresh-like" minimally processed foods. The mode of action that underlies PEF is based on short electric pulses of high voltage applied to a product that is placed between a pair of electrodes, thereby, bringing about electroporation of the bacterial cell wall and its subsequent breakdown [7]. In addition to the type of microorganism and its innate resistance [8], the efficacy of PEF for food preservation depends on processing factors such as electric field strength, number of pulses applied and the treatment time [9], as well as product parameters, including electrical conductivity, viscosity, and pH [10].

MF is a membrane-driven separation process typically employing membranes with a pore size of 1.2 to 1.4 μm that allows removal of bacteria in vegetative and spore forms from milk [11,12]. Commercially available thermal processing methods promise a high degree of microbial safety but can adversely affect other food properties such as color and volatile compounds. Several studies have demonstrated a comparable impact of PEF and MF processes on microbial inactivation in foods [13,14,15,16,17]. Milk has its natural color due to the reflectance of light by dispersed milk fat globules, proteins, and natural milk pigments like riboflavin and carotenoids [18,19]. However, milk color is altered at high processing temperatures as a result of Maillard browning or a temporary increase in lightness due to denaturation of soluble whey proteins [20,21]. The changes in milk compounds that are responsible for change in milk color may, at the same time, affect the perception of flavor in milk due to intensification of some volatile components. It is generally accepted that prolonged or severe heat treatments cause the degradation of the volatile profile of milk, with the increase in concentration of volatile compounds being positively correlated to intensity of the heating [22,23]. The development of such volatile compounds may alter consumer acceptance as suggested by Gandy and others [24]. In contrast, non-thermal processing may have a minimal effect on the concentration of volatiles in foods due to shorter processing times at temperatures below those used for pasteurization. Zhang *et al.* [5] observed increased levels of aldehydes and ketones in pasteurized milk as compared to raw and PEF processed milk. PEF-treated and HHP-treated orange juice-milk beverage showed a reduced loss of volatiles as compared to thermally processed beverage [3]. From a quality and safety perspective, it may be advantageous to use hurdle technology rather than using a single technology to achieve sufficient product safety and maximized quality while maintaining minimal processing. Microfiltration is one of the non-thermal technologies that have been commercially allowed for extending shelf life of milk, although its use as a stand-alone treatment is restricted due to regulatory requirements [25]. Based on microbiological data, previous studies have indicated that the safety of skim milk processed with a combination of PEF and MF is comparable to that achieved by HTST pasteurization [13,17], potentially representing a non-thermal alternative to the well-established heat pasteurization. However, the effect of this novel process on color and volatile compounds has not been determined. Thus, the objective of the present study was to study the effect of different PEF, MF, and hurdle (PEF/MF) processing conditions on the color and flavor profiles of skim milk. In addition, the retention of color and flavor following treatments with these non-thermal technologies was compared to that obtained by heat pasteurization.

2. EXPERIMENTAL MATERIALS AND METHODOLOGY

2.1. Supply, Preparation, and Storage Conditions of Skim Milk

Raw milk obtained from a local dairy processing plant was separated at the Canadian Research Institute for Food Safety at the Department of Food Science, University of Guelph, using a cream separator (STsM-100-18, Motor Sich JSC, Zaporozhye, Ukraine). Skim milk was selected for processing in this study due to its lower fat content, allowing efficient microfiltration without membrane blockage, and because it is common research model for whole milk and has grown in consumer popularity.

2.2. Pulsed Electric Field (PEF) Treatment of Skim Milk

Raw skim milk processing was carried out in a PEF treatment chamber designed at the University of Guelph using an exponential decay pulse generator (PPS 30, University of Waterloo, Waterloo, ON, Canada) to form monopolar pulses with an average width of 1.5 μs. PEF parameters were measured and monitored as described previously [17]. The single PEF chamber used consists of two co-axial stainless steel electrodes encased in an insulating plastic casing. Liquid passes between electrodes through a gap of 0.21 cm. The chamber has a hold up volume of 44 cm^3 of which 25 cm^3 is the volume between the electrodes where milk is treated by means of electric pulses. An external jacket with a volume of 790 cm^3surrounds the outer electrode for additional temperature control, allowing the PEF chamber to be heated or cooled during operation. Skim milk was pumped through Masterflex silicone tubing (size L/S 16) using a peristaltic pump (Masterflex pump drive 7524-40 and pump head 77201-60, Cole Parmer Instrument Co., Vernon Hills, IL, USA). The product was pumped at a flow rate of 20 to 35 mL/min and the temperature at the entry and exit of the PEF chamber was recorded using T-type thermocouples (TMTSS-040G-6, Omega, Stamford, CT, USA) connected to a wireless temperature data logger (OM-SQ2020-2F8, Omega, Stamford, CT, USA). After treatment the outgoing product was cooled by flow through tubing submerged in a cooling water bath (NESLAB RTE 7, Thermo Fisher Scientific Inc., Newlington, NH, USA), set at 2.5 °C. For hurdle treatment, involving subsequent microfiltration, samples were kept in ice in a cold room at 4 °C before further processing to maintain the cold chain and thus, avoid growth of mesophilic microorganisms and potential quality degradation in line with common practice used in the dairy industry. Experiments were conducted at three PEF intensities that reflect different processing demands for skim milk with regard to food safety and energy consumption, which can vary considerably, based on production facilities and locations, and have not been investigated for comparable effects on milk quality parameters, such as color and volatile composition, based on these treatment intensities. The skim milk was processed using electric field strength and treatment time combinations of 28 kV/cm and 2805 μs for low-intensity PEF (PEF-L), 40 kV/cm and 1122 μs for moderate-intensity PEF (PEF-M), and 40 kV/cm and 1571 μs for high-intensity PEF (PEF-H). These processing conditions corresponded to energy densities of 83, 157 and 198 kJ/L, respectively. A pulse frequency of 25 Hz was applied for PEF-L, whereas 17.5 Hz were used for PEF-M and PEF-H treatments. Milk temperature at the inlet of the PEF chamber was 17 °C on average and depending on the treatment intensity applied maximum temperatures of 37 (PEF-L), 56 (PEF-M), and 65 (PEF-H) °C were obtained at the outlet. Following PEF and the cooling step the product temperature was 10 °C and samples were collected at this temperature.

2.3. Microfiltration (MF) of Skim Milk

Bearing in mind that smaller sized membrane pores allow for enhanced microbial reduction in skim milk, the latter was microfiltered at different pore size diameters to compare their effect on the product quality. Moreover, lab and pilot scale systems featuring different membrane materials and designs were used for MF to determine whether or not the processing scale and/or membrane materials and/or membrane designs affect color and volatile compound composition in skim milk. Tangential-flow microfiltration (TFMF) was performed using a laboratory scale Supor tangential-flow filtration system featuring either 0.65 μm (medium screen, 1.4 cm thick) or 1.2 μm (suspended screen, 1.8 cm thick) pore-sized polyethersulfone membranes (Pall Corporation, Port Washington, NY, USA). For laboratory-scale TFMF, a flow rate of 300 mL/min was generated using a Masterflex EW 77200-60 pump (Cole Parmer Instrument Co., Vernon Hills, IL, USA) and the skim milk was delivered by silicone tubing (Masterflex 96400-25, Cole Parmer Instrument Co., Vernon Hills, IL, USA) through either of the TFMF membrane cassettes, with trans-membrane pressures ranging from 0.3 to 0.6 kPa. The permeate flow rate was between 7 and 13 mL/min for milk passing through the TFMF 0.65 μm membrane (MF-0.65), whereas it was between 37 and 53 mL/min for the TFMF 1.2 μm (MF-1.2) membrane. In addition, skim milk processing was also carried out on a pilot scale 1.4 μm pore size, ceramic membrane, cross-flow microfiltration (CFMF) unit (MFS-1, Tetra Pak Filtration Systems, Aarhus, Denmark). The CFMF unit was used at a permeate flow rate (100 L/h) ten times that of retentate (10 L/h), using filter module Type-SCT Membralox (Tetra Pak Filtration Systems). The trans-membrane pressure of the ceramic membrane module was 106 kPa. All microfiltration treatments were carried out at a constant temperature of 35 °C after pre-heating skim milk in a water bath (Isotemp 210, Thermo Fisher Scientific Inc., Newlington, NH, USA) set to this temperature and equipped with a thermometer. In agreement with membrane and equipment manufacturer specifications milk volumes were selected for TFMF and CFMF processing that allowed proper milk circulation through the membranes and process stability before sampling. To avoid membrane clogging over time, laboratory scale membranes were cleaned by recirculating 0.5 N sodium hydroxide followed by 0.1 N nitric acid and 2% bleach at 50 °C, and a final rinse with water immediately after each run. Following the same approach but using different cleaning agents the pilot plant CFMF ceramic membrane was cleaned by recirculation of commercially available mixed acid detergent descaler (Diversey Divos2 (VM13), Diversey, Oakville, ON, Canada) and caustic detergent (Diversey Liquid Bril Tak (VC85)) followed by a final rinse with water.

2.4. PEF/MF Processing of Skim Milk

For hurdle processing, milk was processed using PEF as described above and stored at 4 °C prior to processing with MF. All three PEF treatments: PEF-L, PEF-M and PEF-H were used for hurdle processing followed by MF-0.65 µm and MF-1.2 µm membranes (PEF/MF). However, due to the difficulty in achieving the high sample volumes required for pilot scale microfiltration (MF-1.4) using the laboratory scale PEF unit, only skim milk processed using the PEF-M treatment was coupled with the MF-1.4 µm treatment.

2.5. High-Temperature Short-Time Pasteurization of Skim Milk

High-temperature short-time (HTST) processing of raw skim milk was carried out with a pilot scale, dual stage heat exchanger unit (UHT/HTST Lab-25 EDH, Microthermics Inc., Raleigh, NC, USA) at the Guelph Food Technology Centre (GFTC), Guelph, ON, Canada. Skim milk was kept at 4 °C before processing in the tubular heat exchanger. Milk was processed at two different temperatures of 75 (HTST-75) and 95 °C (HTST-95) for respective holding times of 20 and 45 s, corresponding to flow rates of 0.9 and 0.4 L/min, respectively. The higher intensity HTST pasteurization (95 °C, 45 s) for milk takes into consideration an additional safety margin and extended shelf life in part applied by the industry, but at the same time increased thermal load also renders changes in color and volatile compounds more likely. The heat exchanger pre-heated milk to 55 °C prior to reaching the respective temperatures and cooled the product to 10 °C after the treatments and prior to sampling.

2.6. Color Analysis

Color attributes were measured in Hunter Lab color space using a CM 3500-d spectrophotometer (Konica Minolta Sensing Inc., Mahwah, NJ, USA) equipped with SpectraMagic NX CM-S 100 software (Konica Minolta Sensing Inc.). Reflectance measurements were collected over a wavelength range from 400 to 700 nm. The color values were expressed in Hunter Lab color space as lightness (L), redness (a), and yellowness (b). In order to compare the total color difference (ΔE), between the color properties of untreated samples to those of values obtained after subjecting milk to different treatments, the following equation [26,27] was utilized:

$$\Delta E = \sqrt{\Delta L^2 + \Delta a^2 + \Delta b^2} \quad (1)$$

where, $\Delta L = L_{\text{standard}} - L_{\text{sample}}$, $\Delta a = a_{\text{standard}} - a_{\text{sample}}$, $\Delta b = b_{\text{standard}} - b_{\text{sample}}$.

In addition, the whiteness was calculated for the skim milk samples by converting Hunter Lab to CIE 1931 XYZ color space values using following formulas:

$$Y = \left(\frac{L}{10}\right)^2 \quad (2)$$

$$X = \left(Y + \left(\frac{L}{10} \times \frac{a}{17.5}\right)\right) / 1.02 \quad (3)$$

$$Z = \left(Y - \left(\frac{b}{7}\right) \times \left(\frac{L}{10}\right)\right) / 0.847 \quad (4)$$

and then using American Society for Testing and Materials (ASTM) E313 standard practice for calculating Whiteness index (WI):

$$WI = (3.388 \times Z) - (3 \times Y) \tag{5}$$

For each treatment, samples were collected in triplicate. Color measurements were done in duplicate for each sample.

2.7. Analysis of Volatile Compounds

Headspace volatiles in skim milk were analyzed using SIFT-MS (Selected In Flow Tube-Mass Spectrometry), which is a direct mass spectrometer that applies chemical ionization for the analysis of volatile components, yielding product ions that are analyzed with SIFT-MS technology (Voice200®, Syft Technologies Ltd., Christchurch, New Zealand). The latter identifies and quantifies volatile compounds with a built-in scientific library, that is experimentally determined with product ions and reaction rate coefficients, in combination with the flow tube geometry, the ionic reaction time, the measured flow rates and the pressure applied during sample analysis and requires no chemical standards for calibration. Skim milk samples (35 mL) were placed in 300 mL glass jars with lids closed and then warmed to room temperature before analysis. An empty jar was used as a blank. Twenty-one flavor compounds associated with milk were quantified using selected ion mode with 100 ms time limit, count limit of 10,000, 30 s background scan time and a product scan time of 60 s. Helium was applied as the carrier gas (250 kPa) and the scan was performed following charge transfer from three positively-charged reagent ions (H_3O^+, NO^+, and O_2^+) to the analyte in the flow tube. The temperature of the inlet arm extension was 120 °C and flow tube pressure was 113.6 ± 1.7 mTorr.

Samples were collected in triplicate for each treatment with a duplicate analysis for each sample.

2.8. Statistical Data Analysis

Color and flavor data were statistically analyzed with Sigmaplot version 11.0 (Systat Software Inc., London, UK). One-way analysis of variance (ANOVA) was carried out for three different batches of skim milk ($n = 3$), analyzing all treatments at a confidence interval of 95%. When statistical differences between treatment means were observed, the Holm-Sidak method for multiple pairwise comparisons was used. For adequate reproducibility each treatment was repeated three times and analysis was performed in duplicate.

3. RESULTS

3.1. Color Attributes of Non-Thermally and Thermally Treated Skim Milk

No significant difference in color was found in untreated skim milk and when processed using PEF-L, PEF-M, PEF-H, MF-1.2 and MF-1.4 and different combinations of PEF with MF treatments. The values for the color attributes *L*, *a*, and *b* following various treatments are provided in Table 1.

Total color difference, ΔE, representing an overall color difference of treated samples from the untreated skim milk was calculated using equation 1 and presented in Table 1. Color difference classification was adopted from Cserhalmi and others [27]. Based on this classification system ΔE can be categorized as: 0 to 0.5 = "not noticeable", 0.5 to 1.5 = "slightly noticeable" and >1.5 = "noticeable". Slightly noticeable color differences of 0.51 and 0.79 were observed in skim milk treated with HTST treatment at 75 °C or 95 °C, respectively. However, a noticeable ΔE of 11.03 was obtained in skim milk processed with MF-0.65 ($p < 0.05$). Similar values ($p \geq 0.05$) were obtained when combination treatments were applied consisting of PEF (PEF-L, PEF-M, and PEF-H) and subsequent MF-0.65 for hurdle treatment of skim milk, yielding ΔEs of 11.83, 11.63 and 11.49, respectively. Overall, noticeably better color retention was achieved with PEF-L, PEF-M and PEF-H treatments; 1.2 μm TFMF and 1.4 μm CFMF-based processing, achieving lower ΔE values (in the range between 0.10 and 0.37 ($p < 0.05$)) than by the HTST and 0.65 μm TFMF treatments.

While skim milk pasteurized at 95 °C showed a positive yellowness value, no significant difference in any of the color attributes was observed ($p \geq 0.05$) between the heat-treated skim milk compared to that processed by PEF, MF-1.2 and MF-1.4, and their combinations. Lightness was 16%–18% lower for all treatments that involved skim milk processing through 0.65 μm MF and the corresponding samples also exhibited a higher blueness as indicated by a lower *b* value ($p < 0.05$). As a result significantly lower whiteness index values were obtained for MF-0.65 and hurdle treatments combining PEF of different processing intensities with MF-0.65 as shown in Figure 1 ($p < 0.05$).

Table 1. Comparison of Hunter color space attributes *L* (lightness), *a* (redness) and *b* (yellowness) measured in raw skim milk and after processing with high-temperature short-time pasteurization at 75 (HTST-75) and 95 (HTST-95) °C, pulsed electric field at low (PEF-L), moderate (PEF-M) and high (PEF-H) intensities, 0.65 and 1.2 μm pore size tangential flow microfiltration (0.65 TFMF and 1.2 TFMF, respectively), 1.4 μm-pore size cross-flow microfiltration (1.4 CFMF), and different pulsed electric field-based combinations with the latter three membrane filtration treatments.

Treatments	*L*	*a*	*b*	Δ*E*
Raw Skim Milk	64.50 (±0.87) [a]	−3.92 (±0.08) [a]	−0.15 (±0.15) [a]	-
HTST-75	65.00 (±0.61) [a]	−4.01 (±0.10) [a]	−0.08 (±0.02) [a]	0.51
HTST-95	65.21 (±0.10) [a]	−4.03 (±0.01) [a]	0.19 (±0.04) [a]	0.79
0.65 TFMF	53.90 (±1.25) [b]	−3.92 (±0.03) [a]	−3.21 (±0.28) [b]	11.03
1.2 TFMF	64.22 (±0.05) [a]	−3.94 (±0.05) [a]	−0.28 (±0.17) [a]	0.31
1.4 CFMF	64.71 (±0.08) [a]	−3.89 (±0.03) [a]	−0.33 (±0.04) [a]	0.10
PEF-L	64.26 (±0.68) [a]	−3.88 (±0.02) [a]	−0.20 (±0.17) [a]	0.25
PEF-M	64.58 (±0.51) [a]	−3.88 (±0.01) [a]	−0.16 (±0.26) [a]	0.10
PEF-H	64.85 (±0.60) [a]	−3.96 (±0.05) [a]	−0.07 (±0.16) [a]	0.37
PEF-L/0.65 TFMF	53.07 (±0.65) [b]	−3.99 (±0.04) [a]	−3.12 (±0.11) [b]	11.81
PEF-M/0.65 TFMF	53.37 (±1.23) [b]	−3.99 (±0.04) [a]	−3.52 (±0.03) [b]	11.63
PEF-H/0.65 TFMF	53.44 (±1.38) [b]	−4.02 (±0.04) [a]	−3.28 (±0.24) [b]	11.49
PEF-L/1.2 TFMF	64.51 (±0.50) [a]	−3.92 (±0.02) [a]	−0.15 (±0.22) [a]	0.02
PEF-M/1.2 TFMF	64.36 (±0.42) [a]	−3.92 (±0.01) [a]	−0.23 (±0.03) [a]	0.15
PEF-H/1.2 TFMF	64.65 (±0.32) [a]	−4.00 (±0.02) [a]	−0.20 (±0.06) [a]	0.18
PEF-M/1.4 CFMF	64.57 (±0.18) [a]	−3.91 (±0.02) [a]	−0.34 (±0.07) [a]	0.20
P [1]	*	NS	*	-
SEM [2]	0.539	0.010	0.148	0.711

[a, b] Different superscripted letters in the same column indicate statistical significance between the two means. Values in parentheses following the mean values of the color attributes indicate the standard deviations. [1] *P* stands for the statistical probability. * Refers to a statistical significance of $p < 0.05$. NS indicates no statistical significance. [2] SEM abbreviates the standard error of the mean.

For the remaining non-thermal or thermal treatments no significant differences in whiteness were determined when compared among one another or with raw skim milk ($p \geq 0.05$). Overall, non-thermal treatments using MF-1.2 MF-1.4 and PEF-L, PEF-M, PEF-H, and heat pasteurization did not have a significant impact on the single color properties of skim milk but, with the exception of MF-0.65-based treatments, thermally pasteurized skim milk was in a higher Δ*E* category than non-thermally treated skim milk.

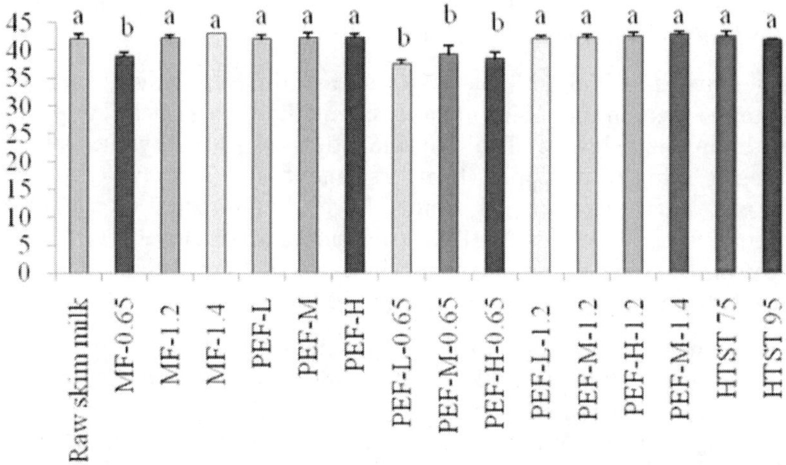

Figure 1. Whiteness obtained in untreated raw skim milk and by processing the latter with low-intensity (PEF-L), medium-intensity (PEF-M) and high-intensity (PEF-H) pulsed electric field, 0.65 (MF-0.65), 1.2 (MF-1.2), and 1.4 (MF-1.4) μm microfiltration, high-temperature short-time (HTST) pasteurization treatments and different pulsed electric field-based combinations with microfiltration treatments (after conversion from Hunter Lab color space data to CIE 1931 XYZ color space values). Different letters (*i.e., a, b*) above bars indicate significant differences between treatment means ($p < 0.05$).

3.2. Flavor Attributes of Non-Thermally and Thermally Treated Skim Milk

The effect of heat, pulsed electric field and microfiltration treatments on the volatile profile of skim milk was measured using SIFT-MS. Volatile compounds generated as a result of heat treatment and non-enzymatic browning reactions in skim milk were of main interest and, therefore, 21 compounds were selected for this study. Quantities of acetic acid and acetaldehyde were not detected by the mass spectrometer, so that data for 19 compounds (Table 2, Table 3 and Table 4) was obtained. Concentrations of ketones were found to be significantly higher in skim milk treated with HTST at 95 °C for a holding time of 45 s when compared to other treatments ($p < 0.05$). The major ketones contributing to this were 2-hexanone, 2-heptanone, 2-octanone, acetone, and butanone. However, there was no marked increase in concentration of these ketones in skim milk pasteurized at 75 °C for 20 s, nor for PEF-treated and MF-treated skim milk ($p \geq 0.05$).

The aldehydes in skim milk were altered in a different way as compared to ketones in skim milk. Thermal treatments had a significant impact on 3-methylbutanal levels. The concentration of 3-methylbutanal was increased significantly upon heat treatment at 95 °C ($p < 0.05$). By contrast, microfiltration did not have an impact on aldehyde concentrations in skim milk. HTST treatments at 75 and 95 °C, PEF and microfiltration treatments did not have a significant impact on the concentration of heptanal or decanal ($p \geq 0.05$).

Short chain fatty acids were the other group of volatiles analyzed in skim milk. Of these, hexanoic acid and butanoic acid were detected in raw skim milk, heat-treated, PEF-treated and microfiltered skim milk. Increased intensity of thermal treatments resulted in an increased concentration of hexanoic acid and the values were significantly higher in skim milk subjected to both HTST-75 and HTST-95 ($p < 0.05$) than in raw skim milk. Pasteurization at the higher temperature (HTST-95) led to a significant increase ($p < 0.05$) in the concentration of butanoic acid, whereas no impact ($p \geq 0.05$) on its concentration was found for skim milk pasteurized at 75 °C. Moreover, no significant difference ($p \geq 0.05$) in hexanoic and butanoic acid concentrations was observed between raw, PEF, MF, and hurdle treated skim milk.

The concentrations of both ethanol and propanol increased in skim milk following heat pasteurization at 95 °C ($p < 0.05$). Contrasting results ($p \geq 0.05$) were obtained for PEF, MF, hurdle processing and HTST-75 treatments, indicating that their concentrations did not change in skim milk subjected to these processing conditions. Heat treatment at 95 °C resulted in increased levels of toluene ($p < 0.05$), while there was no significant impact by HTST-75, PEF, and MF treatments on this volatile compound ($p \geq 0.05$).

The methyl acetate concentration was significantly higher ($p < 0.05$) in skim milk processed at 95 °C. None of the other processes affected the concentration of this volatile component ($p \geq 0.05$). Furthermore, two sulfur compounds, hydrogen sulfide and dimethyl sulfide, were detected by the SIFT-MS equipment. Hydrogen sulfide concentrations were found to be at significantly higher levels in HTST-75 and HTST-95 treated skim milk, reaching the highest levels with the latter treatment ($p < 0.05$). The highest concentration of dimethyl sulfide occurred in skim milk that underwent the HTST-95 treatment. PEF processing and microfiltration did not increase the content of sulfur compounds.

Table 2. Retention of in 3-methyl butanal, heptanal, decanal, butanoic acid, hexanoic acid, dimethyl sulfide, and hydrogen sulfide obtained in skim milk after high-temperature short-time pasteurization ($HTST_{75}$ = 75 °C for 20 s; $HTST_{95}$ = 95 °C for 45 s) and lab-scale tangential-flow microfiltration ($TFMF_{0.65}$ = 0.65 µm; $TFMF_{1.2}$ = 1.2 µm), pilot scale cross-flow microfiltration ($CFMF_{1.4}$ = 1.4 µm), pulsed electric field at low (PEF-L: 28 kV/cm; 2805 µs), moderate (PEF-M: 40 kV/cm, 1122 µs), and high (PEF-H: 40 kV/cm, 1571 µs) processing intensities and selected non-thermal combination treatments using PEF and subsequent microfiltration.

Treatments	3-Methyl butanal [ppb]	Heptanal [ppb]	Decanal [ppb]	Butanoic acid [ppb]	Hexanoic acid [ppb]	Dimethyl sulfide [ppb]	Hydrogen sulfide [ppb]
Raw Skim Milk	3.6 (±1.1) ^{a}	15 (±6.2) ^{a}	2.7 (±0.5) ^{a}	14 (±6.0) ^{a}	2.3 (±0.8) ^{a}	12 (±5.9) ^{a}	1.4 (±0.7) ^{a,b}
$HTST_{75}$	3.3 (±0.1) ^{a}	15 (±1.9) ^{a}	3.3 (±0.0) ^{a}	13 (±1.4) ^{a}	9.7 (±0.3) ^{d}	17 (±2.2) ^{a,b}	4.2 (±0.1) ^{c}
$HTST_{95}$	6.3 (±0.7) ^{b}	24 (±5.2) ^{a}	4.8 (±1.6) ^{a}	44 (±6.0) ^{b}	4.8 (±0.2) ^{c}	25 (±6.1) ^{b}	5.5 (±0.2) ^{d}
$TFMF_{0.65}$	3.9 (±0.0) ^{a}	14 (±10.3) ^{a}	2.2 (±0.3) ^{a}	15 (±5.8) ^{a,b}	1.1 (±0.3) ^{a}	11 (±0.7) ^{a,b}	n.d.
$TFMF_{1.2}$	4.5 (±1.7) ^{a}	17 (±7.3) ^{a}	3.4 (±1.4) ^{a}	13 (±1.8) ^{a}	2.4 (±0.2) ^{a,b}	12 (±0.9) ^{a,b}	1.1 (±0.5) ^{a,b}
$CFMF_{1.4}$	3.5 (±1.5) ^{a}	11 (±1.8) ^{a}	2.3 (±0.8) ^{a}	11 (±3.1) ^{a}	2.5 (±0.3) ^{a,b}	11 (±0.5) ^{a,b}	1.1 (±0.4) ^{a,b}
PEF-L	2.4 (±0.4) ^{a}	10 (±0.7) ^{a}	2.5 (±0.7) ^{a}	12 (±3.3) ^{a}	1.6 (±0.3) ^{a,b}	9.1 (±3.4) ^{a}	0.8 (±0.1) ^{a}
PEF-M	3.0 (±1.7) ^{a}	17 (±2.9) ^{a}	3.4 (±0.8) ^{a}	16 (±7.2) ^{a,b}	1.8 (±0.5) ^{a,b}	15 (±7.0) ^{a,b}	1.3 (±0.2) ^{a,b}
PEF-H	3.9 (±2.8) ^{a}	14 (±2.2) ^{a}	2.6 (±1.1) ^{a}	15 (±5.1) ^{a,b}	2.7 (±0.1) ^{b}	18 (±2.1) ^{a,b}	2.1 (±0.1) ^{b}
PEF-L/$TFMF_{0.65}$	2.5 (±0.4) ^{a}	20 (±0.7) ^{a}	3.9 (±2.3) ^{a}	10 (±3.4) ^{a}	1.7 (±0.2) ^{a,b}	5.0 (±0.8) ^{a}	1.1 (±0.3) ^{a,b}
PEF-M/$TFMF_{0.65}$	3.6 (±2.8) ^{a}	21 (±1.0) ^{a}	3.3 (±1.6) ^{a}	13 (±5.2) ^{a}	1.7 (±0.5) ^{a,b}	7.5 (±1.3) ^{a}	0.9 (±0.3) ^{a,b}
PEF-H/$TFMF_{0.65}$	4.8 (±2.5) ^{a}	16 (±2.1) ^{a}	3.2 (±0.5) ^{a}	9.1 (±2.3) ^{a}	1.5 (±0.5) ^{a,b}	7.9 (±0.7) ^{a}	1.1 (±0.2) ^{a,b}
PEF-L/$TFMF_{1.2}$	2.1 (±0.5) ^{a}	14 (±2.7) ^{a}	2.5 (±0.1) ^{a}	8.6 (±1.9) ^{a}	1.7 (±0.1) ^{a,b}	9.2 (±1.9) ^{a}	0.9 (±0.2) ^{a,b}
PEF-M/$TFMF_{1.2}$	5.6 (±0.2) ^{a}	20 (±0.4) ^{a}	1.5 (±1.2) ^{a}	16 (±0.3) ^{a,b}	2.5 (±0.1) ^{a,b}	17 (±2.7) ^{a,b}	1.6 (±0.2) ^{a,b}
PEF-H/$TFMF_{1.2}$	2.7 (±1.3) ^{a}	13 (±3.0) ^{a}	1.8 (±0.2) ^{a}	8.9 (±3.2) ^{a}	2.1 (±0.5) ^{a,b}	12 (±2.4) ^{a,b}	1.2 (±0.3) ^{a,b}
PEF-M/$CFMF_{1.4}$	2.0 (±0.2) ^{a}	17 (±1.2) ^{a}	3.5 (±0.1) ^{a}	9.1 (±0.7) ^{a}	1.4 (±0.4) ^{a,b}	14 (±2.3) ^{a,b}	1.0 (±0.0) ^{a,b}
P^{1}	*	NS	NS	*	*	*	*
SEM [2]	0.19	0.64	0.12	1.53	0.20	0.87	0.11

[a, b, c] Different superscripted letters in the same column indicate statistical significance between the two means. Values in parentheses following the mean values of compound indicate the standard deviations. [1] P stands for the statistical probability. * Refers to a statistical significance of $p < 0.05$. NS indicates no statistical significance. [2] SEM abbreviates the standard error of the mean.

4. DISCUSSION

For Hunter *L*, *a*, *b* color space values, no significant differences were observed in lightness, redness, and yellowness for skim milk treated with PEF, HTST and, MF-1.2 and MF-1.4 ($p \geq 0.05$). A significant reduction in *L*, *b* and whiteness index values for skim milk that was microfiltered through a 0.65 µm pore size membrane indicates that higher cut-off membranes partially retain suspended milk components [28,29]. Pafylias *et al.* [30] suggested the use of larger pore size membranes of about 1.4 μm to maintain a balance between reduction of bacterial cells and retention of milk components. However, we did not see any significant differences in color attributes for skim milk processed through 1.2 and 1.4 μm pore size MF. Silva *et al.* [31] compared instrumental color differences between microfiltered and pasteurized skim milk and found smaller changes in color coordinates of microfiltered skim milk due to a lack of reactions caused by heating skim milk to pasteurization temperatures. HTST treatments caused a small increase in lightness values for all the samples, which may be due to an increase in number of dispersed components as a result of denaturation of milk proteins at high temperatures [20]. However, the increase in

Table 3. Retention of in ethanol, 1-propanol, 1-octene, toluene, *p*-xylene, and methyl acetate obtained in skim milk after high-temperature short-time pasteurization ($HTST_{75}$ = 75 °C for 20 s; $HTST_{95}$ = 95 °C for 45 s) and lab-scale tangential-flow microfiltration ($TFMF_{0.65}$ = 0.65 μm; $TFMF_{1.2}$ = 1.2 μm), pilot scale cross-flow microfiltration ($CFMF_{1.4}$ = 1.4 μm), pulsed electric field at low (PEF-L: 28 kV/cm; 2805 μs), moderate (PEF-M: 40 kV/cm, 1122 μs), and high (PEF-H: 40 kV/cm, 1571 μs) processing intensities and selected non-thermal combination treatments using PEF and subsequent microfiltration.

Treatments	Ethanol [ppb]	1-Propanol [ppb]	1-Octene [ppb]	Toluene [ppb]	*p*-Xylene [ppb]	Methyl acetate [ppb]
Raw Skim Milk	65 (±18) ^{a}	10 (±2.0) ^{a}	34 (±8.3) ^{a}	4.2 (±0.8) ^{a}	1.6 (±0.9) ^{a}	6.7 (±1.3) ^{a}
$HTST_{75}$	57 (±9.9) ^{a,b}	11 (±1.8) ^{a,b}	31 (±2.1) ^{a}	4.6 (±0.8) ^{a}	1.5 (±0.3) ^{a}	10 (±3.0) ^{a,b}
$HTST_{95}$	131 (±1.8) ^{c}	16 (±2.1) ^{b}	38 (±1.5) ^{a}	6.9 (±1.6) ^{b}	2.0 (±0.8) ^{a}	19 (±4.1) ^{b}
$TFMF_{0.65}$	51 (±6.0) ^{a,b}	11 (±0.8) ^{a,b}	15 (±1.7) ^{a}	4.0 (±0.7) ^{a}	1.6 (±1.1) ^{a}	6.7 (±0.8) ^{a,b}
$TFMF_{1.2}$	60 (±0.9) ^{a,b}	10 (±1.3) ^{a,b}	33 (±8.3) ^{a}	4.6 (±0.6) ^{a}	1.3 (±0.3) ^{a}	6.3 (±0.5) ^{a,b}
$CFMF_{1.4}$	61 (±27) ^{a,b}	9.2 (±3.3) ^{a,b}	22 (±3.7) ^{a}	4.5 (±1.8) ^{a}	2.7 (±0.5) ^{a}	5.6 (±2.3) ^{a,b}
PEF-L	39 (±2.0) ^{a}	7.1 (±1.8) ^{a,b}	28 (±7.8) ^{a}	2.8 (±0.2) ^{a}	0.9 (±0.5) ^{a}	3.5 (±0.1) ^{a}
PEF-M	64 (±16) ^{a,b}	9.3 (±2.8) ^{a,b}	39 (±8.0) ^{a}	5.5 (±0.2) ^{a}	1.4 (±0.4) ^{a}	7.8 (±3.1) ^{a,b}
PEF-H	61 (±7.1) ^{a,b}	9.6 (±4.3) ^{a,b}	29 (±4.7) ^{a}	2.4 (±0.1) ^{a}	2.5 (±0.2) ^{a}	7.5 (±0.7) ^{a,b}
PEF-L/$TFMF_{0.65}$	18 (±13) ^{a}	4.8 (±0.2) ^{a}	31 (±11) ^{a}	2.6 (±0.0) ^{a}	1.9 (±0.0) ^{a}	5.4 (±0.4) ^{a}
PEF-M/$TFMF_{0.65}$	39 (±36) ^{a}	6.1 (±1.7) ^{a,b}	26 (±5.7) ^{a}	3.9 (±0.3) ^{a}	2.1 (±0.6) ^{a}	3.8 (±0.4) ^{a}
PEF-H/$TFMF_{0.65}$	44 (±2.9) ^{a}	5.2 (±2.2) ^{a}	24 (±3.4) ^{a}	2.2 (±0.2) ^{a}	1.2 (±0.5) ^{a}	5.2 (±0.5) ^{a}
PEF-L/$TFMF_{1.2}$	58 (±2.6) ^{a,b}	5.2 (±0.1) ^{a}	28 (±7.7) ^{a}	2.0 (±0.4) ^{a}	0.9 (±0.3) ^{a}	4.1 (±0.6) ^{a}
PEF-M/$TFMF_{1.2}$	84 (±5.3) ^{b,c}	12 (±0.8) ^{a,b}	37 (±1.8) ^{a}	5.6 (±0.7) ^{a}	3.0 (±1.1) ^{a}	6.4 (±0.5) ^{a,b}
PEF-H/$TFMF_{1.2}$	52 (±3.4) ^{a,b}	7.4 (±3.2) ^{a}	29 (±2.2) ^{a}	3.0 (±0.9) ^{a}	1.1 (±0.6) ^{a}	3.8 (±2.2) ^{a}
PEF-M/$CFMF_{1.4}$	42 (±12) ^{a,b}	6.6 (±1.5) ^{a,b}	25 (±5.8) ^{a}	2.2 (±0.4) ^{a}	1.5 (±0.1) ^{a}	4.1 (±0.1) ^{a}
P^{1}	*	*	N.S.	*	N.S.	*
SEM [2]	3.06	0.34	0.95	0.17	0.13	0.37

^{a,b,c} Different superscripted letters in the same column indicate statistical significance between the two means. Values in parentheses following the mean values of compound indicate the standard deviations. [1] *P* stands for the statistical probability. * Refers to a statistical significance of $p < 0.05$. NS indicates no statistical significance. [2] SEM abbreviates the standard error of the mean.

Table 4. Retention of in acetone, butanone, 2-hexanone, 2-heptanone, 2-octanone and 2-nonanone obtained in skim milk after high temperature short time pasteurization ($HTST_{75}$ = 75 °C for 20 s; $HTST_{95}$ = 95 °C for 45 s) and lab-scale tangential-flow microfiltration ($TFMF_{0.65}$ = 0.65 µm; $TFMF_{1.2}$ = 1.2 µm), pilot scale cross-flow microfiltration ($CFMF_{1.4}$ = 1.4 µm), pulsed electric field at low (PEF-L: 28 kV/cm; 2805 µs), moderate (PEF-M: 40 kV/cm, 1122 µs), and high (PEF-H: 40 kV/cm, 1571 µs) processing intensities and selected non-thermal combination treatments using PEF and subsequent the microfiltration.

Treatments	Acetone [ppb]	Butanone [ppm]	2-Hexanone [ppb]	2-Heptanone [ppb]	2-Octanone [ppb]	2-Nonanone [ppb]
Raw Skim Milk	2.7 (± 0.8) ^{a}	174 (±51) ^{a}	6.8 (±3.1) ^{a}	4.9 (±2.5) ^{a}	9.7 (±6.5) ^{a}	4.2 (±1.1) ^{a}
$HTST_{75}$	4.0 (±0.4) ^{a,b}	225 (±23) ^{a,b}	9.0 (±1.1) ^{a,b}	4.8 (±1.7) ^{a,b}	11 (±1.8) ^{a,b}	5.5 (±0.8) ^{a}
$HTST_{95}$	6.2 (±1.0) ^{b}	344 (±0.7) ^{b}	9.3 (±0.3) ^{b}	12 (±2.6) ^{b}	16 (±3.6) ^{b}	5.9 (±0.3) ^{a}
$TFMF_{0.65}$	2.4 (±0.4) ^{a}	169 (±23) ^{a}	2.8 (±1.6) ^{a}	4.6 (±1.1) ^{a,b}	8.9 (±1.4) ^{a,b}	3.3 (±0.1) ^{a}
$TFMF_{1.2}$	2.6 (±0.2) ^{a,b}	188 (±56) ^{a}	3.4 (±1.8) ^{a,b}	4.9 (±4.0) ^{a,b}	9.5 (±3.5) ^{a,b}	3.7 (±2.7) ^{a}
$CFMF_{1.4}$	2.3 (±0.9) ^{a}	140 (±51) ^{a}	7.0 (±1.6) ^{a,b}	3.7 (±0.3) ^{a,b}	6.5 (±0.6) ^{a}	5.0 (±2.4) ^{a}
PEF-L	1.8 (±1.0) ^{a}	120 (±32) ^{a}	5.7 (±1.5) ^{a,b}	3.9 (±0.2) ^{a,b}	7.2 (±1.4) ^{a,b}	2.6 (±0.4) ^{a}
PEF-M	2.6 (±1.9) ^{a,b}	135 (±74) ^{a}	7.0 (±0.7) ^{a,b}	5.1 (±2.8) ^{a,b}	8.8 (±3.5) ^{a,b}	3.5 (±1.9) ^{a}
PEF-H	3.1 (±2.2) ^{a,b}	147 (±58) ^{a}	7.8 (±0.3) ^{a,b}	4.8 (±2.2) ^{a,b}	7.9 (±4.2) ^{a,b}	5.4 (±0.9) ^{a}
PEF-L/$TFMF_{0.65}$	1.5 (±0.9) ^{a}	54 (±8.0) ^{a}	5.0 (±0.1) ^{a,b}	6.0 (±1.2) ^{a,b}	8.7 (±1.8) ^{a,b}	5.2 (±3.8) ^{a}
PEF-M/$TFMF_{0.65}$	1.9 (±1.1) ^{a}	75 (±40) ^{a}	5.7 (±0.2) ^{a,b}	6.5 (±1.4) ^{a,b}	8.8 (±1.8) ^{a,b}	5.4 (±3.6) ^{a}
PEF-H/$TFMF_{0.65}$	1.9 (±1.1) ^{a}	80 (±41) ^{a}	6.9 (±0.6) ^{a,b}	4.7 (±3.1) ^{a,b}	7.4 (±2.8) ^{a,b}	4.5 (±0.9) ^{a}
PEF-L/$TFMF_{1.2}$	1.5 (±0.4) ^{a}	56 (±18) ^{a}	5.0 (±0.1) ^{a,b}	3.5 (±0.3) ^{a}	7.5 (±0.3) ^{a,b}	1.9 (±0.5) ^{a}
PEF-M/$TFMF_{1.2}$	3.0 (±0.2) ^{a,b}	74 (±26) ^{a}	7.6 (±0.4) ^{a,b}	6.6 (±1.0) ^{a,b}	12 (±0.0) ^{a,b}	5.7 (±0.4) ^{a}
PEF-H/$TFMF_{1.2}$	2.2 (±0.6) ^{a}	84 (±60) ^{a}	5.4 (±1.3) ^{a,b}	4.8 (±1.2) ^{a,b}	8.3 (±0.9) ^{a,b}	2.9 (±0.1) ^{a}
PEF-M/$CFMF_{1.4}$	2.1 (±0.4) ^{a}	81 (±11) ^{a}	8.9 (±0.2) ^{b}	3.4 (±0.3) ^{a}	8.8 (±2.4) ^{a}	4.0 (±0.7) ^{a}
p^{1}	*	*	*	*	*	NS
SEM ^{2}	0.12	8.01	0.38	0.36	0.54	0.23

^{a,b,c} Different superscripted letters in the same column indicate statistical significance between the two means. Values in parentheses following the mean values of compound indicate the standard deviations. ^{1} P stands for the statistical probability. * Refers to a statistical significance of $p < 0.05$. NS indicates no statistical significance. ^{2} SEM abbreviates the standard error of the mean.

lightness was not significant ($p \geq 0.05$). The Hunter-*b* value was positive for HTST-95 treatment indicating yellowness in skim milk. This might be attributed to heat induced browning reactions in milk between lactose and amino acids [32,33,34]. However, PEF and MF-treated skim milk retained milk color better than heat treatments as indicated by "not noticeable" total color difference values compared to raw skim milk. As indicated by ΔE values, there was a lower impact on color of skim milk produced with the non-thermal technologies, with the exception of the smallest pore size MF that produced "unacceptable" color values. However, our results are not in agreement with results of Bermúdez-Aguirre *et al.* [35] who observed significant changes in *L, a, b* values of PEF-treated skim milk. The authors, however, did not observe any trends in color change and attributed this change partly to the wearing down of the electrode due to arcing.

Among the 21 volatile compounds analyzed, ketones represented a major class based on their expected contribution to flavor. Ketones in milk can originate from different sources. Some of them may be present in raw milk as a result of feed, e.g., acetone and 2-butanone, while others may be generated or accumulated as a result of β-oxidation of saturated fatty acids followed by decarboxylation or by β-ketoacid decarboxylation [36,37]. A significant increase in concentration of all methyl ketones, except 2-nonanone, was observed for samples subjected to heat treatment ($p < 0.05$). Increased levels of methyl ketones in milk due to thermal treatments have also been reported by other researchers [23,38]. The methyl ketone content of skim milk treated with either PEF or MF alone was similar to that of raw skim milk ($p \geq 0.05$), indicating that they have a minimum effect on the concentration of lipids in skim milk. Comparable results ($p \geq 0.05$) were obtained by hurdle processing of skim milk using both PEF and MF, suggesting a non-significant impact of combined processes on ketones in skim milk.

Another important class of compounds identified and relevant to this study were the sulfur compounds: dimethyl and hydrogen sulfide. These compounds are formed in milk as a result of thermal denaturation of milk whey proteins during processing and are responsible for cooked flavors in milk. Hydrogen sulfide is produced mainly from Strecker degradation of sulfhydryl groups of sulfur-containing amino acids (*i.e.*, cysteine and methionine) in whey proteins. Its production is also attributed to denatured proteins associated with the milk fat globule membrane and the rearrangement of a thiazole group during thermal degradation of thiamine [36,39,40]. In our research, the concentration of hydrogen sulfide in raw skim milk was found to be 1.39 ± 0.68 μg/L. Its concentration was higher when skim milk was pasteurized at 95 °C for 45 s (5.53 μg/L) as compared to skim milk subjected to 75 °C for 20 s (4.21 μg/L). Thus, the hydrogen sulfide concentration was found to increase with the severity of heat treatment. In contrast, PEF treatment applied at the highest intensity, which is known to generate increased temperature in the treatment chamber to about 65 °C by means of Joule heating, however, did not cause any significant increase in hydrogen sulfide concentration (2.05 μg/L) ($p \geq 0.05$). The PEF-L treatment that resulted in the longest exposure (2805 μs) of skim milk to electric pulses also did not cause any significant changes in hydrogen sulfide concentration (0.82 μg/L) when compared to raw skim milk ($p \geq 0.05$). The increase in concentration of sulfur compounds has been associated with growing denaturation of sulfur containing polypeptides affected to a large degree by increasing exposure to heat [39,41]. Dimethyl sulfide has been reported to occur naturally in cow's milk as a result of diet and at low levels it contributes to milk flavor but at high concentrations it may impart an off-flavor to the milk [42,43]. It can also be formed as a result of thermal denaturation of sulfur containing amino acids in milk. Increased dimethyl sulfide content in milk subjected to heat treatment at pasteurization temperatures and ultra-high-temperatures has been reported by different research groups [22,44,45]. In this study, dimethyl sulfide concentration was observed to be significantly higher in skim milk processed at 95 °C for 45 s (23.6 μg/L) as compared to raw skim milk (6.17 to 17.95 μg/L). However, no significant differences in concentrations of dimethyl sulfide from those in raw skim milk were observed for milk that underwent HTST-75, PEF, and MF treatments ($p \geq 0.05$). Hurdle processing of PEF with MF also did not cause any significant changes in concentrations of sulfur compounds in comparison to raw skim milk ($p \geq 0.05$).

Among aldehydes, acetaldehyde concentrations were below the detection limit of SIFT-MS for raw as well as treated skim milk. Aliphatic aldehydes are produced as a result of auto-oxidation of unsaturated fatty acids as well as by breakdown of amino acids in milk [37,40]. No significant differences were found in heptanal and decanal concentrations among raw skim milk, HTST-treated, PEF-treated, and MF-treated skim milk ($p \geq 0.05$). In contrast, Zhang and others [5] observed significantly increased heptanal and decanal levels in pasteurized milk in comparison to raw milk while no significant differences were found in PEF-treated skim milk. However, the increase in aldehyde concentrations in skim milk upon exposure to heat treatment at 95 °C, though not significant ($p \geq 0.05$) could contribute to significant changes in aroma of skim milk due to lower threshold values for some aldehydes [44]. 3-Methyl butanal is a product of Strecker degradation of leucine during the non-enzymatic browning reaction, which occurs during heat treatment of milk [36,37]. No significant increase in its concentration was found after non-thermal PEF and MF treatments ($p \geq 0.05$). Its presence in skim milk treated at 95 °C was found to be significantly higher ($p < 0.05$) than in raw skim milk and in skim milk processed by all other treatments tested. Contarini and Povolo [23] also observed a positive correlation between increased levels of 3-methyl butanal and severity of heat treatment, while no significant differences were found in heptanal concentrations.

Short-chain and medium-chain fatty acids (C-4 to C-14 and some C-16) comprise 45% of total fatty acids in milk fat and are produced in the mammary gland from acetate and β-hydroxybutyrate, the products of bacterial fermentation in the rumen. Activation of acetyl to acetyl CoA and its carboxylation to malonyl CoA results in a stepwise addition of two CH_2groups and hence, increase in fatty acid chain length from short- to medium-chain fatty acids [46]. Out of these straight-chained, even numbered fatty acids, butanoic and hexanoic acids were quantified in this study. An increase in concentration of these two short-chain fatty acids as a result of heat exposure has been reported by Gandy and others [24]. Significantly higher amounts of both butanoic acid and hexanoic acid were observed in skim milk treated at 95 °C. However, HTST pasteurization at 75 °C showed a greater increase in hexanoic acid and, interestingly, the increase was twice the increase observed for the 95 °C treatment. Increased levels of butanoic and hexanoic acid in heat treated milk may impart cheesy, sour, and cream notes to milk as classified by Zhang and others [5] under the category of high olfactometric intensity odorants.

Formation of alcohols by reduction of corresponding carbonyl compounds [29] was observed for milk stored at refrigeration temperatures, where alcohol concentration increased as result of reduction of carbonyl compounds [30,47]. HTST-75, PEF, and MF treatments did not have any significant impact on the concentration of ethanol and 2-propanol in skim milk. Similarly, Zhang *et al.* [5] reported no marked change in concentrations of alcohols subsequent to PEF and HTST processing. However, at a processing temperature of 95 °C, a marked increase in concentration of both these alcohols was observed, which indicates some reducing reactions take place at this temperature. Toluene and *p*-xylene are considered as products of thermal degradation of β-carotene [48] and these compounds have been associated with heat treated milk [24,38,49]. Esters are formed by reaction of carboxylic acids with alcohols. Methyl acetate was the only ester quantified in this study and its concentration increased three-fold in HTST-95 treated skim milk. As observed in similar studies [3,50,51] non-thermal processing had a lesser impact on the concentration of volatile compounds in skim milk than thermal processing and indicates that skim milk of better quality results from non-thermal processing. Last but not least, it is worthy of mention that different scales (lab, pilot plant) of MF processing equipments (1.2 μm TFMF and 1.4 μm CFMF, respectively) used in this study indicated comparable effects on color and composition of volatiles in skim milk, thereby, also substantiating that different designs and materials of the MF membrane modules had no significant effect on these organoleptical quality parameters.

5. CONCLUSIONS

MF (1.2 and 1.4 μm) and PEF alone and in combination (hurdle technology) caused no significant changes to skim milk color and its volatile compound composition, thus, the hurdle approach can be regarded as promising for preservation of skim milk of very good sensory quality. No noticeable color differences from raw skim milk were observed by application of PEF, and MF with a pore size of 1.2 μm or above. Smaller pore size microfiltration (0.65 μm) was found to be unsatisfactory when applied alone or in combination with PEF, resulting in significant changes to skim milk color attributes. Different scale, membrane materials, and membrane designs of the MF systems showed no effect on color and volatile compounds. A significant increase in ketones, fatty acids, hydrocarbons and sulfur compounds was obtained following heat treatment of skim milk at the higher intensity treatment (95 °C, 45 s) applied in this study, while lower intensity heat treatment (75 °C, 20 s) also resulted in increased concentration of selected volatile compounds. However, increased PEF intensity did not significantly alter concentrations of volatile compounds. These findings indicate PEF/MF is an emerging technology with potential to produce skim milk of high sensory quality. Further research on this promising hurdle technology is envisaged such as the study of enzyme activity and vitamin retention in PEF/MF-treated skim milk as well as its application for processing other liquid dairy products.

ACKNOWLEDGMENTS

The authors are thankful for the financial support by the Dairy Farmers of Ontario and the Natural Sciences and Engineering Research Council of Canada that both made the present study possible. Moreover, we are grateful for the supply of raw milk by the Gay Lea Foods Co-operative Limited and for the use of the pilot scale heat exchanger and microfiltration units provided by the Guelph Food Technology Centre.

AUTHOR CONTRIBUTIONS

The present manuscript was written by Anupam Chugh, who carried out the experimental work together with Dipendra Khanal under the guidance and supervision of Markus Walkling-Ribeiro, Milena Corredig, and Mansel W. Griffiths. The experimental design of the study was devised by Walkling-Ribeiro, also responsible for the article correspondence, and Griffiths, the principal investigator of the research project. Due to her expertise in sensory analyses Lisa Duizer, also a co-investigator of the research project, mentored Chugh and Khanal during the course of their volatile compound analyses.

CONFLICTS OF INTEREST

The authors declare no conflict of interest.

REFERENCES

1. Doan, F.J. The color of cow's milk and its value. J. Dairy Sci. 1924, 7, 147–153.
2. Thomas, E.L. Trends in milk flavors. J. Dairy Sci. 1981, 64, 1023–1027.
3. Sampedro, F.; Geveke, D.J.; Fan, X.; Zhang, H.Q. Effect of PEF, HHP and thermal treatment on PME inactivation and volatile compounds concentration of an orange juice-milk based beverage. Innov. Food Sci. Emerg. Technol. 2009, 10, 463–469.
4. Pereda, J.; Jaramillo, D.P.; Quevedo, J.M.; Ferragut, V.; Guamis, B.; Trujillo, A.J. Characterization of volatile compounds in ultra-high-pressure homogenized milk. Int. Dairy J. 2008, 18, 826–834.
5. Zhang, S.; Yang, R.; Zhao, W.; Hua, X.; Zhang, W.; Zhang, Z. Influence of pulsed electric field treatments on the volatile compounds of milk in comparison with pasteurized processing. J. Food Sci. 2011, 76, C127–C132.
6. Chugh, A.; Khanal, D.; Duizer, L.; Walkling-Ribeiro, M.; Griffiths, M.W. Effect of non-thermal pulsed electric fields and tangential-flow microfiltration processing on flavor and color properties of skim milk. In Proceedings of the IFT Annual Meeting and Food Expo, Chicago, IL, USA, 14 July 2013; 14 July 2013. session 074-04.
7. Zimmermann, U. The effect of high intensity electric field pulses on eukaryotic cell membranes: Fundamentals and applications. In Electromanipulation of Cells; Zimmermann, U., Neil, G.A., Eds.; CRC Press: Boca Raton, FL, USA, 1996; pp. 1–106.
8. Waite-Cusic, J.G.; Diono, B.H.S.; Yousef, A.E. Screening for Listeria monocytogenes surrogate strains applicable to food processing by ultra high pressure and pulsed electric field. J. Food Prot. 2011, 74, 1655–1661.
9. Grahl, T.; Markl, H. Killing of microorganisms by pulsed electric fields. Appl. Microbiol. Biotechnol. 1996, 45, 148–157.
10. Griffiths, M.W.; Walkling-Ribeiro, M. Implementation of microbially safe foods with pulsed electric fields. In Food Microbiology: Novel Food Preservation and Microbial Assessment Techniques; Boziaris, I.S., Ed.; Science Publishers—Taylor & Francis CRC Press: Boca Raton, FL, USA, 2014.
11. GEA Filtration. Available online: http://www.geafiltration. com/applications /microfiltration_whole_skim_milk.asp (accessed on 1 September 2013).
12. Gésan-Guiziou, G. Removal of bacteria, spores and somatic cells from milk by centrifugation and microfiltration techniques. In Improving the Safety and Quality of Milk, Volume 1—Milk Production and Processing; Griffiths, M.W., Ed.; CRC Press: Boca Raton, FL, USA, 2010; pp. 356–369.
13. Rodríguez-González, O.; Walkling-Ribeiro, M.; Jayaram, S.; Griffiths, M.W. Factors affecting the inactivation of the natural microbiota of milk processed by pulsed electric fields and cross-flow microfiltration. J. Dairy Res. 2011, 78, 270–278.
14. Sobrino-López, A.; Martín-Belloso, O. Enhancing the lethal effect of high-intensity pulsed electric field in milk by antimicrobial compounds as combined hurdles. J. Dairy Sci. 2008, 91, 1759–1768.
15. Swanson, B.G.; Barbosa-Cánovas, G.V. Inactivation of Listeria innocua in skim milk by pulsed electric fields and nisin. Int. J. Food Microbiol. 1999, 51, 19–30.
16. Odriozola-Serrano, I.; Bendicho-Porta, S.; Martín-Belloso, O. Comparative study on shelf life of whole milk processed by high-intensity pulsed electric field or heat treatment. J. Dairy Sci. 2006, 89, 905–911.

17. Walkling-Ribeiro, M.; Rodríguez-González, O.; Jayaram, S.; Griffiths, M.W. Microbial inactivation and shelf life comparison of "cold" hurdle processing with pulsed electric fields and microfiltration, and conventional thermal pasteurisation in skim milk. Int. J. Food Microbiol. 2011, 144, 379–386.
18. Solah, V.A.; Staines, V.; Honda, S.; Limley, H.A. Measurement of milk colour and composition: Effect of dietary intervention on Western Australian Holstein-Friesian cow's milk quality. J. Food Sci. 2007, 72, S560–S566.
19. Nozière, P.; Graulet, B.; Lucas, A.; Martin, B.; Grolier, P.; Doreau, M. Carotenoids for ruminants: From forages to dairy products. Anim. Feed Sci. Technol. 2006, 131, 418–450.
20. Dunkerley, J.A.; Ganguli, N.C.; Zadow, J.G. Reversible changes in the colour of skim milk on heating suggest rapid reversible heat-induced changes in casein micelle size. Aust. J. Dairy Technol. 1993, 48, 66–70.
21. Browning, E.; Lewis, M.; Macdougall, D. Predicting safety and quality parameters for UHT-processed milks. Int. J. Dairy Technol. 2001, 54, 111–120.
22. Datta, N.; Elliott, A.J.; Perkins, M.L.; Deeth, H.C. Ultra-high-temperature (UHT) treatment of milk: Comparison of direct and indirect modes of heating. Aust. J. Dairy Technol. 2002, 57, 221–227.
23. Contarini, G.; Povolo, M. Volatile fraction of milk: Comparison between purge and trap and solid phase microextraction techniques. J. Agric. Food Chem. 2002, 50, 7350–7355.
24. Gandy, A.L.; Schilling, M.W.; Coggins, P.C.; White, C.H.; Yoon, Y.; Kamadia, V.V. The effect of pasteurisation temperature on consumer acceptability, sensory characteristics, volatile compound composition, and shelf-life of fluid milk. J. Dairy Sci. 2008, 91, 1769–1777.
25. Lewis, M. Improving pasteurized and extended shelf-life milk. In Improving the Safety and Quality of Milk: Volume One, Milk Production and Processing; Griffiths, M.W., Ed.; CRC Press: Boca Raton, FL, USA, 2010; pp. 277–301.
26. Walkling-Ribeiro, M.; Noci, F.; Riener, J.; Cronin, D.A.; Lyng, J.G.; Morgan, D.J. The impact of thermosonication and pulsed electric fields on Staphylococcus aureus inactivation and selected quality parameters in orange juice. Food Bioprocess Technol. 2009, 2, 422–430.
27. Cserhalmi, Z.; Sass-Kiss, Á.; Tóth-Markus, M.; Lechner, N. Study of pulsed electric field treated citrus juices. Innov. Food Sci. Emerg. Technol. 2006, 7, 49–54.
28. Sachdeva, S.; Buchheim, W. Separation of native casein and whey proteins using micrfiltration. Aust. J. Dairy Technol.1997, 52, 92–97.
29. Vivekanand, V.; Kentish, S.E.; Connor, O.; O'Connor, A.J.; Barber, A.R.; Stevens, G.W. Microfiltration offers environmentally friendly fractionation of milk proteins. Aust. J. Dairy Technol. 2004, 59, 186–188.
30. Pafylias, I.; Cheryan, M.; Mehaiab, M.A.; Saglam, N. Microfiltration of milk with ceramic membranes. Food Res. Int.1996, 29, 141–146.
31. Silva, R.C.S.N.; Vasconcelos, C.M.; Suda, J.Y.; Minim, V.P.R.; Pires, A.C.S.; Carvalho, A.F. Acceptance of microfiltered milk by consumers aged from 7 to 70 years. Rev. Inst. Adolfo Lutz 2012, 71, 481–487.
32. Pagliarini, E.; Vernile, M.; Peri, C. Kinetic study on colour changes in milk due to heat. J. Food Sci. 1990, 55, 1766–1767.
33. Patton, S. Browning and associated changes in milk and its products: A review. J. Dairy Sci. 1955, 38, 457–478.
34. Boekel, M.A. Effect of heating on Maillard reactions in milk. Food Chem. 1998, 62, 403–414.
35. Bermúdez-Aguirre, D.; Fernández, S; Esquivel, H.; Dunne, P.C.; Barbosa-Cánovas, G.V. Milk processed by pulsed electric fields: Evaluation of microbial quality, physicochemical characteristics, and selected nutrients at different storage conditions. J. Food Sci. 2011, 76, S289–S299.
36. Fennema, O.R. Food Chemistry, 3rd ed.; Fennema, O.R., Ed.; CRC Press: New York, NY, USA, 1996; pp. 254–273.

37. Calvo, M.M.; de la Hoz, L. Flavour of heated milks: A review. Int. Dairy J. 1992, 2, 69–81.
38. Contarini, G.; Povolo, M.; Leardi, R.; Toppino, P.M.; Lattiero-Casealio, I.S.; Lombardo, V.; Salerno, V.B. Influence of heat treatment on the volatile vompounds of milk. J. Agric. Food Chem. 1997, 45, 3171–3177.
39. Hutton, J.T.; Patton, S. The origin of sulfhydrylgroups in milk proteins and their contributions to "cooked" flavour. J. Dairy Sci. 1952, 35, 699–705.
40. Zabbia, A.; Buys, E.M.; de Kock, H.L. Undesirable sulphur and carbonyl flavour compounds in UHT milk: A review. Crit. Rev. Food Sci. Nutr. 2012, 52, 21–30.
41. Al-Attabi, Z.; D'Arcy, B.R.; Deeth, H.C. Volatile sulphur compounds in UHT milk. Crit. Rev. Food Sci. Nutr. 2009, 49, 28–47.
42. Toso, B.; Procida, G.; Stefanon, B. Determination of volatile compounds in cows' milk using headspace GC-MS. J. Dairy Res. 2002, 69, 569–577.
43. Patton, S.; Forss, D.A.; Day, E.A. Methyl sulfide and the flavour of milk. J. Dairy Sci. 1956, 39, 1469–1470.
44. Vazquez-Landaverde, P.A.; Velazquez, G.; Torres, J.A; Qian, M.C. Quantitative determination of thermally derived off-flavour compounds in milk using solid-phase microextraction and gas chromatography. J. Dairy Sci. 2005, 88, 3764–3772.
45. Vazquez-Landaverde, P.A.; Torres, J.A.; Qian, M.C. Quantification of trace volatile sulfur compounds in milk by solid-phase microextraction and gas chromatography-pulsed flame photometric detection. J. Dairy Sci. 2006, 89, 2919–2927.
46. Advanced Dairy Chemistry, 3rd ed.; Fox, P.F.; McSweeney, P.L. (Eds.) Kluwer Academic/Plenum: New York, NY, USA, 2003–2006; p. 4.
47. Nursten, H.E. The flavour of milk and dairy products: I. Milk of different kinds, milk powder, butter and cream. Int. J. Dairy Technol. 1997, 50, 48–56.
48. Mader, I. Beta-Carotene: Thermal degradation. Science 1964, 144, 533–534.
49. Valero, E.; Villamiel, M.; Miralles, B.; Sanz, J.; Martínez-Castro, I. Changes in flavour and volatile components during storage of whole and skimmed UHT milk. Food Chem. 2001, 72, 51–58.
50. Vallverdú-Queralt, A.; Bendini, A.; Barbieri, S.; di Lecce, G.; Martin-Belloso, O.; Toschi, T.G. Volatile profile and sensory evaluation of tomato juices treated with pulsed electric fields. J. Agric. Food Chem. 2013, 61, 1977–1984.
51. Sampedro, F.; Geveke, D.J.; Fan, X.; Rodrigo, D.; Zhang, Q.H. Shelf-life study of an orange juice-milk based beverage after PEF and thermal processing. J. Food Sci. 2009, 74, S107–S112.

CHAPTER 12

Biosorption of silver cations onto *Lactococcus lactis* and *Lactobacillus casei* isolated from dairy products

Maciej Milanowski[1,2], Paweł Pomastowski[1,2], Viorica Railean-Plugaru[1,2], Katarzyna Rafińska[1,2], Tomasz Ligor[1,2], Bogusław Buszewski[1,2]*

[1] Department of Environmental Chemistry and Bioanalytics, Faculty of Chemistry, Nicolaus Copernicus University, Toruń, Poland,

[2] Interdisciplinary Centre of Modern Technologies, Nicolaus Copernicus University, Toruń, Poland

ABSTRACT

The current work deals with the phenomenon of silver cations uptake by two kinds of bacteria isolated from dairy products. The mechanism of sorption of silver cations by *Lactococcus lactis* and *Lactobacillus casei* bacteria was investigated. Inductively coupled plasma–mass spectrometry (ICP-MS) was used for determination of silver concentration sorbed by bacteria. Analysis of charge distribution was conducted by diffraction light scattering method. Changes in the ultrastructure of *Lactococcus lactis* and *Lactobacillus casei* cells after treatment with silver cations were investigated using transmission electron microscopy observation. Molecular spectroscopy methods, namely Fourier transform-infrared spectroscopy (FT-IR) and matrix-assisted laser desorption/ionization mass spectrometry (MALDI MS) were employed for description of the sorption mechanism. Moreover, an analysis of volatile organic compounds (VOCs) extracted from bacterial cells was performed.

INTRODUCTION

Lactic acid bacteria (LAB) are gram-positive, facultatively anaerobic, usually nonmotile and nonsporulating bacteria, whose characteristic feature is production of lactic acid from carbohydrates by means of fermentation as their major end product [1, 2]. This capability and other features have led to the wide use of LAB in numerous industrial applications. The most important ones are: production of fermented foods using starter cultures and utilization as probiotics [3–6]. Moreover, LAB have been recognized in the field of bioremediation, especially when bacterial strains act as sorbents [7–9]. *Lactobacillus acidophilus* has been employed for arsenic(III) removal from waste water [10]. Cadmium and lead ions can be bound by certain species of *Lactobacillus* and *Bifidobacterium* [11]. Chromium resistant *Lactobacillus* strains have displayed their ability to reduce chromium(VI) to chromium(III) [12]. Furthermore, *Lactobacillus casei* was found to be the most effective binding agent of Cu^{2+} amongst several tested microorganisms [13]. Also, biosorption mechanism of silver cations by *Lactobacillus* sp. strain A09 was investigated in the study of Lin et al. [14].

Nowadays, there are broad industrial and medical applications of silver nanoparticles (AgNPs), including electronics, food industry, clothing, medical devices and cosmetics [15]. This indicates continuously increasing and widespread usage of silver nanoparticles. AgNPs are in electrochemical equilibrium with silver cations [16], hence their high demand causes increased amount of this element being released into the environment, especially into the aquatic one. Therefore, the problem of its removal is a top issue [17–21] and LAB are microorganisms which can be used for uptake of silver cations [22]. In this field it is crucial to understand the mechanisms of biosorption of Ag by bacteria species capable of binding silver cations.

The aim of this study was to develop a methodology for biosorption of silver cations by/onto isolates of lactic acid bacteria. For this purpose, (i) isolation of selected LAB obtained from milk products (e.g. milk, cheese) was conducted; (ii) identification of isolated bacteria by sequencing 16S rRNA genes, as well as intact cell MALDI-TOF MS was performed; (iii) microbiological profile description of LAB was carried out; (iv) biosorption of silver ions on selected LAB was employed and (v) determination of silver binding mechanisms was realized with the use of multiple techniques.

MATERIALS AND METHODS

Instrumentation

NanoDrop 2000c (Thermo Fisher Scientific, Wilmington, DE, USA) was used to measure concentration of DNA. For polymerase chain reaction (PCR) amplification we employed Mastercycler® pro thermocycler (Eppendorf, Hamburg, Germany) and electrophoresis was carried out with the use of PowerPac™ Universal Power Supply (Bio-RAD Laboratories, Hercules, CA, USA). Identification of the isolated bacteria and investigation of silver biosorption were both performed with the use of ultrafleXtreme MALDI-TOF/TOF mass spectrometer (Bruker Daltonik, Bremen, Germany). Concentration of silver was determined with the use of CX 7500 ICP-MS spectrometer (Agilent Technologies, Santa Clara, CA, USA). FT-IR SPECTRUM 2000 used for acquisition of IR spectra was purchased from PerkinElmer (Waltham, MA, USA). The zeta potential was analyzed by Zetasizer Nano Series analyzer (Malvern Instruments, Malvern, United Kingdom). For transmission electron microscopy we used Tecnai F20 X-Twin system (FEI Europe, Eindhoven, The Netherlands). The employed ultramicrotome was Leica EM UC7 (Leica Microsystems, Wetzlar, Germany). We also used Axio Observer D1 fluorescence microscope (Carl Zeiss, Oberkochen, Germany). GC/MS analyses were carried out using an Agilent 6890A gas chromatograph (Agilent Technologies, Santa Clara, CA, USA) coupled with an Agilent 5975 Inert XL MSD mass spectrometer (Agilent Technologies). The system was equipped with a CP-Porabond-Q 25 m × 0.25 mm × 3 μm column (Agilent Technologies). Extractions of volatile organic compounds (VOCs) were performed using 75 μm Carboxen/polydimethylsiloxane (PDMS) fiber (Supelco, Bellefonte, PA, USA).

Materials

Various combinations of media were employed for bacteria cultivation, namely: Tryptic Soy Broth (TSB; Soybean-Casein Digest Medium; Bacto, Sydney, Australia), Tryptic Soy Agar (TSA; Soybean-Casein Digest Agar Medium; Bacto), de Man, Rogosa and Sharpe Agar (MRSA; Sigma Aldrich, Steinheim, Germany), de Man, Rogosa and Sharpe (MRS) Bulion and M17 Bulion (Merck, Darmstadt, Germany). EXTRACTME DNA BACTERIA kit (version: 1.2014; DNA-Gdańsk, BLIRT, Gdańsk, Poland) was used to isolate LAB. Taq PCR Master Mix Kit (Qiagen, Hilden, Germany) together with Ideal Marker DNA M10Kpz (DNA-Gdańsk, BLIRT, Gdańsk, Poland) were applied for PCR reaction. Electrophoresis was conducted using RNase-Free Water (Qiagen), agarose (MAXIMUS, Łódź, Poland), 1× TAE buffer (Bioline, London, United Kingdom) and ethidium bromide (AppliChem, Darmstadt, Germany). QIAquick PCR Purification Kit (Qiagen) was applied for PCR products purification. For transmission electron microscopy we used 1× phosphate-buffered saline (PBS) buffer, glutaraldehyde, ethanol, uranyl acetate and lead citrate purchased from Sigma Aldrich. MTP Anchor Chip 384 target (Bruker Daltonik, Bremen, Germany) was used in MALDI-TOF MS experiments, as well as chemicals from Sigma Aldrich: LC-MS grade Chromasolv water, ethanol, acetonitrile, trifluoroacetic acid (TFA), isopropanol, and products from Avantor Performance Materials (Gliwice, Poland): acetone, hydrochloric acid and methanol. Matrices for MALDI analyses were α-cyano-4-hydroxycinnamic acid (HCCA) and 2,5-dihydroxybenzoic acid (DHB) (both from Sigma Aldrich). KCl and NaCl buffers (both from Avantor Performance Materials) were purchased for zeta potential measurements. Other chemicals and consumables such as: Ag solution, KBr, acridine orange and CaF_2 plate were acquired from Sigma Aldrich. 22 mL headspace crimp top vials and polytetrafluoroethylene (PTFE)/butyl septa for HS-SPME-GC/MS experiments were purchased from PerkinElmer (Waltham, MA, USA). Ultra-pure water from a Milli-Q water system (Millipore, Bedford, MS, USA) was used throughout the work.

Milk products (milk, cheese, cottage cheese, whey, powdered milk, powdered whey) were supplied by Dairy Factory in Drzycim and Piotrków Kujawski, Poland.

Isolation of bacteria from milk products

Samples of milk, cheese, cottage cheese, whey and powdered milk were plated on three culture media, namely TSA, MRSA and M17, and incubated at 37°C for 24 h. Then, the same combination of media was streaked with the obtained biological material using a sterile inoculation loop and incubated at 37°C for 24 h. Subsequently, the grown colonies were applied for preparation of dilutions in the range of 10^{-1} to 10^{-8} using sterilized 0.87% KCl and double distilled water (H_2O_{dd}). All dilutions were plated (1 mL of inoculum) onto Petri dishes with culture media, MRSA and M17, respectively. The plates were placed into an incubator for 24 h at 37°C.

Microbiological characteristics of LAB

Enumeration of the total number of microorganisms was performed according to the method described in the Polish norm PN-A-86034-04:1993. A series of sequential dilutions (1 mL of material: 9 mL of H_2O/KCl) was prepared from the initial dilution (10 mL of milk: 90 mL of H_2O/KCl). The first dilution was made by transferring 1 mL of a sample to 9 mL of H_2O/KCl which made it 10^{-2} dilution of the original sample. Immediately after the 10^{-2} dilution was shaken, the vessel was uncapped and 1 mL was aseptically transferred to a second 9 mL-portion of H_2O/KCl solution. As a consequence, the second blank represented a 10^{-3} dilution of the original sample. Shaking the 10^{-3} dilution vigorously and transferring 1 mL of it to the third 9 mL-portion of water/KCl resulted in obtaining 10^{-4} dilution blank. The process was continuously repeated until the dilution of 10^{-7} was reached. All samples (1 µL of suspension) were plated in duplicate onto Petri dishes filled with M17 and MRSA media and then incubated at 37°C.

Identification of the isolated bacteria by 16S PCR

Isolation of DNA from 27 selected bacteria was conducted using the EXTRACTME DNA BACTERIA kit. Samples were prepared according to the manufacturer's protocol. Concentration of DNA was measured using NanoDrop 2000c within the range of 10.8 to 22.8 ng μL^{-1}, and, subsequently, 1 µL of each DNA sample was transferred into a vial. 2 µL of Taq PCR Master Mix Kit was added to each sample together with 1 µL of F1 primers (5′-GAG TTT GAT CCT GGC TCA G-3′) and R12 primers (5′-ACG GCT ACC TTG TTA CGA CTT-3′). Finally, the samples were placed in Mastercycler® pro thermocycler with the following program: initial denaturation at 95°C for 2 min followed by 30 cycles of denaturation at 94°C for 1 min, annealing at 55°C for 1 min and extension at 72°C for 2 min with the final extension at 72°C for 5 min. The PCR amplification fragments were resolved by agarose (1.0%) gel electrophoresis at 100 V for 30 min. The gel was stained with ethidium bromide, and the bands were visualized under UV light. Then, the PCR products were purified and sequenced according to Hrynkiewicz et al. [23]. At first, the PCR products were purified using the QIAquick PCR Purification Kit. Then, the PCR primers acted as sequencing primers in the process of direct sequencing of PCR products. The sequences were aligned manually aided by the Sequencher system (TW Version 5.1, Gene Codes, Ann Arbor, MI, USA). Comparison of the obtained sequences of all isolates was performed using the BLAST database of the National Center for Biotechnology Information (NCBI).

Identification of the isolated bacteria by intact cell MALDI-TOF MS

The matrices: HCCA (10 mg mL^{-1}) and DHB (50 mg mL^{-1}) were prepared in Bacterial Solution (EtOH/ACN/H_2O, 1:1:1 (v/v/v)). Trifluoroacetic acid (TFA) solution was added to the Bacterial Solution to reach the final concentration of 2.5% v/v. Then, under sterile conditions, two loopfuls (approximately 10 mg) of bacterial cells were suspended in 5 μL of Bacterial Solution and thoroughly vortexed for 30 s. 2 μL of bacterial suspension was mixed with 2 μL of a matrix and then 1 μL of the mixture was spotted on a ground steel MALDI target. After 30 min, when all spots had dried, the target was placed in the ultrafleXtreme MALDI-TOF/TOF mass spectrometer for a measurement according to Pomastowski et al. [24]. The ultrafleXtreme mass spectrometer was equipped with a modified Nd:YAG laser (smartbeam II™) operating at the wavelength of 355 nm and the frequency of 2 kHz. Intact cell (IC) MALDI TOF MS spectra were recorded manually in linear positive mode within *m/z* range of 300–30000 and applying the acceleration voltage of 25 kV. All mass spectra were acquired and processed with the dedicated software: flexControl and flexAnalysis, respectively (both from Bruker).

Biosorption of silver onto *L. lactis* and *L. casei*

10 mL of M17 and MRSA was inoculated with *L. lactis* and *L. casei*, respectively. Three solutions of silver cations at concentrations of 1 mg L^{-1} (1 ppm), 0.1 mg L^{-1} (0.1 ppm) and 0.01 mg L^{-1} (0.01 ppm) were made from a stock solution of silver (60 mg L^{-1} $AgNO_3$),. After addition of silver cations, each of the inoculated solutions was measured for optical density (OD) with incubation lasting for 24 h at 37°C. The following MALDI-TOF MS experiments were conducted in order to determine changes in spectra between bacteria alone (control) and bacteria treated with silver cations (1 ppm, 0.1 ppm and 0.01 ppm). The settings and conditions of the measurements using the ultrafleXtreme mass spectrometer were identical as described above.

Independently, 5 mL (at 10^6 cells/mL) of inoculum was prepared. Then, 1 mL of inoculum was transferred to three flasks with TSB medium. One of them served as a control. 2.5 mL of a stock solution of Ag^+ was added to the second flask (OPTION 1) at the time of inoculation. All flasks were then incubated in a water bath (37°C) with shaking (150 rpm) for 24 h. After this period 2.5 mL of Ag^+ stock solution was transferred into the third flask (OPTION 2) and all flasks were subsequently incubated under the same conditions for 48 h. Finally, all the samples were centrifuged for 15 min

(14500 rpm). The separated supernatants and the pellets were stored in a refrigerator (4°C) and used for further analyses.

Inductively Coupled Plasma—Mass Spectrometry (ICP-MS)

The obtained supernatants from OPTION 1 & 2 were transferred into sterile Eppendorf tubes (500 µL). The pellets from OPTION 1 & 2 were suspended in 5 mL of sterile H_2O and centrifuged again for 15 min (14500 rpm) to purify the biomass from the medium. Next, the supernatant was removed and the pellet was suspended in 500 µL of H_2O. 300 µL of $HNO_3$and 100 µL of HCl were added to the suspension and incubated in a thermomixer (90°C) with shaking for 4 h until the precipitate was completely dissolved. Each sample was examined for the presence of silver using inductively coupled plasma—mass spectrometry (ICP-MS) with a quadrupole mass analyzer.

Fourier Transform Infrared spectroscopy (FT-IR)

The bacterial pellets from OPTION 1 & 2 were suspended in 250 µL of H_2O and thoroughly centrifuged. One drop of the suspension (120 µL) was transferred onto acetone-cleaned plate of CaF_2 and dried at 105°C for approximately 2 min. IR spectra were acquired in the range of 4000–400 cm^{-1} and 20 scans were carried out for each sample. The measurements were performed using FT-IR SPECTRUM 2000.

Zeta (ζ) potential measurement

The measurement of zeta potential was carried out with different solvents: 0.7% KCl (pH 2–11) and 0.7% NaCl (pH 3–9). The bacterial pellet from OPTION 2 was suspended in 1 mL of H_2O and vortexed thoroughly until obtaining a homogeneous suspension of bacteria. Then, 980 µL of a buffer with a specific pH was added to 20 µL of bacteria solution and mixed by pipetting. The whole sample was vortexed for 30 s and incubated for 1 min in an ultrasonic bath. Thus prepared samples were placed directly in a cuvette and all measurements were recorded using the Zetasizer Nano Series analyzer. The ζ-potential values were provided directly by the instrument. The measurements were carried out at 25°C in triplicate for 10 cycles. The obtained results were shown in diagrams using the sigmoidal model available in the CurveExpert Professional 2.0 software (Hyams Development, USA).

Transmission Electron Microscopy (TEM)

Selected samples from the control and OPTION 2 were measured using Tecnai F20 X-Twin electron microscope coupled with Energy Dispersive X-ray (EDX) detector. First, a portion of 10 mL of culture broth was centrifuged, washed twice with sterilized phosphate-buffered saline, and the biomass was subjected to transmission electron microscopy. Bacterial cells were fixed with 2% glutaraldehyde for 4 h. After eliminating the remaining glutaraldehyde, dehydration process was conducted with 20, 30, 40, 50, 60, 70, 80, 90, and 100% ethyl alcohol. Fixed cells were then infiltrated and embedded in LR Gold resin. The embedded material was cut on Leica EM UC7 ultramicrotome into ultra-thin sections (50 nm) and placed on grids coated with formvar. Ultrathin sections were subsequently stained with 2.5% uranyl acetate and 0.4% lead citrate solutions, and examined using Tecnai F20 X-Twin system.

Fluorescence microscopy analysis

Fluorescence microscopy examination was performed via acridine orange/ethidium bromide (AO/EB) staining in order to discriminate living and dead cells. Membrane integrity visualization was detected through the filter set at 43 He (excitation 550/25, emission 605/70) and 38 (excitation 470/40, emission 520/50) using Axio Observer D1 fluorescence microscope. Bacteria cells of *L. casei* and *L. lactis* grown in TSB medium were treated with $AgNO_3$ (1 ppm). Two methods (OPTION 1 & 2) were applied for both strains. In the first method (OPTION 1) $AgNO_3$ (1 ppm) was added to cell culture medium (1×10^6 of cells) and analysis was performed after 1 h and 24 h of incubation. In the second method (OPTION 2) the inoculum of bacteria (1×10^6 of cells) was first incubated at 37°C for 24 h, then treated with $AgNO_3$ (1 ppm) and analyzed after 1 h and 24 h of incubation at 37°C. In both cases cells were stained with AO/EB with the final concentration of 0.12/0.4 $\mu g\ mL^{-1}$ and incubated at room temperature for 5 min in the dark. Cells were then centrifuged and the remaining supernatant was discarded to eliminate the unbound dyes. Next, the cell pellet was resuspended in a small volume of PBS-1X. Control samples were prepared basing on cells untreated with $AgNO_3$ (1 ppm). All experiments were performed in a dark room to avoid photobleaching of the dyes.

Gas Chromatography—Mass Spectrometry (GC/MS)

Empty glass vials were stored at 60°C for 2–3 days, and then removed from a dryer and left to stand for about 15 minutes to equilibrate with the surroundings. 2 g of the bacterial pellet from OPTION 2 was weighed, transferred into a vial and capped. The headspace air was pumped out twice through a syringe to generate underpressure. VOCs were extracted and analyzed using headspace solid phase microextraction gas chromatography—mass spectrometry (HS-SPME-GC/MS). Volatile compounds were collected using 75 μm Carboxen/PDMS fiber at 50°C for 15 min. Gas chromatography was performed with helium as the carrier gas with the flow rate of 1.4 mL min^{-1}; the temperature of the split-splitless injector was set at 200°C. The oven temperature program was as follows: the initial 40°C were kept for 2 min, and ramped at 10°C min^{-1} to 140°C, then again ramped at 5°C min^{-1} to 270°C and kept for 5 min. Acquisition was performed at the mass range of 30–300 m/z. Spectra were collected for ions generated by electron ionization (EI) at 70 eV; both the ion source and the transfer line temperature was set at 200°C.

RESULTS

Isolation and characteristics of bacteria isolated from milk products

Bacteria were isolated from the investigated milk products and were described by visible morphological features. Isolate colonies appearing in plates were classified according to their shape, color, texture, size, etc. S1 Table shows selected 27 bacteria strains along with a description of the applied medium, diluent, and a brief characterization. M17 and MRSA proved to be the best media for the culture of lactic acid bacteria owing to the greatest number of grown colonies on the plates. On the other hand, the type of a diluent (H_2O or KCl) used for preparation of inoculum had no significant effect on the growth of the bacteria.

Identification of the obtained LAB by 16S PCR and intact cell MALDI-TOF MS

Sequences obtained from 16S rRNA gene sequence analysis were used to search for similarities using the BLAST (basic local alignment search tool; http://blast.ncbi.nlm.nih.gov/Blast.cgi) program. Identity of the representative isolates were determined on the basis of the highest scores (typically ≥ 95%). The results of this study showed that analysis with 16S rRNA gene sequencing could provide successful identification for the investigated LAB species, *L. lactis* (100%) and *L. casei* (100%) (Table 1).

Table 1. Results of 16S rRNA sequencing of *L. lactis* and *L. casei* isolated from milk products.

Name	Identified strain	BLAST	Overlap (%)
Cheese isolated from MRSA HOH yellow dark, medium: M17	*Lactobacillus casei*	*Lactobacillus casei* [KF673500]	1430/1430 (100%)
		Lactobacillus paracasei subsp. *paracasei* [KF418817]	1430/1430 (100%)
		Lactobacillus casei [JN560891]	1430/1430 (100%)
Cow milk from Świecie district, Poland	*Lactococcus lactis*	*Lactococcus lactis* [JN7852396]	1433/1433 (100%)

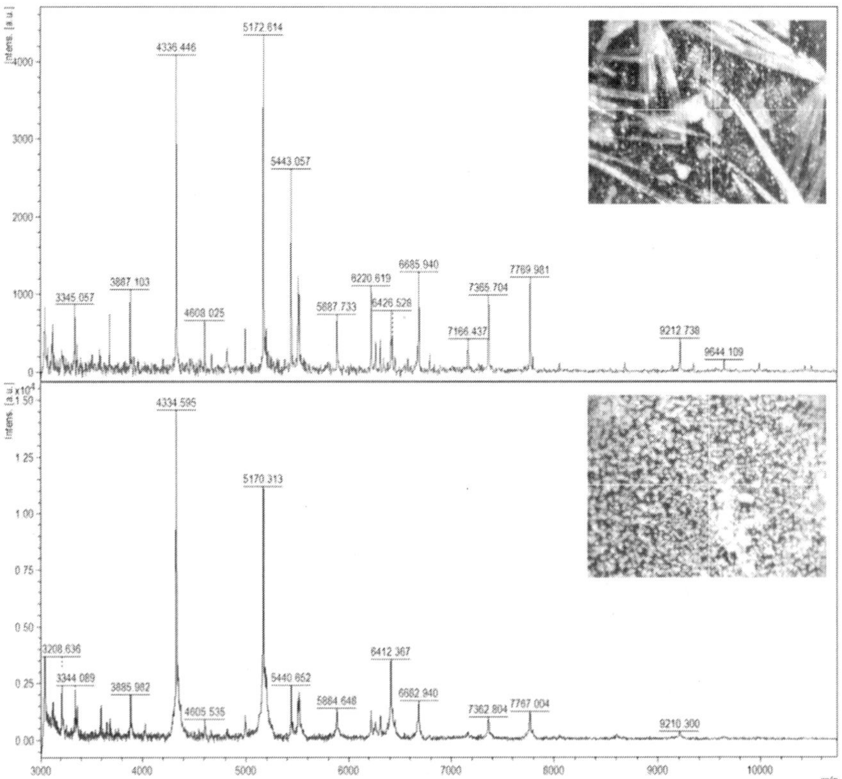

Fig 1. A spectrum of *L. lactis* recorded using MALDI-TOF MS technique with the use of DHB (top) and HCCA (bottom) matrix; photos of crystals of the matrices are shown in the corner.

Two bacterial strains: *L. lactis* and *L. casei*, were chosen for identification using the ultrafleXtreme MALDI-TOF/TOF mass spectrometer. The acquired spectra were compared with MALDI spectra of reference bacterial strains: *L. lactis* ATCC 49032 and *L. casei* ATCC 334 and the coverage of 99%– 100% was obtained. Moreover, in order to select appropriate conditions for the analysis, we tested and compared two different matrices: DHB and HCCA. As portrayed in Fig 1 a greater number of signals (i.e. 70) was recorded for *L. lactis* when α-cyano-4-hydroxycinnamic acid (HCCA) was used as a matrix, in comparison to application of 2,5-dihydroxybenzoic acid (DHB) where only 43 ions were detected.

In the case of *L. casei*, we recorded 78 signals for DHB matrix. HCCA gave only a slightly smaller number of signals (74) on the obtained spectra (Fig 2).

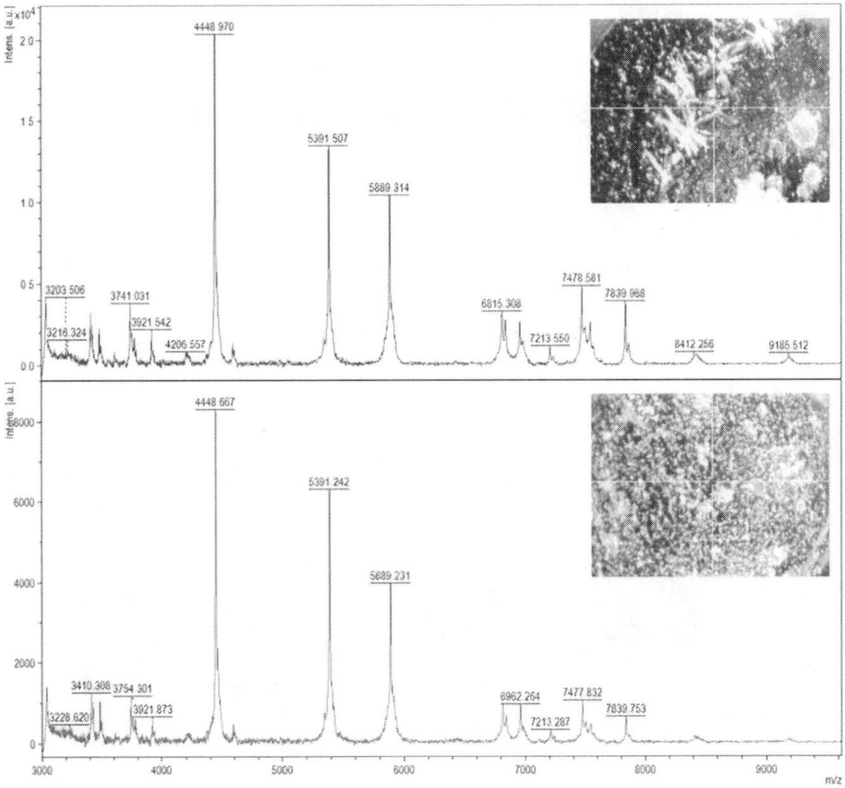

Fig 2. A spectrum of *L. casei* recorded using MALDI-TOF MS technique on DHB (top) and HCCA (bottom) matrix; photos of crystals of the matrices are shown in the corner.

Based on the value of the signal to noise ratio and intensity of individual peaks, it can be concluded that HCCA matrix is a better choice in the case of bacteria *L. lactis.* On the contrary, better quality of spectra was obtained using DHB matrix for *L. casei.* The optimal starting optical density of microbial cells was in the range of 0.3–0.4. Among the dilutions– 1:1, 1:10 and 1:100 the best mass spectra were recorded for 1:1 dilution. MS spectra obtained for the studied species at this dilution factor were characterized by the highest intensity, high signal to noise ratio and the largest number of registered peaks. Therefore, in order to prepare reference spectra we used the 1:1 dilution of bacterial cells in HCCA matrix.

Biosorption of silver onto selected LAB evaluated by MALDI-TOF MS

Fig 3 shows OD values obtained after 24 h of incubation of *L. lactis* and *L. casei* with 1 mg L^{-1}(1 ppm), 0.1 mg L^{-1} (0.1 ppm) and 0.01 mg L^{-1} (0.01 ppm) of silver cations.

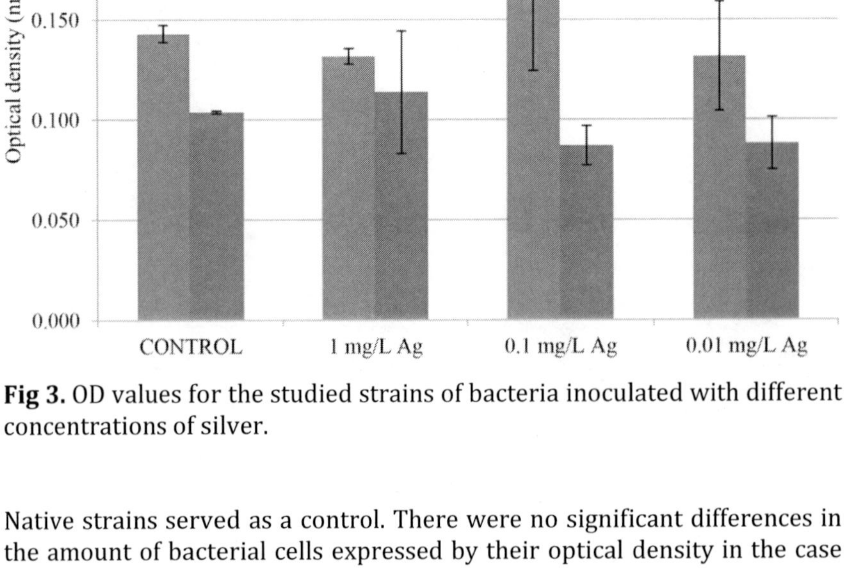

Fig 3. OD values for the studied strains of bacteria inoculated with different concentrations of silver.

Native strains served as a control. There were no significant differences in the amount of bacterial cells expressed by their optical density in the case of both *L. lactis* and *L. casei.* None of the applied concentrations of Ag^+ affected the final OD, moreover the number of cells remained relatively constant in comparison with the control. Hence, in this study the concentration of silver cations was specially selected so as not to be cytotoxic to the isolated lactic acid bacteria and not to inhibit their growth.

In order to determine changes in the cells of LAB after silver binding MALDI-TOF MS spectra from bacteria inoculated with silver were compared with previously obtained reference spectra of *L. lactis* and *L. casei.* Fig 4 illustrates that MALDI-TOF MS spectra derived from silver-enhanced *L. lactis* were significantly different than those originating from a native strain.

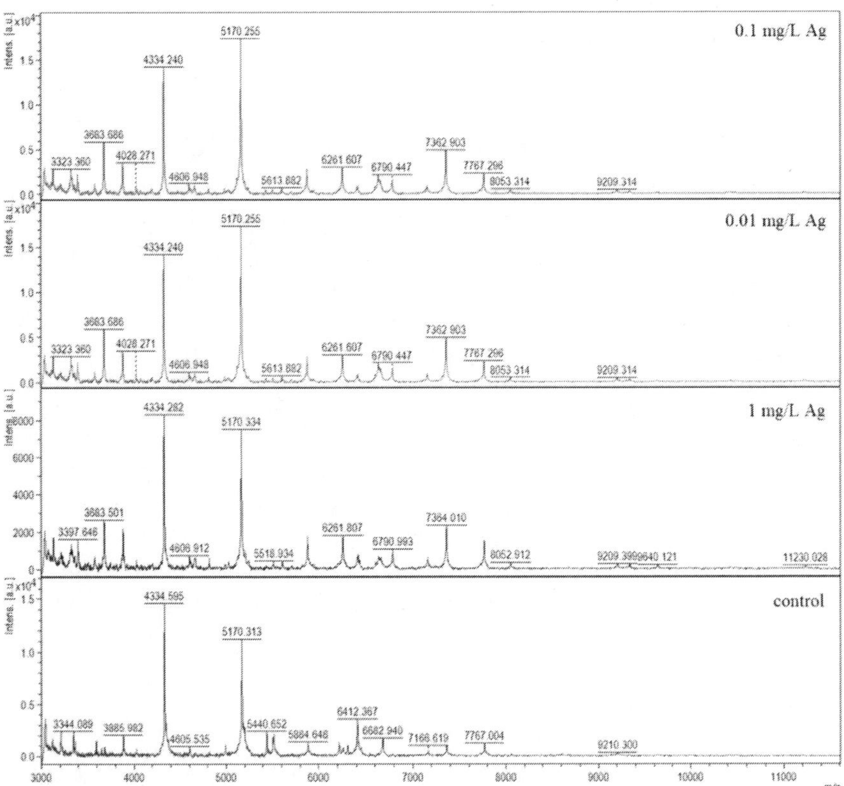

Fig 4. MALDI-TOF MS spectra of *L. lactis* with different concentrations of silver compared to a reference spectrum of these bacteria.

Seven signals appeared to have significantly increased intensity in samples containing silver. This effect was observed for the peaks at 3683.607, 3885.089, 5170.334, 5884.648, 7166.619, 7362.804 and 7767.004 *m/z*. Moreover, intensity of the peaks at 5440.652 and 5518.837 *m/z* were increased at the control spectrum. Fourteen signals were recorded only for samples with silver cations. They were: 3133.449, 3212.085, 3322.476, 3397.050, 3583.491, 4820.593, 6261.807, 6424.261, 6640.434, 6790.993, 8053.314, 9338.727, 9640.121 and 11230.028 *m/z*. On the other hand, eight signals were present exclusively for *L. lactis* without any modifications: 3119.572, 3208.636, 3344.089, 3594.997, 6217.919, 6308.829, 6412.367 and 6682.940 *m/z*. These peaks were absent from MS spectra of the samples with the addition of silver. For *L. casei* no changes in MALDI-TOF MS spectra were observed.

ICP-MS measurement

Fig 5 illustrates differences in the content of silver for particular variants (OPTION 1 & 2) in the tested bacterial cells before and after inoculation with silver.

In the case of *L. lactis*, the content of silver in both variants (OPTION 1 & 2) remained at a similar level of about 79% which means that *L. lactis* bacteria bound similar amount of silver regardless of the time of Ag^+ addition. This observation confirms the thesis that Ag^+ in the concentration of 1 ppm has no significant effect on the physiology of *L. lactis* and the ability of this strain to adopt Ag^+. In the case of *L. casei*, in turn, significant differences in the processes of Ag^+ binding were observed. Bacteria supplemented with silver cations 24 h after inoculation (OPTION 2) bound 70.77% ± 3.21% of the element as compared to bacteria incubated with silver at the time of inoculation (OPTION 1) where sorption was at the level of 95.96% ± 2.89%. The results indicate an ability for adaptation of *L. casei* grown in an environment abundant in silver cations. Moreover, the performed study demonstrated higher adsorption efficiency of bacterial cells inoculated at the time of addition of silver cations.

FT-IR spectroscopy

Spectral characteristics of native and modified LAB was conducted to find active chemical groups involved in the silver-binding process. S1 and S2 Figs show FT-IR spectra of native strains and strains after modification resulting from addition of silver. In the case of *L. lactis* a broad peak localized in the region of 1530–1630 cm^{-1} was observed upon the addition of silver to the sample (S1 Fig), which is characteristic for samples with Ag addition. The absorption band at 1456 cm^{-1} was flattened in OPTION 2. Moreover, signals at 1059 and 1236 cm^{-1}became deeper in the case of silver-treated *L. lactis*. For *L. casei*, in turn, the only observed change was a broad peak in the region of 1530–1630 cm^{-1} in OPTION 1 (S2 Fig).

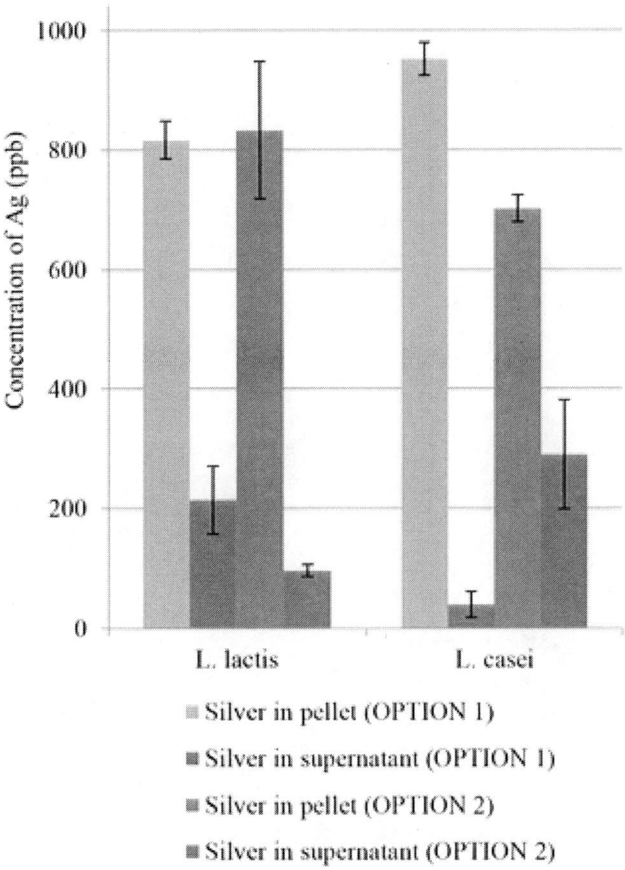

Fig 5. Concentration of Ag in OPTION 1 & 2 measured using ICP-MS.

Determination of zeta (ζ) potential

S3(A) Fig for *L. lactis* shows the size of a bacterial cell within 1000–5000 nm in each of the used buffers (KCl and NaCl) and at all tested pH values. The graph (B) displays the same peak with reduced intensity and an additional weak signal from a particle of about 5000–10000 nm in size. The following charts show changes in the value of the zeta potential resulting from changes in pH of a buffer (Fig 6).

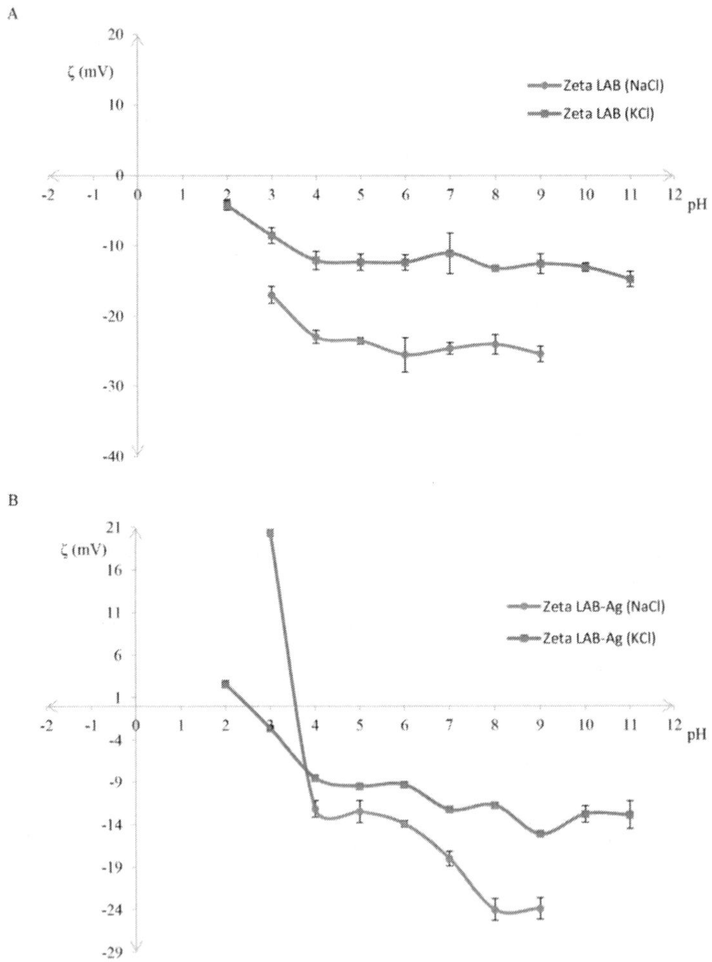

Fig 6. Zeta potential of bacterial cells of *L. lactis* in different solutions, and obtained before (A) and after (B) silver addition.

Fig 6(A) illustrates charge distribution of native *L. lactis*. The zeta potential of *L. lactis* suspended in 0.7% KCl in pH range of 2–11 oscillated between −4 and −15 mV. In the case of 0.7% NaCl, a decrease of zeta potential was observed (ranging from −17 to −25 mV) in pH range of 2–11. Fig 6(B) shows zeta potential values for *L. lactis* after sorption of Ag^{+}. Depending on the buffer applied, the zeta potential for *L. lactis* with sorbed Ag^+ ranged from +20 to −24 mV for NaCl buffer, and from +3 to −16 mV for KCl buffer. There was a sharp decrease in the potential values registered for both buffers: from +20 mV to −12 mV in the case of NaCl buffer and from +3 mV to −3 mV for KCl buffer. Values of the isoelectric point of *L. lactis* with sorbed Ag^+ were approximately 3.5 and 2.5 for NaCl and KCl buffers, respectively. After the drop, constant values of ζ potential for KCl and its continuous decrease until −24 mV for NaCl buffer were observed.

TEM imaging

In order to investigate changes in the ultrastructure of *L. lactis* and *L. casei* cells after treatment with Ag^+, a transmission electron microscopy observation was performed (Fig 7).

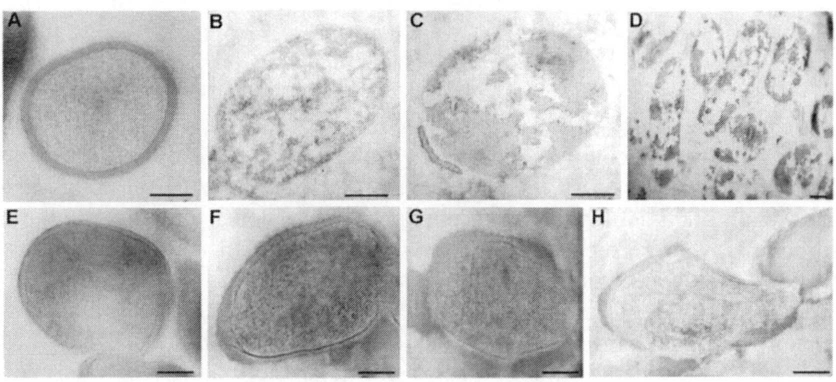

Fig 7. Ultrastructure of bacterial cells of *L. lactis* and *L. casei*..

(A)–native *L. lactis,* (B), (C), (D)–*L. lactis* cells after 48 hours of incubation with silver cations (OPTION 2), (E)–native *L. casei,* (F), (G), (H)–*L. casei* cells after 48 hours of incubation with silver cations (OPTION 2). Bar: 200 nm.

Control cells of *L. lactis* not treated with $AgNO_3$ (Fig 7(A)) have an oval shape, homogenous cytoplasm and a relatively thick cell wall. After 48 hours of incubation with Ag^+, the ultrastructure underwent characteristic changes (Fig 7(B), 7(C) and 7(D)). Electron light bands were visible on the area of the cytoplasm. The cell wall was present only in the form of individual pieces on single cells, whereas most of the observed cells were completely deprived of it. There was no continuity of the cell membrane, either.

Cells of *L. casei* are oval in shape and have a homogenous cytoplasm and a smooth surface (Fig 7(E)). After incubation with $AgNO_3$ their ultrastructure did not undergo such significant changes as in the case of *L. lactis*. Thickening and uneven surface of cell walls was a characteristic change (Fig 7(F), 7(G) and 7(H)). Their cytoplasm was still electron dense, although much less homogeneous. Only in a few cells of *L. casei* changes in the shape and separation of cell membrane from the cell wall were observed. These results suggest that $AgNO_3$ has a limited effect on the ultrastructure of *L. casei* cells.

Fluorescence microscopy examinations with $AgNO_3$

For visualization of *L. lactis* and *L. casei* viability after the treatment with $AgNO_3$ (1 ppm) we performed staining with nucleic acid dyes (acridine orange and ethidium bromide). The living cells that showed only green fluorescence and the dead ones which displayed red color are shown in Fig 8.

In the case of the first method (OPTION 1), only viable cells displaying green fluorescence were visible in the samples of *L. lactis* and *L. casei* at the time of inoculation. After 1 h of incubation with $AgNO_3$, dead cells for both bacterial species could be observed. However, specific differences were noticeable after 24 h of incubation. In the case of *L. casei*, the number of dead cells displaying red fluorescence was much higher than the number of those displaying green fluorescence which suggests that *L. casei* is not as resistant to the used concentration of silver cations as *L. lactis*. An ability for adaptation of cells to Ag cations was recorded for *L. lactis*, which was manifested by a high number of cells displaying green fluorescence, which in turn indicated that under these conditions the majority of cells were alive. For the second method (OPTION 2) (Ag^+ added 24 h after inoculation) the results were comparable regardless of the chosen bacteria. After 1 h and 24 h of incubation with silver cations a growing number of dead cells could be noticed. The number of dead cells surpassed the number of live cells, hence the toxic effect of $AgNO_3$ (1 ppm) was clearly visible.

HS-SPME-GC/MS experiment

HS-SPME-GC/MS analysis of bacterial headspace resulted in identification of 13 volatiles. The chromatogram acquired for *L. lactis* is illustrated in Fig 9.

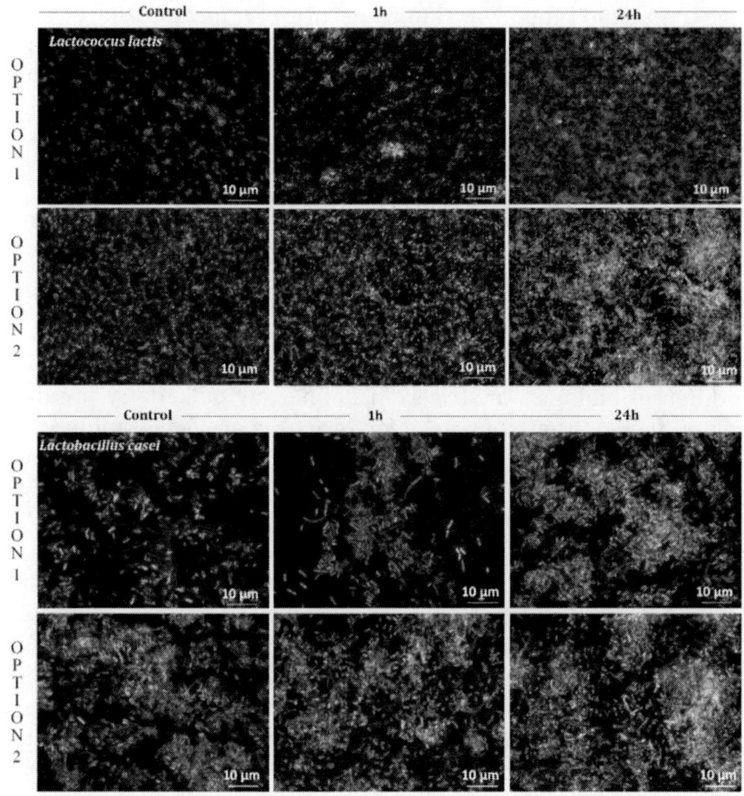

Fig 8. Comparative viability of *L. lactis* and *L. casei* cells after the treatment with $AgNO_3$ (1 ppm) regarding the first (OPTION 1) and the second method (OPTION 2).

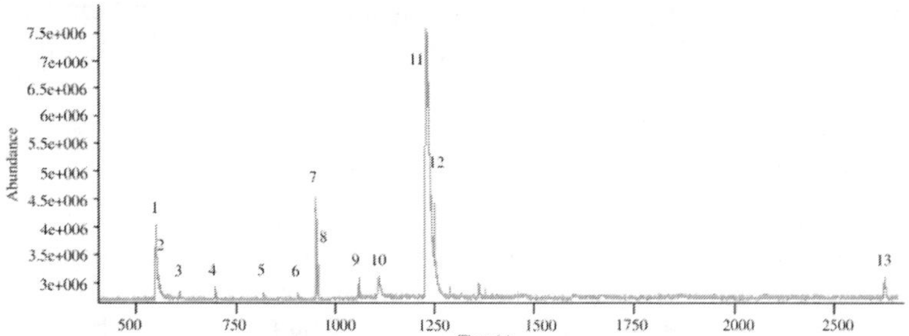

Fig 9. Chromatogram of VOCs from *L. lactis* obtained using HS-SPME-GC/MS, where: (1) acetaldehyde, (2) ethanol, (3) butane, (4) acetone, (5) 2-methylpropanal, (6) ethenyl acetate, (7) 2-butanone, (8) ethyl acetate, (9) hexane, (10) 3-methylbutanal, (11) 1-pentanol, (12) 3-methyl-1-butanol, (13) 2-phenylethyl acetate.

Among compounds found in the headspace of *L. lactis* the most abundant were: 1-pentanol, 2-phenylethyl acetate and 2-butanone. Other compounds, such as: ethyl acetate, hexane, acetaldehyde and 2-methylpropanal could also be distinguished.

DISCUSSION

Isolation and identification of bacteria from milk products

Due to their widespread distribution, lactic acid bacteria can be isolated from different biological matrices, including raw milk, dairy products and fermented foods. Other sources of lactobacteria are fecal samples, soil and plants [25, 26]. In particular, samples of human and animal milk are habitats rich in nutrients and they are often used for isolation of a diverse spectrum of LAB [27–29]. In this field, growth media such as MRSA and M17 are often employed for cultivation of LAB [30, 31]. In our study, identification of *L. lactis* and *L. casei* was carried out by two different techniques, namely: 16S rRNA gene sequencing and intact cell MALDI-TOF MS. The first technique is commonly used for identification of LAB [32, 33]. However, a growing interest of researchers in methods utilizing MALDI-TOF MS has been recently observed, due to high sample throughput, repeatability and precision of this technique. Currently, MALDI-TOF MS profiling of whole bacterial cells is used more frequently for bacterial identification [34, 35]. In order to select optimal conditions for analysis we tested and compared two matrices: HCCA and DHB. The former one proved to be more appropriate for *L. lactis*, whereas the latter one—for *L. casei*. This could result from different nature of the matrices. HCCA forms fine crystals, whereas DHB is characterized by large needles (Figs 1 and 2). This feature causes specific alterations in the extraction mechanism of bacterial components [24].

Biosorption of silver

In general, biosorption is a process of removal of substances from a solution by biological material [36]. It is a widely used technique for elimination of heavy metals (e.g. Cd, Pb, Cu, Cr) from waters by live or dead biomass and cellular products [37, 38]. The processes of biosorption involve, inter alia, absorption, adsorption, ion exchange, surface complexation and precipitation [36]. There are various types of biosorbents including bacteria, fungi, yeasts, algae, as well as industrial and agricultural waste [37, 39]. For example, bacteria of *Lactobacillus* sp. strain A09 show silver binding properties [14]. Recently, brewery fermentation industry waste yeast has been employed to adsorb Ag(I) from an aqueous solution [40]. Our study revealed that also *L. lactis* and *L. casei* isolated from milk and its products are good biosorbents of silver cations as they absorbed about 70–96% of silver from 1 ppm solution. The rate of silver cations uptake was the highest for *L. casei* when Ag^+ was added at the time of inoculation. There is a report that bacteria originating from metal-containing waters or sediments show greater tolerance to heavy metals than those from uncontaminated environments [41]. The obtained results indicate that this strain of bacteria very quickly adapts to the presence of silver cations in the environment when the culture is at the stationary growth phase. However, when silver cations were added to bacteria at the logarithmic phase of growth the rate of uptake was significantly reduced. It suggests that cells in this phase of growth are very susceptible to the toxic effect of silver.

Furthermore, we observed several changes in MALDI-TOF MS spectra recorded for silver-treated *L. lactis* in comparison to a reference spectrum of a native strain. We registered increased intensity of some ions and additional signals for samples containing silver. These alterations are an obvious result of the presence of a stressing agent i.e. silver cations. They are a well-known bactericide and their activity against multidrug-resistant, highly pathogenic or probiotic bacteria has been extensively described [42, 43]. The statement regarding stressing influence of silver on bacteria is supported by a recent work of Chudobova et al. who investigated the effect of addition of metal ions (including Ag^+) on acquired MALDI-TOF MS spectra of bacterial proteome [44]. They studied modifications in *Staphylococcus aureus* protein composition and registered additional peaks in the obtained spectra suggesting metabolic changes of heavy metal resistant *S. aureus*. These signals were connected with changes in protein morphology due to the metal influence. Such alterations in the metabolic profile of bacteria were also observed in the study of Gopal et al. This group proposed a mechanism for interaction of five nanoparticles (of Ag, NiO, Pt, TiO_2 and ZnO) with *S. aureus* and *Pseudomonas aeruginosa*. Significant changes in the number and intensity of signals on MALDI-MS spectra of both bacterial strains were reported after a treatment with AgNPs, and in comparison with a control sample of native bacteria [45]. Antibacterial properties and influence of graphene-based silver nanoparticles (AgNPs—GE) on *P. aeruginosa* were investigated in the work of He et al. These researchers performed MALDI—TOF/TOF MS analysis to find changes in proteomic profile induced by AgNPs—GE and $AgNO_3$. Untreated bacteria served as a control. They found the total of 28 proteins which were altered in expression following the exposure of *P. aeruginosa* to AgNPs—GE and $AgNO_3$ [46]. This is another confirmation of the thesis that addition of silver influences metabolic and proteomic profiles of bacteria, which can be monitored using MALDI-TOF MS technique.

Further analyses of LAB-Ag

FT-IR spectroscopy is as an advanced tool for characterization of molecular composition which allows characterization of conformationally distinct structures in biological molecules [47]. S1 Fig shows a broad absorption peak formed after addition of silver, which is assigned to amide II and amide I vibrations of β-pleated sheet structures of proteins [48, 49]. The disappearing peak at 1456 cm^{-1} in OPTION 2 may indicate C-H deformation of >CH_2 group in lipid proteins [47–49]. Two bands at 1059 and 1236 cm^{-1} in silver-treated *L. lactis* resulted from the presence of C-O-C and C-O groups dominated by ring vibrations in various polysaccharides and components of proteins contributing to amide III band, respectively [47, 50]. Hence, it is possible that carboxyl groups of proteins are involved in the process of biosorption of silver. As can be seen in S2 Fig, addition of Ag to *L. casei* caused formation of a broad peak in the region of 1530–1630 cm^{-1}, which could be attributed to amide II vibrations [48, 49].

S3(A) Fig demonstrates that regardless of the used buffers and the applied pH values the size of bacterial cell of *L. lactis* was registered in the range of 1000–5000 nm. Broadening of the base of the peak and a new signal observed in the range of 5000–10000 nm (S3(B) Fig) indicate formation of aggregates of bacterial cells resulting from the presence of electrostatic interactions on the surface or attachment of an additional molecule [51, 52]. This is directly connected with the obtained results from the zeta potential measurements illustrated in Fig 6(B) for LAB-Ag. Bacteria suspended in NaCl had a more negative potential value as compared to bacteria suspended in KCl. Hence, bacteria suspended in NaCl showed greater dispersion stability. This phenomenon is caused by repulsive forces between solvated biocolloids of bacterial cells generated by sodium ions [51, 53]. Potassium ions have a greater ionic radius than sodium ions, which reduces the shielding effect between biocolloids [54]. When analyzing influence of pH of the system a sharp decrease of zeta potential was noticed for pH 2–4 which was a result of progressive deprotonation of carboxyl groups of bacterial structural components. For pH 5–8 a constant value of zeta potential could be observed which reflected the proteolytic balance [53, 55]. Bacteria after biosorption of Ag^+ show higher electric charge. In the case of NaCl, bacteria at pH 2–3 acquired charge of about +20 mV; for KCl this charge was +3 mV. The increased value of ζ potential indicated reduced dispersion stability and destabilization of the biocolloidal system. LAB-Ag demonstrated reduced stability of biocolloids as compared to native cells. This was a consequence and also an evidence of the biosorption processes of silver cations on the surface of bacterial cells [22, 53, 55, 56].

Ultrastructural TEM analysis of bacterial strains *L. lactis* and *L. casei* treated with $AgNO_3$ is portrayed in Fig 7. The changes ((B), (C), (D)) prove that the applied silver concentration had destructive influence on the ultrastructure of *L. lactis* cells and led to their death. In the case of *L. casei* ((F), (G), (H)), the observed increase in wall thickness can suggest a mechanism preventing penetration of silver cations into the cell and the site of immobilization. These processes are probably utilized to protect the fragile protoplast from the toxic effects of silver, and thus they can increase the ability of cell sorption. Thickening of the cell walls is a phenomenon that has already been observed in gram-positive bacterium *S. aureus* in response to the presence of vancomycin. It is suggested that this process prevents penetration of antibiotic molecules into cells [57]. The changes in ultrastructure of LAB cells suggest the uptake of silver cations and no reduction of silver cations at the concentration of 1 ppm, which is illustrated in Fig 3.

Fig 8 demonstrates viability of *L. lactis* and *L. casei* after treatment with $AgNO_3$ (1 ppm). At the stationary phase of growth (OPTION 1) *L. lactis* appeared to be more resistant to the applied concentration of silver than *L. casei*, which suggest adaptation to this specific environment. Silver cations added to *L. lactis* and *L. casei* cells, which were at the logarithmic growth phase, induced a toxic effect—the number of dead cells was significantly higher than the one of live cells. It can be found in the literature that *L. lactis* was tested in terms of its ability to adapt to different conditions, such as heat shock, osmotic stress, low pH, starvation [58], and other stressing factors [59–62]. These alterations often changed the physiology of bacteria, hence it is possible that the presence of silver influences molecular and metabolic processes of the tested *L. lactis*, as portrayed in Fig 8.

Fig 9 shows a profile of volatile organic compounds for *L. lactis* obtained using HS-SPME-GC/MS. This technique is widely used for extraction of VOCs from different matrices, including headspace of bacterial strains. For example, it was employed to absorb the volatile compounds from headspace of *Helicobacter pylori* [63], *Clostridium difficile* [64], and many others [65, 66]. Volatile organic compounds were extracted by analyzing food-related products such as milk [67], cheese [68], kefir-like beverages [69] and bushera (Ugandan traditional fermented beverage) [70] inoculated with *L. lactis* and other bacterial strains. In the headspace of *L. lactis* we found volatile compounds originating from natural metabolism of bacteria. VOCs such as acetone, butane, ethanol, acetaldehyde and 3-methylbutanal were detected in numerous studies using HS-SPME-GC/MS technique [71–74]. Hence, our methodology seems to be appropriate for observation of changes of bacterial metabolism.

CONCLUSIONS

Milk and its products are sources of LAB resistant to the presence of silver below 100 ppm and capable of its biotransformation and assimilation. We found that two investigated lactic acid bacteria, *L. lactis* and *L. casei*, had the ability to sorb silver cations which was most likely associated with the cell surface of bacterial strain evaluated using TEM examinations. In our study we demonstrated changes in spectra obtained from MALDI-TOF MS profiling of whole bacterial cells when using different concentrations of silver. ICP-MS analyses proved that *L. casei* could grow in the environment of silver cations and that these bacteria could bind more of this element if it was added at the time of inoculation. Such dependence was not observed in the case of *L. lactis*. The bacteria bound nearly the same amount of silver during inoculation and upon addition after 24 hours. FT-IR spectroscopy and ζ potential measurements showed that several groups, mostly carboxyl groups, were involved in binding and biotransformation of silver cations. Also profiling of volatile organic compounds performed for *L. lactis* demonstrated a variety of compounds in bacterial headspace which were typical products of bacterial metabolism. Finally, it can be stated that the employed combination of various instrumental methods provided new possibilities for interpretation of physicochemical phenomena occurring at the interface of bacteria—bacteria and bacteria—metal—bacteria.

ACKNOWLEDGMENTS

The authors are thankful to Dr. J. Rudnicka for her technical assistance in GC/MS analyses. This work was supported by Symfonia I grant No. 2013/08/W/N28/00701, Maestro-6 No. 2014/14/A/ST4/00641 (2015–2018), Preludium grant No.: 2014/15/N/ST4/03702 from the National Science Centre, Poland. The work was also financially supported by the National Science Centre, Poland in the frame of the Opus 11 project No. 2016/21/B/ST4/02130 (2016–2019).

AUTHOR CONTRIBUTIONS

1. **Conceptualization:** BB MM PP VRP KR.
2. **Funding acquisition:** BB TL MM.
3. **Investigation:** MM PP VRP KR.

4. **Methodology:** TL MM PP VRP KR.
5. **Project administration:** BB TL MM.
6. **Resources:** PP VRP KR.
7. **Software:** MM VRP.
8. **Supervision:** BB TL.
9. **Validation:** MM.
10. **Visualization:** PP VRP KR.
11. **Writing – original draft:** MM VRP.
12. **Writing – review & editing:** BB TL MM PP VRP KR.

REFERENCES

1. Pfeiler EA, Klaenhammer TR. The genomics of lactic acid bacteria. Trends Microbiol. 2007; 15(12): 546–553. pmid:18024129
2. Khalid K. An overview of lactic acid bacteria. Int J Biosci. 2011; 1(3): 1–13.
3. Leroy F, De Vuyst L. Lactic acid bacteria as functional starter cultures for the food fermentation industry. Trends Food Sci Tech. 2004; 15(2): 67–78.
4. Ljungh A, Wadström T. Lactic acid bacteria as probiotics. Curr Issues Intest Microbiol. 2006; 7(2): 73–89. pmid:16875422
5. Giraffa G. Selection and design of lactic acid bacteria probiotic cultures. Eng Life Sci. 2012; 12(4): 391–398.
6. Douillard FP, de Vos WM. Functional genomics of lactic acid bacteria: from food to health. Microb Cell Fact. 2014; 13(Suppl 1): S8.
7. Monachese M, Burton JP, Reid G. Bioremediation and tolerance of humans to heavy metals through microbial processes: a potential role for probiotics? Appl Environ Microbiol. 2012; 78(18): 6397–6404. pmid:22798364
8. Mrvčić J, Stanzer D, Solić E, Stehlik-Tomas V. Interaction of lactic acid bacteria with metal ions: opportunities for improving food safety and quality. World J Microbiol Biotechnol. 2012; 28(9): 2771–2782. pmid:22806724
9. Bhakta JN, Ohnishi K, Munekage Y, Iwasaki K, Wei MQ. Characterization of lactic acid bacteria-based probiotics as potential heavy metal sorbents. J Appl Microbiol. 2012; 112(6): 1193–1206. pmid:22404232
10. Singh AL, Sarma PN. Removal of arsenic(III) from waste water using Lactobacillus acidophilus. Bioremediat J. 2010; 14(2): 92–97.
11. Halttunen T, Salminen S, Tahvonen R. Rapid removal of lead and cadmium from water by specific lactic acid bacteria. Int J Food Microbiol. 2007; 114(1): 30–35. pmid:17184867
12. Mishra R, Sinha V, Kannan A, Upreti RK. Reduction of Chromium-VI by Chromium Resistant Lactobacilli: A Prospective Bacterium for Bioremediation. Toxicol Int. 2012; 19(1): 25–30. pmid:22736899
13. Mosier AP, Behnke J, Jin ET, Cady NC. Microbial biofilms for the removal of Cu2+from CMP wastewater. J Environ Manage. 2015; 160: 67–72. pmid:26093466
14. Lin Z, Zhou C, Wu J, Zhou J, Wang L. A further insight into the mechanism of Ag+biosorption by Lactobacillus sp. strain A 09. Spectrochim Acta Part A. 2005; 61(6): 1195–1200.

15. Ahamed M, Alsalhi MS, Siddiqui MK. Silver nanoparticle applications and human health. Clin Chim Acta. 2010; 411(23–24): 1841–1848. pmid:20719239
16. Pomastowski P, Sprynskyy M, Žuvela P, Rafińska K, Milanowski M, Liu JJ, et al. Silver-Lactoferrin Nanocomplexes as a Potent Antimicrobial Agent. J Am Chem Soc. 2016; 138(25): 7899–7909. pmid:27263865
17. Chae YJ, Pham CH, Lee J, Bae E, Yi J, Gu MB. Evaluation of the toxic impact of silver nanoparticles on Japanese medaka (Oryzias latipes). Aquat Toxicol. 2009; 94(4): 320–327. pmid:19699002
18. Bondarenko O, Juganson K, Ivask A, Kasemets K, Mortimer M, Kahru A. Toxicity of Ag, CuO and ZnO nanoparticles to selected environmentally relevant test organisms and mammalian cells in vitro: a critical review. Arch Toxicol. 2013; 87(7): 1181–1200. pmid:23728526
19. Asztemborska M, Jakubiak M, Książyk M, Stęborowski R, Polkowska-Motrenko H, Bystrzejewska-Piotrowska G. Silver nanoparticle accumulation by aquatic organisms—neutron activation as a tool for the environmental fate of nanoparticles tracing. Nukleonika. 2014; 59(4): 169–173.
20. Taju G, Abdul Majeed S, Nambi KS, Sahul Hameed AS. In vitro assay for the toxicity of silver nanoparticles using heart and gill cell lines of Catla catla and gill cell line of Labeo rohita. Comp Biochem Physiol C Toxicol Pharmacol. 2014; 161(1): 41–52.
21. Khan MS, Jabeen F, Qureshi NA, Asghar MS, Shakeel M, Noureen A. Toxicity of silver nanoparticles in fish: a critical review. J Biodivers Environ Sci. 2015; 6(5): 211–227.
22. Sintubin L, De Windt W, Dick J, Mast J, van der Ha D, Verstraete W, et al. Lactic acid bacteria as reducing and capping agent for the fast and efficient production of silver nanoparticles. Appl Microbiol Biotechnol. 2009; 84(4): 741–749. pmid:19488750
23. Hrynkiewicz K, Złoch M, Kowalkowski T, Baum C, Niedojadło K, Buszewski B. Strain-specific bioaccumulation and intracellular distribution of Cd2+ in bacteria isolated from the rhizosphere, ectomycorrhizae, and fruitbodies of ectomycorrhizal fungi. Environ Sci Pollut Res. 2015; 22(4): 3055–3067.
24. Pomastowski P, Szultka M, Kupczyk W, Jackowski M, Buszewski B. Evaluation of Intact Cell Matrix-Assisted Laser Desorption/Ionization Time-of-Flight Mass Spectrometry for Capillary Electrophoresis Detection of Controlled Bacterial Clumping. J Anal Bioanal Tech. 2015; S13: 008.
25. Abdullah SA, Osman MM. Isolation and identification of lactic acid bacteria from raw cow milk, white cheese and Rob in Sudan. Pak J Nutr. 2010; 9(12): 1203–1206.
26. Teneva-Angelova T, Beshkova D. Non-traditional sources for isolation of lactic acid bacteria. Ann Microbiol. 2012; 66(1): 449–459.
27. Martín R, Heilig HG, Zoetendal EG, Jiménez E, Fernández L, Smidt H, et al. Cultivation-independent assessment of the bacterial diversity of breast milk among healthy women. Res Microbiol. 2007; 158(1): 31–37. pmid:17224259
28. Carminati D, Tidona F, Fornasari ME, Rossetti L, Meucci A, Giraffa G. Biotyping of cultivable lactic acid bacteria isolated from donkey milk. Lett Appl Microbiol. 2014; 59(3): 299–305. pmid:24749686
29. Nejati F, Babaei M, Taghi-Zadeh A. Characterisation of Lactobacillus helveticus strains isolated from home-made dairy products in Iran. Int J Dairy Technol. 2016; 69(1): 89–95.
30. Perin LM, Nero LA. Antagonistic lactic acid bacteria isolated from goat milk and identification of a novel nisin variant Lactococcus lactis. BMC Microbiol. 2014; 14(1): 36.
31. Davati N, Tabatabaee Yazdi F, Zibaee S, Shahidi F, Edalatian MR. Study of Lactic Acid Bacteria Community From Raw Milk of Iranian One Humped Camel and Evaluation of Their Probiotic Properties. Jundishapur J Microbiol. 2015; 8(5): e16750. pmid:26060561
32. Delgado S, Rachid CT, Fernández E, Rychlik T, Alegría Á, Peixoto RS, et al. Diversity of thermophilic bacteria in raw, pasteurized and selectively-cultured milk, as assessed by culturing, PCR-DGGE and pyrosequencing. Food Microbiol. 2013; 36(1): 103–111. pmid:23764225

33. von Neubeck M, Baur C, Krewinkel M, Stoeckel M, Kranz B, Stressler T, et al. Biodiversity of refrigerated raw milk microbiota and their enzymatic spoilage potential. Int J Food Microbiol. 2015; 211: 57–65. pmid:26173200

34. Dec M, Urban-Chmiel R, Gnat S, Puchalski A, Wernicki A. Identification of Lactobacillus strains of goose origin using MALDI-TOF mass spectrometry and 16S-23S rDNA intergenic spacer PCR analysis. Res Microbiol. 2014; 165(3): 190–201. pmid:24607713

35. Hong Y, Li J, Qin P, Lee SY, Kim HY. Predominant lactic acid bacteria in mukeunji, a long-term-aged kimchi, for different aging periods. Food Sci Biotechnol. 2015; 24(2): 545–550.

36. Gadd GM. Biosorption: critical review of scientific rationale, environmental importance and significance for pollution treatment. J Chem Technol Biotechnol. 2009; 84(1): 13–28.

37. Lesmana SO, Febriana N, Soetaredjo FE, Sunarso J, Ismadji S. Studies on potential applications of biomass for the separation of heavy metals from water and wastewater. Biochem Eng J. 2009; 44(1): 19–41.

38. Wang J, Chen C. Biosorbents for heavy metals removal and their future. Biotechnol Adv. 2009; 27(2): 195–226. pmid:19103274

39. Vijayaraghavan K, Yun YS. Bacterial biosorbents and biosorption. Biotechnol Adv. 2008; 26(3): 266–291. pmid:18353595

40. Zhao Y, Wang D, Xie H, Won SW, Cui L, Wu G. Adsorption of Ag (I) from aqueous solution by waste yeast: kinetic, equilibrium and mechanism studies. Bioprocess Biosyst Eng. 2015; 38(1): 69–77. pmid:24996651

41. Kim SJ. Effect of heavy metals on natural populations of bacteria from surface microlayers and subsurface water. Mar Ecol Prog Ser. 1985; 26: 203–206.

42. Kim JS, Kuk E, Yu KN, Kim JH, Park SJ, Lee HJ, et al. Antimicrobial effects of silver nanoparticles. Nanomedicine. 2007; 3(1): 95–101. pmid:17379174

43. Mikiciuk J, Mikiciuk E, Wrońska A, Szterk A. Antimicrobial potential of commercial silver nanoparticles and the characterization of their physical properties toward probiotic bacteria isolated from fermented milk products. J Environ Sci Health B. 2016; 51(4): 222–229. pmid:26766747

44. Chudobova D, Dostalova S, Ruttkay-Nedecky B, Guran R, Rodrigo MA, Tmejova K, et al. The effect of metal ions on Staphylococcus aureus revealed by biochemical and mass spectrometric analyses. Microbiol Res. 2015; 170: 147–156. pmid:25189671

45. Gopal J, Manikandan M, Hasan N, Lee CH, Wu HF. A comparative study on the mode of interaction of different nanoparticles during MALDI-MS of bacterial cells. J Mass Spectrom. 2013; 48(1): 119–127. pmid:23303755

46. He T, Liu H, Zhou Y, Yang J, Cheng X, Shi H. Antibacterial effect and proteomic analysis of graphene-based silver nanoparticles on a pathogenic bacterium Pseudomonas aeruginosa. Biometals. 2014; 27(4): 673–682. pmid:24961696

47. Beekes M, Lasch P, Naumann D. Analytical applications of Fourier transform-infrared (FT-IR) spectroscopy in microbiology and prion research. Vet Microbiol. 2007; 123(4): 305–319. pmid:17540519

48. Davis R, Mauer LJ. Fourier transform infrared (FT-IR) spectroscopy: A rapid tool for detection and analysis of foodborne pathogenic bacteria. In: Mendez-Vilas A. Current research, technology and education topics in Applied Microbiology and Microbial Biotechnology Volume II. Formatex Research Center; 2010. pp. 1582–1594.

49. Alvarez-Ordóñez A, Mouwen DJ, López M, Prieto M. Fourier transform infrared spectroscopy as a tool to characterize molecular composition and stress response in foodborne pathogenic bacteria. J Microbiol Methods. 2011; 84(3): 369–378. pmid:21256893

50. Santos MI, Gerbino E, Tymczyszyn E, Gomez-Zavaglia A. Applications of Infrared and Raman Spectroscopies to Probiotic Investigation. Foods. 2015; 4(3): 283–305. pmid:28231205

51. Zhang Y, Yang M, Portney NG, Cui D, Budak G, Ozbay E, et al. Zeta potential: a surface electrical characteristic to probe the interaction of nanoparticles with normal and cancer human breast epithelial cells. Biomed Microdevices. 2008; 10(2): 321–328. pmid:18165903
52. Kłodzińska E, Szumski M, Dziubakiewicz E, Hrynkiewicz K, Skwarek E, Janusz W, et al. Effect of zeta potential value on bacterial behavior during electrophoretic separation. Electrophoresis. 2010; 31(9): 1590–1596. pmid:20422634
53. Railean-Plugaru V, Pomastowski P, Wypij M, Szultka-Mlynska M, Rafinska K, Golinska P, et al. Study of silver nanoparticles synthesized by acidophilic strain of Actinobacteria isolated from the of Picea sitchensis forest soil. J Appl Microbiol. 2016; 120(5): 1250–1263. pmid:26864807
54. Zhang C, Shen L, Yin H, Wang A, Yan X. Catalytic performances of potassium and sodium hydroxides/carbonates and calcium and magnesium oxides on hydrolysis of α-chloropropionic acid to lactic acid. Reac Kinet Mech Cat. 2014; 113(1): 201–210.
55. Buszewski B, Railean-Plugaru V, Pomastowski P, Rafińska K, Szultka-Mlynska M, Golinska P, et al. Antimicrobial activity of biosilver nanoparticles produced by a novel Streptacidiphilus durhamensis strain. J Microbiol Immunol Infect. 2016; pii: S1684-1182(16)30023-8.
56. Pomastowski P, Sprynskyy M, Buszewski B. The study of zinc ions binding to casein. Colloids Surf B Biointerfaces. 2014; 120: 21–27. pmid:24905676
57. Cui L, Ma X, Sato K, Okuma K, Tenover FC, Mamizuka EM, et al. Cell wall thickening is a common feature of vancomycin resistance in Staphylococcus aureus. J Clin Microbiol. 2003; 41(1): 5–14. pmid:12517819
58. Sanders JW, Venema G, Kok J. Environmental stress responses in Lactococcus lactis. FEMS Microbiol Rev. 1999; 23(4): 483–501.
59. Raynaud S, Perrin R, Cocaign-Bousquet M, Loubiere P. Metabolic and transcriptomic adaptation of Lactococcus lactis subsp. lactis Biovar diacetylactis in response to autoacidification and temperature downshift in skim milk. Appl Environ Microbiol. 2005; 71(12): 8016–8023. pmid:16332781
60. Chen J, Shen J, Solem C, Jensen PR. Oxidative stress at high temperatures in Lactococcus lactis due to an insufficient supply of Riboflavin. Appl Environ Microbiol. 2013; 79(19): 6140–6147. pmid:23913422
61. Chen J, Shen J, Ingvar Hellgren L, Ruhdal Jensen P, Solem C. Adaptation of Lactococcus lactis to high growth temperature leads to a dramatic increase in acidification rate. Sci Rep. 2015; 5: 14199. pmid:26388459
62. Ercan O, Wels M, Smid EJ, Kleerebezem M. Molecular and metabolic adaptations of Lactococcus lactis at near-zero growth rates. Appl Environ Microbiol. 2015; 81(1): 320–331. pmid:25344239
63. Ulanowska A, Kowalkowski T, Hrynkiewicz K, Jackowski M, Buszewski B. Determination of volatile organic compounds in human breath for Helicobacter pylori detection by SPME-GC/MS. Biomed Chromatogr. 2011; 25(3): 391–397. pmid:21321973
64. Tait E, Hill KA, Perry JD, Stanforth SP, Dean JR. Development of a novel method for detection of Clostridium difficile using HS-SPME-GC-MS. J Appl Microbiol. 2014; 116(4): 1010–1019. pmid:24320174
65. Arnold JW, Senter SD. Use of digital aroma technology and SPME GC-MS to compare volatile compounds produced by bacteria isolated from processed poultry. J Sci Food Agric. 1998; 78(3): 343–348.
66. Papaleo MC, Romoli R, Bartolucci G, Maida I, Perrin E, Fondi M, et al. Bioactive volatile organic compounds from Antarctic (sponges) bacteria. N Biotechnol. 2013; 30(6): 824–838. pmid:23619351
67. Alemayehu D, Hannon JA, McAuliffe O, Ross RP. Characterization of plant-derived lactococci on the basis of their volatile compounds profile when grown in milk. Int J Food Microbiol. 2014; 172: 57–61. pmid:24361833

68. Gaglio R, Scatassa ML, Cruciata M, Miraglia V, Corona O, Di Gerlando R, et al. In vivo application and dynamics of lactic acid bacteria for the four-season production of Vastedda-like cheese. Int J Food Microbiol. 2014; 177: 37–48. pmid:24598514
69. Nambou K, Gao C, Zhou F, Guo B, Ai L, Wu ZJ. A novel approach of direct formulation of defined starter cultures for different kefir-like beverage production. Int Dairy J. 2014; 34(2): 237–246.
70. Muyanja CM, Narvhus JA, Treimo J, Langsrud T. Isolation, characterisation and identification of lactic acid bacteria from bushera: a Ugandan traditional fermented beverage. Int J Food Microbiol. 2003; 80(3): 201–210. pmid:12494920
71. Buszewski B, Ulanowska A, Ligor T, Jackowski M, Kłodzińska E, Szeliga J. Identification of volatile organic compounds secreted from cancer tissues and bacterial cultures. J Chromatogr B. 2008; 868(1–2): 88–94.
72. Romoli R, Papaleo MC, de Pascale D, Tutino ML, Michaud L, LoGiudice A, et al. Characterization of the volatile profile of Antarctic bacteria by using solid-phase microextraction-gas chromatography-mass spectrometry. J Mass Spectrom. 2011; 46(10): 1051–1059. pmid:22012672
73. Tait E, Perry JD, Stanforth SP, Dean JR. Identification of volatile organic compounds produced by bacteria using HS-SPME-GC-MS. J Chromatogr Sci. 2014; 52(4): 363–373. pmid:23661670
74. Xu YY, Lu H, Wang X, Zhang KQ, Li GH. Effect of volatile organic compounds from bacteria on nematodes. Chem Biodivers. 2015; 12(9): 1415–1421. pmid:26363885

CHAPTER 13

Crystal Fingerprinting: Elucidating The Crystals of Cheddar, Parmigiano-Reggiano, Gouda, and Soft Washed-Rind Cheeses using Powder X-Ray Diffractometry

G. F. Tansman[1] & P. S. Kindstedt [1] & J. M. Hughes[2]

[1.]*Department of Nutrition and Food SciencesUniversity of VermontBurlingtonUSA*

[2.]*Department of GeologyUniversity of VermontBurlingtonUSA*

ABSTRACT

Crystals in cheese may be considered defects or positive features, depending on the variety and mode of production (industrial, artisanal). Powder x-ray diffractometry (PXRD) offers a simple means to identify and resolve complex combinations of crystals that contribute to cheese characteristics. The objective of the present research was to demonstrate the application of PXRD to study crystals from a range of different cheese types, specifically Cheddar, Parmigiano-Reggiano, Gouda, and soft washed-rind (smear ripened) cheeses. In studies of Parmigiano-Reggiano and long-aged Gouda, PXRD has confirmed that hard (crunchy) crystals that form abundantly within these cheeses consist of tyrosine. Furthermore, PXRD has tentatively identified the presence of an unusual form of crystalline leucine in large (up to 6 mm in diameter) spherical entities, or "pearls", that occur abundantly in 2-year-old Parmigiano Reggiano and long-aged Gouda cheeses, and on the surface of rindless hard Italian-type cheese. Ongoing investigations into the nature of these "pearls" are providing new insight into the roles that crystals play in the visual appearance and texture of long-aged cheeses. Crystals also sometimes develop profusely in the eyes of long-aged Gouda, which have been shown by PXRD to consist of tyrosine and the aforementioned presumptive form of crystalline leucine. Finally, crystals have been shown by PXRD to form in the smears of soft washed-rind cheeses. These crystals may be associated in some cheeses with gritty mouth feel and with zonal body softening that occurs during ripening. Heightened interest in artisanal cheeses highlights the need to better understand crystals and their contributions to cheese characteristics.

Keywords: X-ray diffraction Crystals Cheese Tyrosine Leucine

This paper is part of the Special issue dedicated to the 9th International Cheese Symposium held in Cork, Ireland and organized by Teagasc in collaboration with University College Cork and INRA, 12th & 13th November 2014.

1 INTRODUCTION

Crystals that form in cheese have been the subject of scientific inquiry for over a century, and x-ray diffraction (XRD) has been marshaled as a direct analytical approach to identify cheese crystals for almost as long, the earliest studies dating back to the 1930s (Tuckey and Ruehe 1938; Tuckey et al. 1938). Most XRD investigations of cheese crystals have been restricted to Cheddar cheese and have focused primarily on calcium lactate pentahydrate (CLP) crystals (Agarwal et al. 2006; Chou et al. 2003; Conchie et al. 1960; Dybing et al. 1988; Harper et al. 1953; Tuckey et al. 1938; Washam et al. 1985). A few studies of Cheddar also identified the occurrence of crystals consisting of tyrosine (Conchie et al. 1960; Harper et al. 1953; Shock et al. 1948), cysteine (Harper et al. 1953; Shock et al. 1948), and calcium phosphate (Conchie and Sutherland 1965). In the United States, visible crystals are often viewed by the Cheddar industry as defects and sources of confusion for consumers (Johnson 2014). Therefore, there has been considerable interest to understand the factors that promote crystal formation and develop measures to prevent their occurrence. In contrast, in some traditional cheeses such as Parmigiano-Reggiano and hand-crafted artisanal cheeses, crystals are viewed as important contributors to cheese character and quality (Noël et al. 1996; Zannoni et al. 1994). Growing consumer appreciation of traditional and artisanal cheeses is also fueling interest in understanding the nature and origins of crystals and their contributions to cheese quality and character.

Studies of crystals in cheeses other than Cheddar, such as tyrosine crystals in Roquefort (Dox 1911), CLP, calcium phosphate and tyrosine crystals in Grana cheeses (Bottazzi et al. 1982, 1994), calcium phosphate crystals in soft white mold surface-ripened (bloomy rind) cheeses (Boutrou et al. 1999; Brooker 1987; Gaucheron et al. 1999; Karahadrian and Lindsay 1987), and CLP and calcium phosphate crystals in Serra cheese (Parker et al. 1998), have generally employed indirect methods such as chemical analyses or microscopy coupled with differential staining and x-ray microanalysis to identify presumed crystalline species. Although indirect approaches may provide useful presumptive identification of crystalline species, only XRD can furnish direct and definitive results because the XRD pattern of a specific crystal is determined by its three-dimensional atomic arrangement, which is unique to that crystal and thus analogous to a fingerprint. Consequently, the XRD pattern of an unknown species can be compared to those of known crystals; a perfect match with one of the known patterns provides definitive identification. Today, a growing database of over 250,000 known XRD patterns is accessible through the International Centre for Diffraction Data (ICDD).

Powder x-ray diffractometry (PXRD), which utilizes crystalline specimens that ideally have been pulverized to a fine powder, is a versatile and user-friendly application of XRD. Major enhancements in the computing power of PXRD instrumentation, and advances in the software algorithms for data analysis, have opened up new opportunities for the study of cheese crystals. For example, the authors recently demonstrated that the two different enantiomeric forms of CLP that were presumed to occur in Cheddar cheese (i.e., the L(+) and D(−)/L(+) forms) can be identified and differentiated using PXRD, based on subtle differences in the XRD patterns of the two enantiomeric configurations (Tansman et al. 2014). The application of PXRD to study crystals in Cheddar and other cheese varieties could offer new insights into the factors that govern crystallization phenomena in cheese and the roles that crystals play in determining cheese character and quality. Therefore, the objective of the present research was to demonstrate the application of PXRD to study crystals from a range of different cheese types, specifically Cheddar, Parmigiano-Reggiano, Gouda, and soft washed-rind (smear ripened) cheeses.

2 MATERIALS AND METHODS

2.1 Cheese samples and crystal collection

All cheese samples were commercially produced and purchased from local retail outlets, and crystals were harvested for analysis as described below, except for Cheddar cheese and a hard Italian-style cheese. In the case of Cheddar, the manufacturer harvested large (up to ca. 5 mm) internal crystals from the body of a Cheddar cheese sample (aged 2 years before sale) and delivered the crystals to the University of Vermont for identification after the manufacturer received a consumer complaint about unknown objects in their cheese. Crystals from a hard Italian-style cheese (obtained courtesy of Mark Johnson, Center for Dairy Research, University of Wisconsin) were scraped from the cheese surface using a spatula. Small (ca. 1–2 mm) dense crystals embedded in the body of Parmigiano-Reggiano cheese samples (aged 2 years) and Gouda cheese samples (aged 2 years) were physically excised from the surrounding cheese matrix using a dissecting needle and tweezers. The isolated crystals were brushed free of adhering cheese matrix. Parmigiano-Reggiano samples also contained white spherical entities that ranged from barely visible to around 6 mm in diameter, which will be referred to as "pearls" in this publication. Pearls were separated from the surrounding cheese matrix by slicing the sample with a wire cutter into 1-cm sections and gently applying pressure to the cheese surrounding the pearl using thumb and index finger. Gentle pressure caused the surrounding matrix to fracture and separate from the pearls, enabling the pearls to be isolated and brushed free of adhering cheese matrix. Similar pearls, though smaller in size distribution, also were harvested from the Gouda samples. Gouda samples also displayed extensive crystal formation along the surfaces of internal eyes. Crystals were scraped free from the surfaces of eyes using a spatula. The six soft washed-rind (smear ripened) cheeses that were examined in this study were produced by different companies, four located in the United States and two in Italy.

2.2 Analytical methods

Pearls collected from Parmigiano-Reggiano samples were evaluated for size distribution by placing them on a vibrating stack of four sieves of successively smaller mesh size (5.6, 4.00, 3.35, and 2.00 mm) and counting the number of pearls retained by each sieve. The pearls were then pooled to form a composite sample, grated to fine particles in a blender, and analyzed in duplicate for total solids content by drying in a forced-draft oven at 100 °C for 24 h, fat content by a modified Babcock method (Atherton and Newlander 1977), and for soluble leucine (i.e., free amino acid) content by a modified gas chromatography–mass spectrometry method as follows. Thirteen milligrams of pearls was homogenized in 100 mL of deionized water adjusted to a pH of 1 with HCl. The suspension was vortexed for 10 min and centrifuged at 5000 rpm for an additional 5 min to remove particulate material from the solution. A 1-mL aliquot of the mixture was removed for analysis and to this was added 0.1 mL of 8.37 mM 555 d_3 leucine (Isotec 486825 Lot TPI633-SP). The sample was mixed with 1 mL of 50% acetic acid and passed through cation exchange resin (Biorad AG50W-x8 cation exchange resin 100–200 mesh). The resin was then rinsed three times with 2 mL of deionized water. Amino acids were eluted off the resin by the addition of 2 mL of 4 N ammonium hydroxide. The eluate was collected and evaporated under nitrogen gas. To the dry vial was added 200 μL of a 50:50 mixture of *N*-methyl-*N*-(*t*-butyldimethylsilyl) trifluoro-acetamide + 1% *t*-butyldimethylchlorosilane (MtBSTFA + 1% t-BDMCS)/acetonitrile. The solution was heated for 30 min in a heating block at 80 °C and then transferred to an autosampler vial. A 1-μL aliquot was injected into a ZB1 gas chromatographer (GC) column (30 m in length; 0.25 mm inner diameter with 0.25 μm film). The GC was run isothermally at 165 °C with a column flow rate of 1 mL helium per minute. The mass spectrometer (MS) was run in selected ion monitoring mode and monitored for a mass to charge of 200 *m/z* for leucine and 203 *m/z* for labeled leucine.

Pooled samples of the cheese matrix surrounding the pearls were similarly grated and analyzed for total solids and fat contents. Gouda samples that displayed extensive crystal formation along the surfaces of internal eyes were examined by stereoscopic microscopy before the crystals were harvested for analysis. Smear material was scraped from the surface of soft washed-rind (smear ripened) cheeses employing a spatula, using care not to remove the underlying cheese matrix along with the smear. Smear material was analyzed for selected mineral content by inductively coupled atomic emission spectrometry as follows. A scraping of surface smear from one Vermont washed-rind cheese was dried and defatted in acetone and analyzed for Ca, Mg, K, and P by ICP-OES (Optima 3000DV; Perkin Elmer Corp, Norwalk, CT, USA). Calibration standards were prepared according to the instrument manufacturer's suggested guidelines, to cover the range of concentrations in the sample set. Two-point calibrations (plus a Calibration Blank) were used for ICP analysis. Continuing Calibration Verification samples, prepared from an independent source, were used to check the calibration periodically.

In preparation for PXRD, dense internal crystals obtained from Cheddar, Parmigiano-Reggiano, and Gouda cheeses, as well as crystals scraped from the surface of the hard Italian-style cheese and from the eyes of Gouda, were dried and defatted in acetone, finely ground using a mortar and pestle, and transferred to a glass diffraction slide with a well. The powder in the slide well was leveled with the surface of the diffraction slide using a microscope slide. Diffraction patterns were generated from cheese crystals using a MiniFlex II powder x-ray diffractometer (Rigaku, The Woodlands, TX). Diffractograms were generated at a speed of 2° 2θ.min^{-1} between 5 and 50° 2θ. Diffraction patterns were compared to existing entries archived in the ICDD database to determine matches using a proprietary calculation called a "figure of merit". The figure of merit was calculated automatically using PDXL, the diffractometer software package (Rigaku, The Woodlands, TX). An acceptance threshold of 1.0 was set for the figure of merit. Diffraction patterns were also generated from finely grated samples of pearls, and samples of the surrounding cheese matrix, from Parmigiano-Reggiano and Gouda cheeses as described. Samples of smears scraped from the surfaces of soft washed-rind (smear ripened) cheeses were loaded directly onto the slide well and immediately analyzed without any additional sample preparation.

3 RESULTS AND DISCUSSION

3.1 Cheddar

Cheddar cheese crystals in this study were large internal crystals (Fig. 1), not surface crystals, which were of particular interest because their presence triggered a consumer complaint to the cheese manufacturer about possible foreign objects in the cheese. The x-ray diffraction pattern, shown in Fig. 2, identified the crystals as CLP in the D(−)/L(+) enantiomeric configuration based on different XRD patterns of the two enantiomeric forms reported previously (Tansman et al. 2014). CLP crystals at the surface of Cheddar cheese have been studied extensively. However, little work has been conducted on internal crystals, and only one previous report identified large internal CLP crystals as occurring in the D(−)/L(+) form (Tansman et al. 2014). In the present study, the large internal D(−)/L(+) CLP crystals suggest a microbial origin involving nonstarter lactic acid bacteria (NSLAB) because conventional starter cultures for Cheddar cheese are comprised of strains of *Lactococcus lactis* ssp. *lactis* and *cremoris*, which ferment lactose to L(+) lactate exclusively and which lack the capacity to convert L(+) lactic acid to the D(−) form. In contrast, various NSLAB, including several strains of *Pediococcus pentosaceus* (Thomas and Crow 1983) and *Lactobacillus curvatus* (Somers et al. 2001), are able to racemize L(+) lactate to D(−)

lactate and promote D(−)/L(+) CLP crystallization. Furthermore, racemizing NSLAB have been shown to produce biofilms on equipment surfaces, which can be a source of contamination leading to the formation of D(−)/L(+) CLP crystals in Cheddar (Somers et al. 2001). Thus, a possible strategy to reduce or prevent the occurrence of large internal D(−)/L(+) CLP crystals might involve identifying and eradicating the offending source of NSLAB contamination. Alternatively, D(−) lactate in Cheddar cheese could also be produced if *Lactobacillus helveticus*, which ferments lactose to D(−) lactate, were included in the starter culture, as is sometimes practiced in Cheddar manufacture to enhance flavor (Johnson 2014). In this scenario, corrective action might involve removing *Lb. helveticus* from the starter. In either case, determining the enantiomeric form of CLP by PXRD can provide useful diagnostic information.

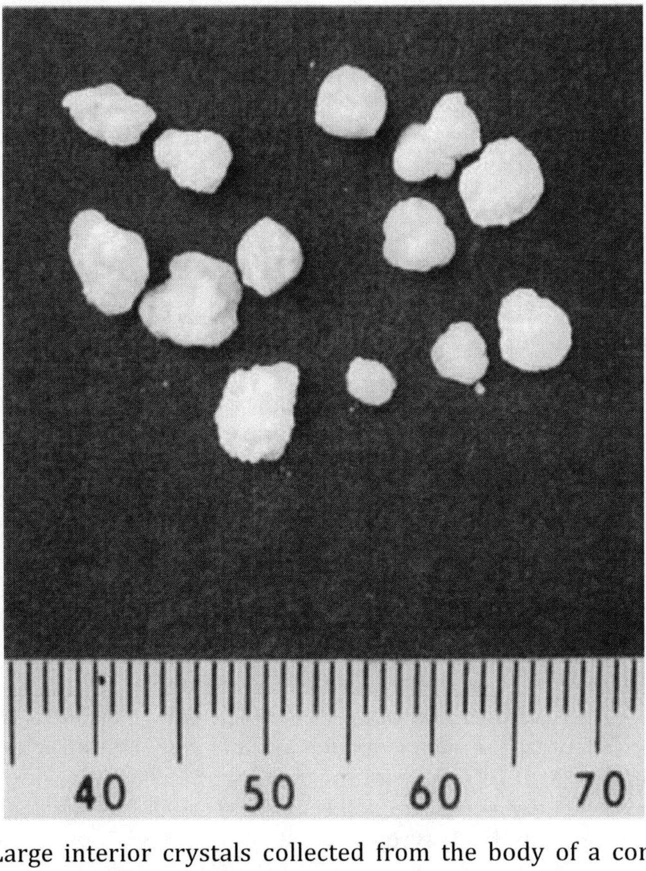

Fig. 1. Large interior crystals collected from the body of a commercial Cheddar cheese that was aged for 2 years before distribution for retail sale. The scale is in millimeters

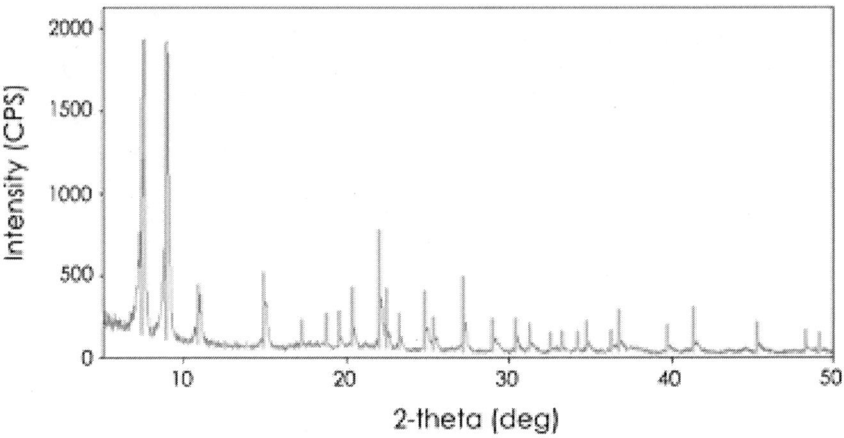

Fig. 2. X-ray diffraction pattern (*in red*) from large interior crystals collected from the body of a commercial Cheddar cheese that was aged for 2 years before distribution for retail sale (see Fig. 1). The *green bars* represent the reference card (ICDD card number 00-029-1596) labeled "calcium lactate pentahydrate"

3.2 **Parmigiano-Reggiano**

In traditional practice, Grana-type cheeses such as Parmigiano-Reggiano are pierced and pried open, which results in very rough and irregular internal cheese surfaces. In contrast, Parmigiano-Reggiano cheese for retail sale in the United States is sometimes cut into wedge-shaped slices with very smooth surfaces and then vacuum packaged in clear film. Smooth surfaces accentuate visible heterogeneities within the cheese body, such as small white crystals (Fig. 3, solid arrows) and spherical pearls that appear as pale white circles against the darker cheese matrix (Fig. 3, dashed arrows). Thus, these heterogeneous elements strongly influence the visual appearance of cheese samples packaged in this manner for the American market. Small white entities in Parmigiano-Reggiano (Bottazzi et al. 1982) and Grana Padano (Bianchi et al. 1974) cheese were presumptively identified as tyrosine crystals based on analyses of amino acid content, but studies using XRD to confirm the crystalline identities of such elements are lacking. The diffraction pattern generated by PXRD (Fig. 4) from the small white elements collected from Parmigiano-Reggiano (Fig. 3, solid arrows) confirms that they were tyrosine crystals, and only tyrosine crystals.

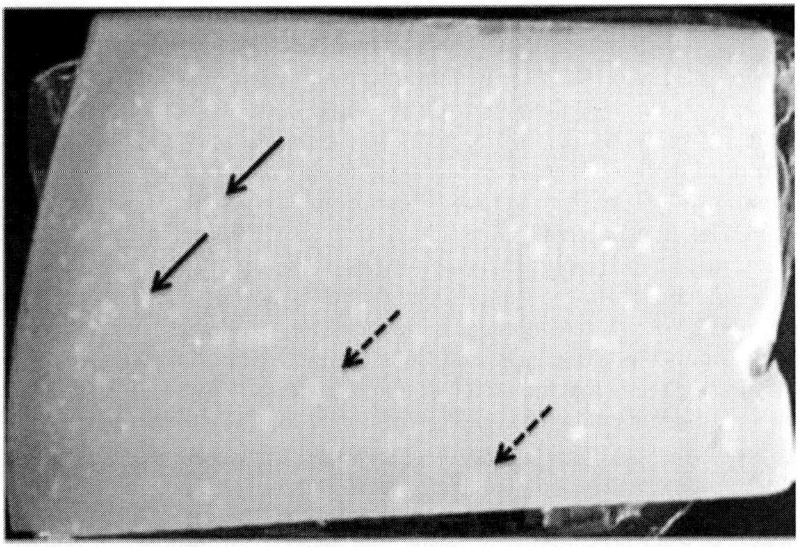

Fig. 3. Visual appearance of a wedge sample of commercially produced Parmigiano-Reggiano cheese that was aged for 2 years before retail distribution. Small white crystals (*solid arrows*) and pale white spherical "pearls" (*dashed arrows*) are noted

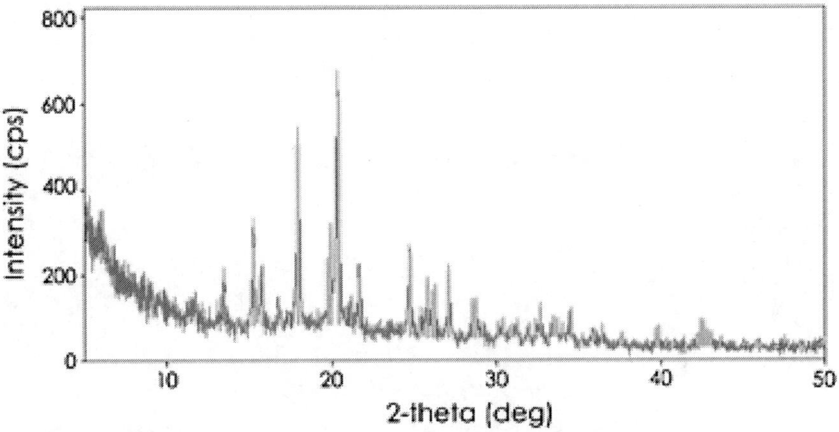

Fig. 4. X-ray diffraction pattern (*in red*) from small white interior crystals collected from the body of a commercial Parmigiano-Reggiano cheese that was aged for 2 years before distribution for retail sale (see Fig. 3). The *green bars* represent the reference card (ICDD card number 00-039-1840) labeled "tyrosine"

The striking abundance of visible pearls in the Parmigiano-Reggiano samples, which the authors have also observed in similarly packaged samples of Grana Padano, prompted closer examination. Pearls having diameters ≥2 mm, collected from a single cheese sample weighing ca. 680 g, are shown in Fig. 5. The pearls are grouped according to size based on retention by sieves of successive mesh sizes ranging from 5.60 to 2.00 mm. The pearls displayed remarkably consistent near spherical geometries irrespective of size (Fig. 5). Approximately 67 pearls per 100 g of cheese, which accounted for about 2% of the total weight of the cheese, were collected from the sample. The volume fraction of the cheese matrix occupied by pearls in this cheese appeared to be quite considerable. Given the abundance of the pearls, and their distinctly different body relative to that of the cheese matrix, which enabled the pearls to be easily harvested by hand, it seems likely that pearls influence the fracture properties of the cheese. The size distribution of pearls harvested from five different commercial samples of Parmigiano-Reggiano, all weighing about 680 g, is presented in Table 1. All five samples contained a few pearls greater than 5.6 mm in diameter, many more between 4 and 5.6 mm, and the greatest number between 2 and 4 mm. Assuming that all pearls grow at the same rate, the smallest pearls harvested were presumably the most recent to form during aging. Continued aging of the cheese, therefore, would be expected to result in nonlinear increases in the volume fraction occupied by pearls as the large number of smaller pearls increase in diameter. Such a bloom in pearl volume is likely to exert an increasingly greater effect on cheese fracture properties as aging progresses.

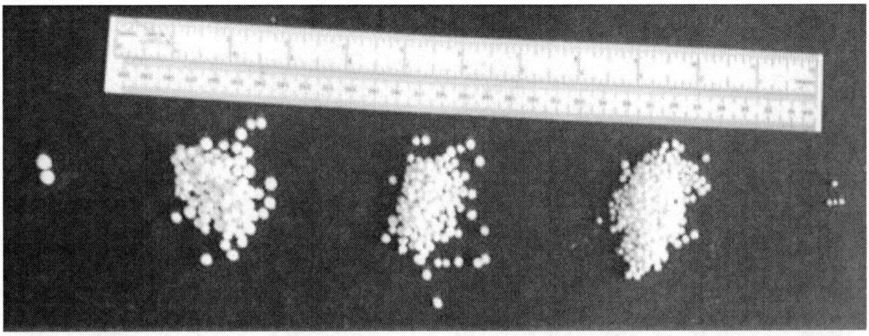

Fig. 5. "Pearls" having diameters ≥2 mm, which were collected from a single sample of Parmigiano-Reggiano cheese weighing ca. 680 g. The pearls are grouped according to retention on sieves of successive mesh size. The ruler is 300 mm long

Table 1. Average number of "pearls" collected from five different commercial samples of Parmigiano-Reggiano cheese

Sieve mesh size	Mean[a]	Standard deviation[a]
5.60	7	8
4.00	127	46
2.00	321	84

Each cheese sample weighed approximately 680 g. Means and standard deviations correspond to the numbers of pearls retained on stacked sieves of decreasing mesh size from 5.60 to 2.00 mm

[a] $n=5$

The total solids and fat contents of composite samples of the pearls and surrounding cheese matrix are shown in Table 2. The total solids content was 8.72% higher in the pearls than in the cheese matrix, but the fat content was 4% lower, indicating that the higher solids content of the pearls was not caused simply by greater loss of moisture from the pearls relative to losses from the surrounding matrix during aging. Instead, the accumulation of solids other than fat contributed to higher total solids, and lower fat content, in the pearls relative to the matrix; presumably, this increase in solids was associated with the growth in scaffolding that enabled the pearls to expand spherically, and which contributed to the firmer texture of the pearls relative to the surrounding matrix. The high fat content of the pearls (28%) indicates that pearls' scaffolding engulfed and occluded fat globules as the pearls increased in diameter.

Table 2. Compositional analyses of a composite sample of "pearls", and the cheese matrix surrounding the pearls, collected from five different commercial samples of Parmigiano-Reggiano cheese

	Pearls	Cheese matrix
Total solids (%)	79.09	70.37
Fat (%)	28	32
FDM (%)	35.4	45.5
Free leucine (%)	1.0	–

The chemical makeup of the pearl scaffolding and its structural organization are unknown but may be related to crystallization phenomena. For example, Bianchi et al. (1974) observed large (up to 6 mm) entities in Grana Padano cheese, apparently similar to the pearls observed in the present study, which they referred to as "spots". Spots collected from cheeses aged for 18 and 25 months contained around 9.9 and 7.7% leucine, respectively (Bianchi et al. 1974). Such extraordinarily high leucine levels suggest a possible structural role of crystalline leucine in the Grana Padano spots. However, the composite sample of pearls from Parmigiano-Reggiano in the present study contained only 1.0% leucine present as free amino acids. Currently, the authors have no explanation for this discrepancy, but it is worth noting that analysis of the Parmigiano-Reggiano pearls by PXRD gave a diffraction pattern, shown in Fig. 6, that corresponded to a partial diffraction pattern for crystalline leucine, which is characterized by a very strong peak at 6.00° 2θ. In contrast, a composite sample of the cheese matrix surrounding the pearls did not diffract x-rays (data not shown). Sou et al. (2013) recently observed a similar partial diffraction pattern for leucine, which they postulated corresponds to an alternative crystalline form of leucine. Further studies will be needed to determine whether this presumptive alternative leucine crystal contributes to the structural scaffolding of pearls in Parmigiano-Reggiano and Grana Padano cheeses. However, it is interesting to note that crystals collected from the surface of the hard Italian-style cheese, shown in Fig. 7, also yielded the same partial diffraction pattern for leucine (data not shown), indicating that the same presumptive crystals can occur in both pearls located internally in cheese and at the surface of cheese when the right conditions are present.

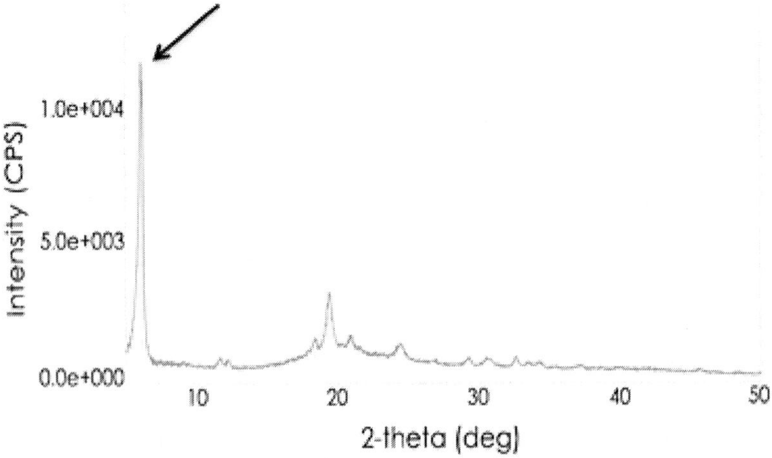

Fig. 6. X-ray diffraction pattern from a composite sample of interior "pearls" collected from five commercial Parmigiano-Reggiano cheese samples that were aged for 2 years before distribution for retail sale. The pattern displays a prominent peak at 6.00° 2θ (see *arrow*)

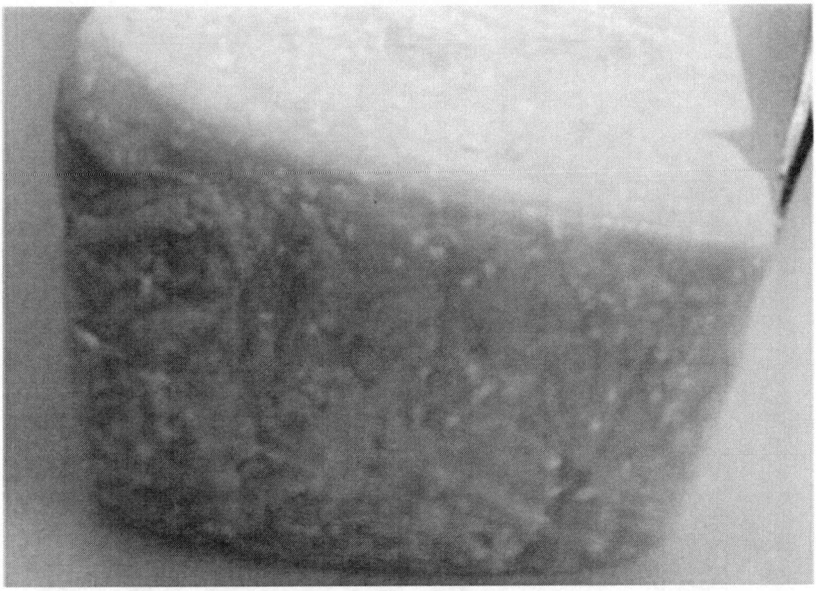

Fig. 7. Surface of a hard Italian-style cheese from the United States displaying extensive crystal coverage

3.3 **Gouda**

Like Parmigiano-Reggiano, the 2-year aged Gouda samples in this study displayed visible heterogeneities within the cheese body, including small white crystals (Fig. 8a, solid arrows) and spherical pearls that appear as pale white circles against the darker cheese matrix (Fig. 8a, dashed arrows). Profuse crystal blooms also were observed in the eyes of the cheese (Fig. 8b, solid arrow). Analysis by PXRD confirmed that the small white crystals embedded in the cheese body consisted of tyrosine (data not shown), and that the pearls contained the presumptive alternative form of crystalline leucine characterized by an intense XRD peak at 6.00° 2θ (data not shown), similar to the pattern obtained with Parmigiano-Reggiano pearls (Fig. 6). This is perhaps surprising because the starter cultures used for Gouda traditionally do not include the thermophilic strains used in Grana type cheeses, although the inclusion of thermophilic adjuncts in Gouda manufacture is becoming more common (Düsterhöft et al. 2011). Again, further studies will be needed to determine whether the presumed crystalline leucine contributes to the structural scaffolding of pearls.

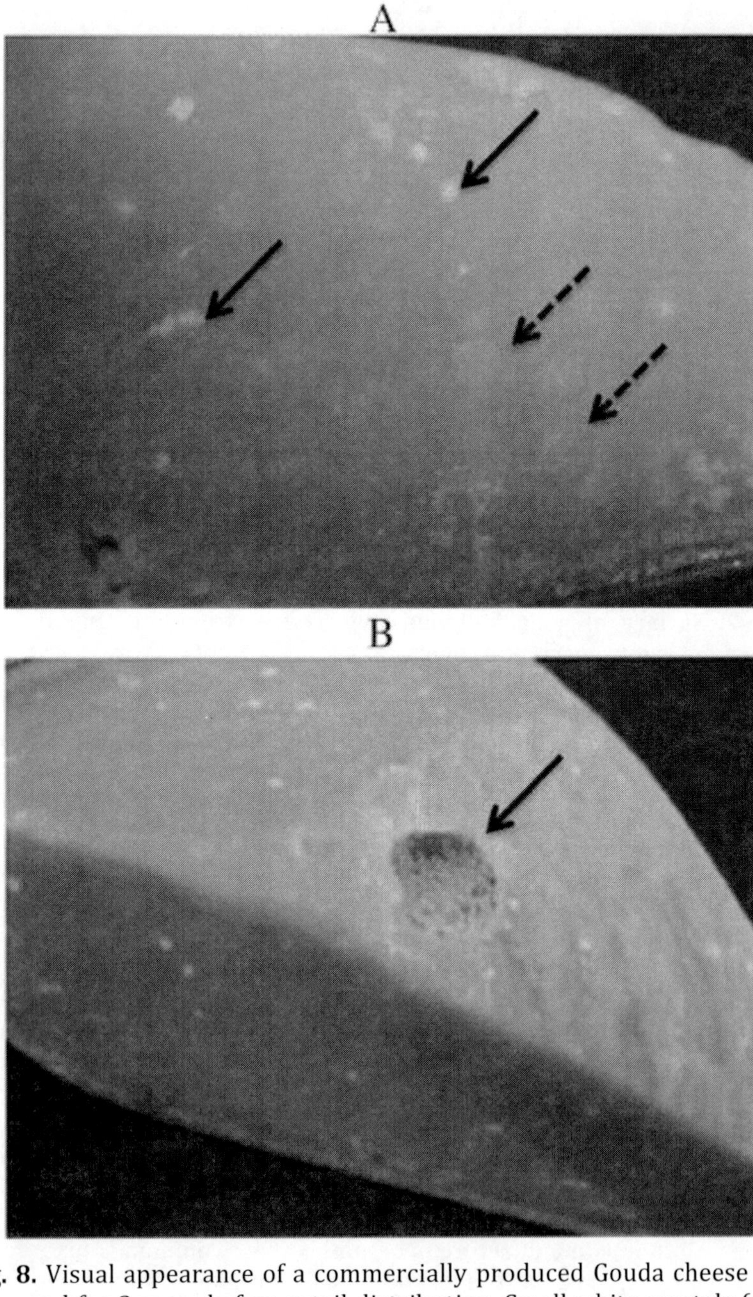

Fig. 8. Visual appearance of a commercially produced Gouda cheese that was aged for 2 years before retail distribution. Small white crystals (*solid arrows*, **A**) and pale white spherical “pearls” (*dashed arrows*, **A**) are noted. Also, eyes within the cheese were lined with crystals (*solid arrow*, **B**)

When the crystals that lined the eyes of Gouda were observed using stereophase contrast microscopy, the visual appearance suggested the presence of two different crystal morphologies, one characterized by an open structure (Fig. 9, solid arrows) and the other by a compact structure (Fig. 9, dashed arrows). The diffraction pattern of the crystals generated by PXRD confirmed the presence of two different crystalline species: tyrosine crystals plus leucine in the presumed alternative configuration, as indicated by the intense peak at 6.00° 2θ (Fig. 10). Thus, the same presumptive crystals can occur in both pearls located internally in cheese and on a cheese surface, in this case the surface of eyes.

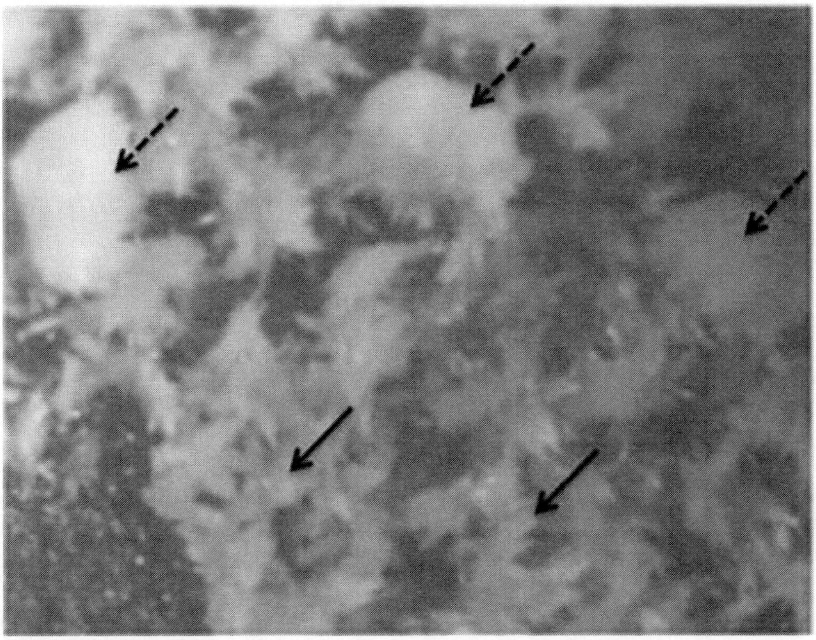

Fig. 9. Phase contrast photomicrograph (×11.25 magnification) of crystals lining the surface of an eye in a commercially produced Gouda cheese that was aged for 2 years before retail distribution (see Fig. 8b). Two different crystal morphologies, one characterized by an open structure (*solid arrows*) and the other by a compact structure (*dashed arrows*), appear to be present

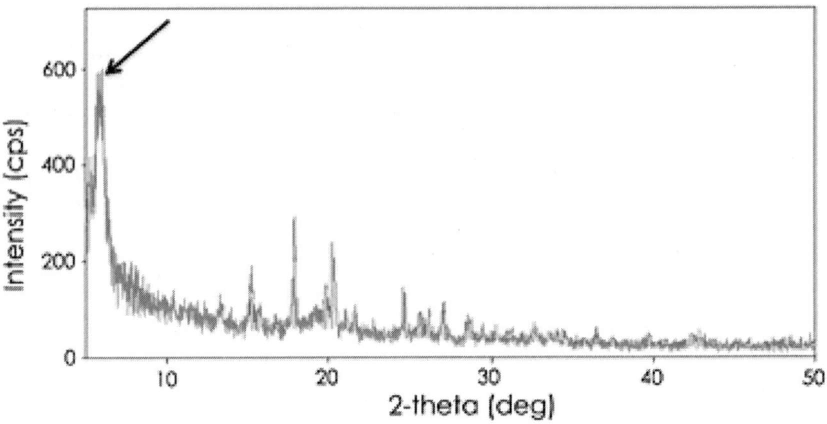

Fig. 10. X-ray diffraction pattern (*in red*) from crystals collected from the surfaces of eyes in a commercial Gouda cheese sample that was aged for 2 years before distribution for retail sale (see Fig. 8b). The *green bars* represent the reference card (ICDD card number 00-039-1840) labeled "tyrosine". Also evident is a prominent peak at 6.00° 2θ (see *arrow*)

3.4 **Soft washed-rind (smear ripened) varieties**

The mineral content of a surface smear collected from one of the soft washed-rind (smear ripened) cheeses is shown in Table 3. Calcium and P were present in the smear at concentrations around ten times higher than that of other minerals measured, which suggests that calcium and phosphate ions migrated from the cheese body to the surface in a manner similar to what is believed to occur in bloomy rind cheeses, i.e., as a result of high surface pH that develops during ripening, which triggers surface crystallization of calcium phosphate (Boutrou et al. 1999; Brooker 1987; Gaucheron et al. 1999; Karahadrian and Lindsay 1987). Therefore, the authors expected to find one or more forms of crystalline calcium phosphate in the smears of the six soft washed-rind cheeses that were analyzed by PXRD. However, smears from all six cheeses each produced a unique XRD pattern that could not be matched to any pattern in the ICDD database. Diffraction of x-rays indicates the presence of crystals in a sample, thus the XRD patterns confirmed that crystalline species were present in the smears of all six cheeses but none could be identified. A possible explanation for the unknown XRD patterns became evident during the course of the analyses when the authors noted the samples dried out extensively during the diffraction procedure, dehydrating the extant phases. This strongly suggested that the PXRD results contained diffraction artifacts of the experimental conditions. Consequently, the preparation of smear samples for PXRD was modified to include the addition of a thin layer mineral oil over the sample well to prevent sample desiccation. This modification produced reproducible data, thereby preempting artifacts obtained previously. When the modified sample preparation procedure was applied to smear material from two of the washed-rind cheeses analyzed previously, the XRD pattern for ikaite, which consists of calcium carbonate in an unusual hexahydrate form, was identified in the smears (Fig. 11). Additional analyses of other washed-rind cheeses are underway and will be presented in a subsequent report.

Table 3. Concentrations (g.kg^{-1}) of selected minerals in dried and defatted smear material collected from the surface of a soft washed-rind (smear ripened) cheese

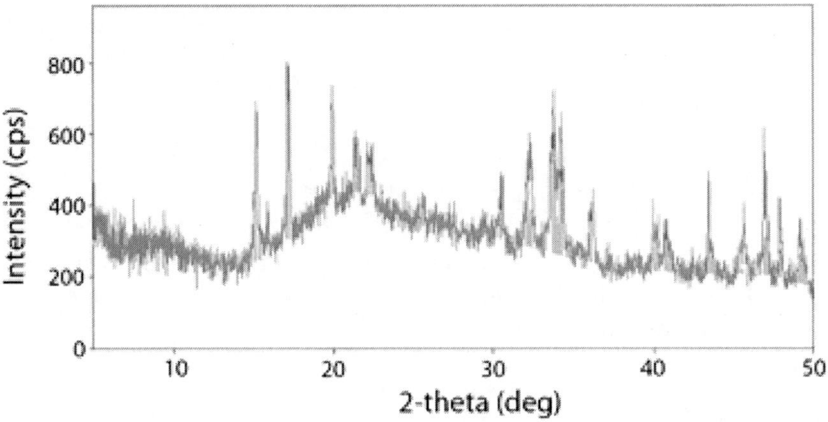

Fig. 11. X-ray diffraction pattern (*in red*) from smear material collected from surface of a commercial soft washed-rind cheese. The *green bars* represent the reference card (ICDD card number 01-074-7174) labeled "ikaite"

The formation of surface crystals, whether of calcium phosphate or other species, is of great interest because of the possible implications for zonal softening, which has been studied extensively in bloomy rind cheeses but not well characterized in soft washed-rind cheeses. Surface crystal formation also has implications for the development of sandiness or grittiness, which the authors have heard anecdotally from cheesemakers and cheesemongers may be considered desirable or undesirable depending on the market.

4 CONCLUSION

Powder x-ray diffractometry is a versatile and powerful analytical approach to identify and differentiate crystalline species in cheese. Crystals influence the quality and character of many cheeses, ranging from extra hard, long-aged types, such as grana cheeses, to soft, rapidly ripened cheeses such as surface ripened washed-rind and bloomy-rind types. This study confirmed the presence of tyrosine crystals and a crystalline species presumptively identified as leucine in both Parmigiano-Reggiano and Gouda cheeses, as well as large interior crystals of D(−)/L(+) calcium lactate pentahydrate in Cheddar cheese and ikaite crystals at the surface of a soft washed-rind cheese. Advancements in PXRD technology have facilitated new avenues of research relating to structure-function implications of crystallization phenomena, such as the role of leucine crystals in the formation and growth of pearls in Parmigiano-Reggiano and Gouda cheeses, and the possible role of surface crystals in the zonal softening of soft surface ripened cheeses.

ACKNOWLEDGMENTS

This study was funded by the United States Department of Agriculture Hatch Project VT-H01905. Purchase of the x-ray diffractometer was funded by National Science Foundation Grant EAR-0922961.

CONFLICT OF INTEREST

Author Tansman, author Kindstedt, and author Hughes declare that they have no conflict of interest.

REFERENCES

1. Agarwal S, Sharma K, Swanson BG, Yüksel GÜ, Clark S (2006) Nonstarter lactic acid bacteria biofilms and calcium lactate crystals in Cheddar cheese. J Dairy Sci 89:1452–1466
2. Atherton HV, Newlander JA (1977) Chemistry and testing of dairy products, 4th ed. Avi Pub Co, Westport, CT
3. Bianchi A, Beretta G, Caserio G, Giolitti G (1974) Amino acid composition of granules and spots in Grana Padano cheeses. J Dairy Sci 57:1504–1508

4. Bottazzi V, Battistotto B, Bianchi F (1982) The microscopic crystalline inclusions in Grana cheese and their x-ray microanalysis. Milchwissenschaft Milk Sci Internat 37(10):577–580
5. Bottazzi V, Lucchini F, Rebecchi A, Scolari GL (1994) I cristalli del formaggio grana (crystals present in Grana cheese). Scienza e Tecnica Lattiero-Casearia 45(1):7–14
6. Boutrou R, Gaucheron F, Piot M, Michel F, Maubois J-L, Léonil J (1999) Changes in the composition of juice expressed from Camembert cheese during ripening. Lait 79:503–513
7. Brooker BE (1987) The crystallization of calcium phosphate at the surface of mould-ripened cheeses. Food Microstruct 6:25–33
8. Chou Y-E, Edwards CG, Luedecke LO, Bates MP, Clark S (2003) Nonstarter lactic acid bacteria and aging temperature affect calcium lactate crystallization in Cheddar cheese. J Dairy Sci 86:2516–2524
9. Conchie J, Sutherland BJ (1965) The nature of steaminess in Cheddar cheese. J Dairy Res 32:35–44
10. Conchie A, Lawrence J, Czulak J, Cole WF (1960) Aust J Dairy Technol 15:120
11. Dox AW (1911) The occurrence of tyrosine crystals in Roquefort cheese. J Amer Chem Soc 33:423–425
12. Düsterhöft EM, Engels W, van den Berg G (2011) Dutch-type cheeses. In Encylopedia of dairy sciences, 2nd edn. Academic Press/Elsevier, London
13. Dybing ST, Wiegand JA, Brudvig SA, Huang EA, Chandan RC (1988) Effect on processing variables on the formation of calcium lactate crystals on Cheddar cheese. J Dairy Sci 71:1701–1710
14. Gaucheron F, Le Graët Y, Michel F, Briard V, Piot M (1999) Evolution of various salt concentrations in the moisture and in the outer layer and centre of a model cheese during its brining and storage in an ammoniacal atmosphere. Lait 79:553–566
15. Harper WJ, Swanson AM, Sommer HH (1953) Observations on the chemical composition of white particles in several lots of Cheddar cheese. J Dairy Sci 36:368–372
16. Johnson M (2014) Crystallization in cheese. Dairy Pipeline, Wisconsin Center for Dairy Research, 26(3): 1–5
17. Karahadrian C, Lindsay RC (1987) Integrated roles of lactate, ammonia, and calcium in texture development of mod surface-ripened cheese. J Dairy Sci 70:909–918
18. Noël Y, Zannoni M, Hunter EA (1996) Texture of Parmigiano Reggiano cheese: statistical relationship between rheological and sensory variates. Lait 76:243–254
19. Parker ML, Gunning PA, Macedo AC, Malcata FX, Brocklehurst TF (1998) The microstructure and distribution of micro-organisms within mature Serra cheese. J Appl Microbiol 84:523–530
20. Shock AA, Harper WJ, Swanson AM, Sommer HH (1948) What's in those "white specks" on Cheddar. Wisconsin Agric Experiment Station Bull 474:31–32
21. Somers EB, Johnson ME, Wong ACL (2001) Biofilm formation and contamination of cheese by nonstarter lactic acid bacteria in the dairy environment. J Dairy Sci 84:1926–1936
22. Sou T, Kaminskas LM, Nguyen T-H, Calberg R, McIntosh MP, Morton DAV (2013) The effect of amino acid excipients on morphology and solid-state properties of multi-component spray dried formulations for pulmonary delivery of biomacromolecules. Eur J Pharm Biopharm 83:234–243
23. Tansman G, Kindstedt PS, Hughes JM (2014) Powder x-ray diffraction can differentiate between enantiomeric variants of calcium lactate pentahydrate crystal in cheese. J Dairy Sci 97:7354–7362
24. Thomas TD, Crow VL (1983) Mechanism of D(−)-lactic acid formation in Cheddar cheese. N Z J Dairy Sci Technol 18:131–141
25. Tuckey SL, Ruehe HA (1938) An x-ray diffraction analysis of Cheddar cheese. J Dairy Sci 21:777–789
26. Tuckey SL, Ruehe HA, Clark GL (1938) X-ray diffraction analysis of white specks in Cheddar cheese. J Dairy Sci 21:161

27. Washam CJ, Kerr TJ, Hurst VJ, Rigsby WE (1985) A scanning electron microscopy study of crystalline structures in commercial cheese. Dev Ind Microbiol 26:749–761
28. Zannoni M, Bertozzi L, Hunter EA (1994) Comparison of Parmigiano-Reggiano and American Parmesan cheeses by sensory analysis of texture. Scienza E Technica Lattiero-Casearia 45(6):505–518

CHAPTER 14

DairyMGT: A Suite of Decision Support Systems in Dairy Farm Management

Victor E. Cabrera[1]

[1] Department of Dairy Science, University of Wisconsin-Madison, U.S.A.

1. INTRODUCTION

Dairy farming is a highly dynamic and integrated production system that requires continuous and intense decision-making. Several dairy farm components that include 1) cattle, 2) crops, 3) soils, 4) weather, 5) management, 6) economics, and 7) environment are extremely interrelated [1]. These components and their sub-components dynamically affect and are affected among them. Therefore, an efficient decision support system (DSS) framework within an integrated systems approach is critical for successful dairy farming management and decision-making [2-5].

This chapter describes the development, application, and adoption of a suite of more than 30 computerized DSS or decision support tools aimed to assist dairy farm managers and dairy farm advisors to improve their continuous decision-making and problem solving abilities. These DSS emerged in response of dairy farm managers' needs and were shaped with their input and feedback [6-7]. No single or special methodology was used to develop each or all of these DSS, but instead a combination and adaptation of methods and empirical techniques with the overarching goal that these DSS were: 1) highly user-friendly, 2) farm and user specific, 3) grounded on the best scientific information available, 4) remaining relevant throughout time, and 5) providing fast, concrete, and simple answer to complex farmers' questions [2, 8-11]. After all, these DSS became innovative tools converting expert information into useful and farm-specific management decisions taking advantage of latest software and computer technologies.

All the DSS object of this chapter are hosted at http://DairyMGT.info, *Tools* section and are categorized within dairy farming management and decision making such as: 1) nutrition and feeding, 2) reproductive efficiency, 3) heifer management and cow replacement, 4) production and productivity, 5) price risk management and financial analysis, and 6) environmental stewardship. Depending on the complexity, the specific purpose, and the requirements of dairy farm decision makers, some DSS are completely online applications, others are Macromedia Flash tools, others are Spreadsheets, and others are self-extractable and installable programs.

This chapter discusses the challenges on the development of these DSS with respect to the trade-offs among user-friendly design, computational detail, accuracy of calculations, and bottom line efficiency performance and effective decision-making. It portrays DSS development strategies, within the computational resources available, that succeeded in their primary objective of providing dairy farm mangers fast and reliable responses to perform efficient and effective decision-making.

The chapter reveals practical and real-life applications of a number of these DSS to demonstrate satisfactory system assessment, acceptable future predictability, adequate scenario evaluation, and, consequently, satisfactory decision-making.

The chapter also covers aspects of DSS dissemination and adoption evaluation, including the inception and development of a dedicated webpage; local, national and international usage, requested presentations, and academic publications.

The chapter also infers the possible role of emerging and evolving new technologies such as smart phones and tablets in the intersection of DSS, real-time applications, and mobile devices, which is a fast growing area of development within the dairy farming industry.

2. DESCRIPTION OF DAIRYMGT.INFO DECISION SUPPORT TOOLS

This section lists and describes the DSS object of this chapter. These DSS are categorized in main areas of dairy farm management, as they appear in the DairyMGT.info: *Tools* webpage.

2.1. Nutrition and Feeding (Dairymgt.Info → Tools→ Feeding)

Dairy farmers recognize that the largest item cost in a dairy farm system is feed, whether purchased or farm-grown. Obviously the major source of income in a dairy farm operation is the milk sale. Consequently, managing and optimizing the milk income over feed cost is a critical decision that affects not only economic sustainability, but also has large impacts regarding environmental stewardship[12]. Farmers also recognize that every farm is completely different and that market conditions are constantly changing. Therefore, beyond established farm feeding rations, there is a need for tools to permanently adjust strategic feeding decisions. Take as an example corn grain and its highly volatile price. Corn is a staple feed commodity for dairy farm feeding and consequently its price influences largely diet costs. With sudden corn price swings farmers confront permanently the question of re-considering the amount of corn in the diet. This question can be responded by estimating the marginal value of milk (also depending on highly volatile prices) to corn according to lactation stage and current amount of corn in the diet. The optimal use of corn would occur when the marginal value of milk equals the marginal value of corn, which at research-based feed efficiency levels [13], would solely depend on the ever-changing price relationship of milk and corn. The tool "*Corn Feeding Strategies*" shows these relationships in a graphical, dynamic, and interactive way so dairy farmers can optimize the amount of corn grain in each farm feeding group according to ever-changing market price conditions.

Take as another example the price of the main dairy cattle feed commodities and their relationship with milk price according to feed efficiency changes throughout lactation states. Research data indicate that the use of concentrates (i.e., corn, soybean meal) have a substantially higher impact on milk production during early or mid-lactation than in late lactation [14]. Under this premise, increased use of forages is justified in late lactation to maximize the overall milk income over feed cost, which however depends on ever-changing feed commodity prices. The tool "*Income Over Feed Cost*" graphs interactively the milk income over feed cost weekly for entire lactations and shows the impact of feed commodity prices on the dynamic milk income over feed cost value. Therefore, dairy farmers can fine-tune their feeding strategies to maximize their milk income over feed cost according to lactation states and feed prices swings.

Sometimes dairy farmers need additional help on formulating their diets to optimize feed concentrate supplementation. Research trails indicate that the optimal level of concentrate supplements in a diet could be achieved by using milk production response to crude protein (CP) and its components of rumen un-degradable protein (RUP), and rumen degradable protein (RDP), according to particular cow-group rations [15]. The tool "*Income over Feed Supplement Cost*" performs an optimization according to defined feed ingredients, prices, and CP (RUP, RDP) restrictions to maximize the net return. The tool helps dairy farm decision makers to select the most cost effective concentrate supplements in the diet, especially from the point of view of providing adequate amounts of RUP and RDP, which not only optimizes the net return, but also reduces the amount of nitrogen excretion and hence environmental impacts.

Dairy farmers also want to know what are the best-priced feed ingredient choices in the market. This information would drive farmer feed purchase decisions. The tool called "*FeedVal 2012*" is a dynamic and interactive matrix that finds the estimated price of a feed as an aggregated sum of its individual nutrients values according to the nutrient content and prices of a set of defined feed ingredients available in the market. The tool then compares the actual price of a feed ingredient with its calculated price. The result is a list of ingredients with their relative prices, indicating if an ingredient is a bargain or an expensive proposition.

Another critical factor in the quest for feed efficiency and maximum milk income over feed cost is the analysis of "benchmarking" with respect to feed efficiency, milk income, and feed costs [16]. Results from surveying dairy farm rations and farm prices reveals an impressive difference regarding to feed costs, feed consumption, and overall milk income over feed cost among otherwise similar dairy farms. A large and important opportunity exists then to improve the milk value net of the feed costs by comparing performance among farms. Therefore an online database structure and DSS was developed: "*Dairy Extension Feed Cost Evaluator*," Figure 1. This tool performs advanced benchmarking analyses for a group of users within a region, state, or country throughout a defined timeline by querying an online database, which is permanently being updated by the users. The tool allows users to "drill-down" the analysis and find out the driving factors for differences, an important step toward improving dairy farm feed efficiency and income over feed cost.

Dairy farmers also require some simpler evaluation tools for feed additives. The tool "*Optigen® Evaluator*" analyzes the economic value of including this slow release urea additive while maintaining diets at the same level of protein and dry matter intake. The tool "*Dairy Ration Feed Additive Break-Even Analysis*" determines any additive's additional milk production needed to justify its economic inclusion in the diet.

Figure 1. Screen snapshot of DSS Dairy Extension Feed Cost Evaluator.

Finally, regarding nutrition and diets, there is some evidence that dairy farmers might be over-feeding a large proportion of lactating cows when they feed the same diet ration to a large group of animals. Diets are normally formulated to provide enough nutrients to the most productive animals, which in turn gives extra nutrients to the less productive animals within the same group. Therefore, splitting lactating cows in smaller groups and offering group-specific feeding rations provide more precise nutrient requirements, increase herd's income over feed cost, and decrease nutrient excretion [17]. The tool "*Grouping Strategies for Feeding Lactating Dairy Cattle*" calculates dynamically individual cow nutrient requirements and optimizes cow grouping feeding strategies within particular farm constraints.

2.2. Reproductive Efficiency (Dairymgt.Info → Tools→ Reproduction)

Reproductive efficiency plays a critical role in the economics of dairy farming. However, assess the economic value of it is extremely difficult and complex [5]. A first step on understanding the economic impact of reproductive programs is to demonstrate the milk value net of feed cost dependent on the pregnancy time. The tool "*Exploring Pregnancy Timing Impact on Income over Feed Cost*" shows interactively and dynamically a cow's total milk income net of feed costs to a fixed lactation's pregnancy time and defined lactation curves. The tool illustratesand quantifies the economic value of having cows pregnant at the right time.

Dairy farmers are also required to do complex decisions regarding the best reproductive programs for the lactating herd population. New reproductive management strategies, whether they use hormonal synchronization technologies, heat detection methods, or a combination of both, are continuously and permanently evolving. Dairy farmers need not only to keep up-to-date with all these technologies, but also make the best decisions according to their own conditions [5]. Dairy farmers usually know which reproductive programs are more efficient from the reproductive point of view of getting more cows pregnant. Farmers also have a good handle on costs incurred according to reproductive programs. Nonetheless, dairy farmers have difficulty assessing the overall profitability of reproductive programs. Not surprisingly, they have long demanded for a systematic economic analysis to analyze reproductive programs. The tool "*UW-DairyRepro$Plus*" is a complex, still user-friendly, decision support systems that assess the economic value of farm-defined alternative reproductive programs for a particular farm according to prevalent market conditions. These tools allow farmers to be highly specific regarding their current or alternative reproductive programs. Besides reporting the most important reproductive parameters for each alternative program, the tools find the reproductive program with the best economic outcome and calculates the difference in net returns a farm would have when using alternative reproductive programs.

Sex-sorted semen that increases the chance of female offspring is a relatively new technology being widely adopted in the dairy industry. Farm-specific sexed semen's economic value and, moreover, when and how to use it, are critical. The tool "*Economic Value of Sexed Semen for Dairy Heifers*" (Figure 2) finds interactively the gain (or loss) of different reproductive program management strategies that include sexed semen compared with solely using conventional semen [18].

As important as to find out the value of specific-defined reproductive programs is to explore the value of improving the overall reproductive efficiency. The tool "*Dairy Reproductive Economic Analysis*" is a Markov-chain stochastic dynamic model packed in a simple to use online application. This tool integrates detailed parameters of pregnancy, abortion, and culling risks to perform iterations during 9 lactations until a herd reaches a steady state [19]. Then, the economic value of a reproductive program is determined by using predicted milk production curves, calve value, replacement costs, and other economic figures. The end result is a net return tied to a reproductive performance.

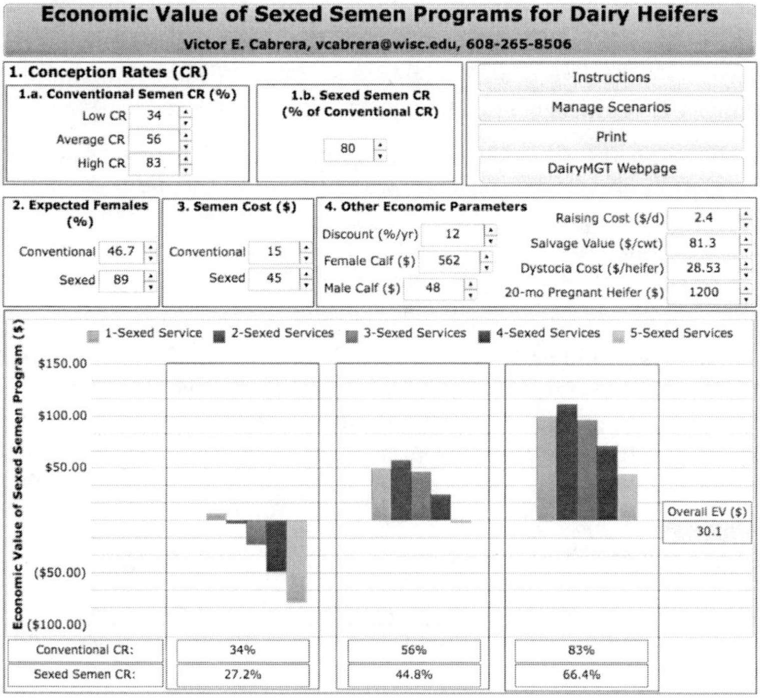

Figure 2. Screen snapshot of DSS Economic Value of Sexed Semen for Dairy Heifers.

2.3. Heifer Management And Cow Replacement (Dairymgt.Info → Tools→ Heifers / Replacement)

Whether farmers raise replacement heifers or not, they benefit from decisions related to this dairy farming enterprise. One first step on the economic decision about heifers is to determine the overall cost associated with rearing heifers according to estimated time to first calving. The tool "*Heifer Break-Even*" calculate the daily and accumulated cost for rearing heifers up to 12 months, 24 months, and beyond 24 months according to farm-defined prices for forages, corn, and soybean meal. Farmers use this tool to decide if to raise their own heifers, use custom-raising heifer services, or simply buy heifer replacements, according to market prices.

When farmers raise heifers on-farm, another decision comes along: to use or not to use accelerated feeding programs for boosting the early development of calves. The tool "*Cost-Benefit of Accelerated Feeding Programs*" gives dairy farmers the opportunity to compare hand-by-hand their current heifers' feeding program with an alternative accelerated feeding program within farm defined conditions. This tool shows economic differences at weaning and calving and calculates the amount of milk amount that would be needed to pay for heifer rearing costs.

In addition to the decisions of raising heifers and if to use accelerated feeding programs, dairy farmers want to know the number of heifers needed to maintain (or increase) the herd size according to farm long-term goals, reproductive efficiency, and heifers' culling rates. The tool "*Heifer Replacement*" calculates the number of replacement animals needed (springer heifers) responding to farm specific data inputs.

Dairy farmers would need to buy (or sell) springing heifers if the number of he replacements is fewer (or greater) than the required number to achieve the goal of maintaining or expand the herd size. Consequently, they need support on estimating the right price to pay (or to sell) springing heifers. The tool "*Value of a Springer*" performs a projection of the net return an animal would have under farm specific conditions. This value indicates the value of a replacement to break-even its costs. Because of the uncertainty in the milk price, milk production, and the productive lifetime, the model presents outcomes under different price and lifetime scenarios, so farmers can make decisions based on their assertion of the future prices and their risk preferences.

Furthermore, dairy farmers need to make critical decisions if to keep or replace a cow from the herd. The optimal decision will depend on which alternative would bring a greater net return in the future. The tool "*The Economic Value of a Dairy Cow*" (Figure 3) is a complex Markov-chain simulation model, still a user-friendly application that calculates interactively the economic value of a cow (or the value of each single cow in a herd) compared with its replacement [20]. Farmers use this value to make more informed decisions if to keep or replace cows. This tool, in addition, calculates the expected herd demographics and the average herd net return for better and additional dairy farm management and decision-making.

2.4. Production And Productivity (Dairymgt.Info → Tools→ Production)

Dairy farmers face several decisions regarding production-related issues. In order to make best decisions, they would like to know how their farm milk production profile compares to other similar farms. Besides milk amount produced per animal, the shape of the herd's lactation curves is critical to pinpoint management weaknesses and strengths of a particular farm. The tool "*Lactation Benchmark Curves for Wisconsin*" displays different parity lactation curves for different production levels herds obtained by processing 3.6 million lactation records. Dairy farmers can define their own lactation curves to assess their production performance compared with the benchmarked records. Similarly, farmers find great benefit of projecting their own lactation curves and compare specific dairy herd cows to the standards of the whole herd, which can be accomplished by using the tool "*Milk Curve Fitter,*" Figure 4.

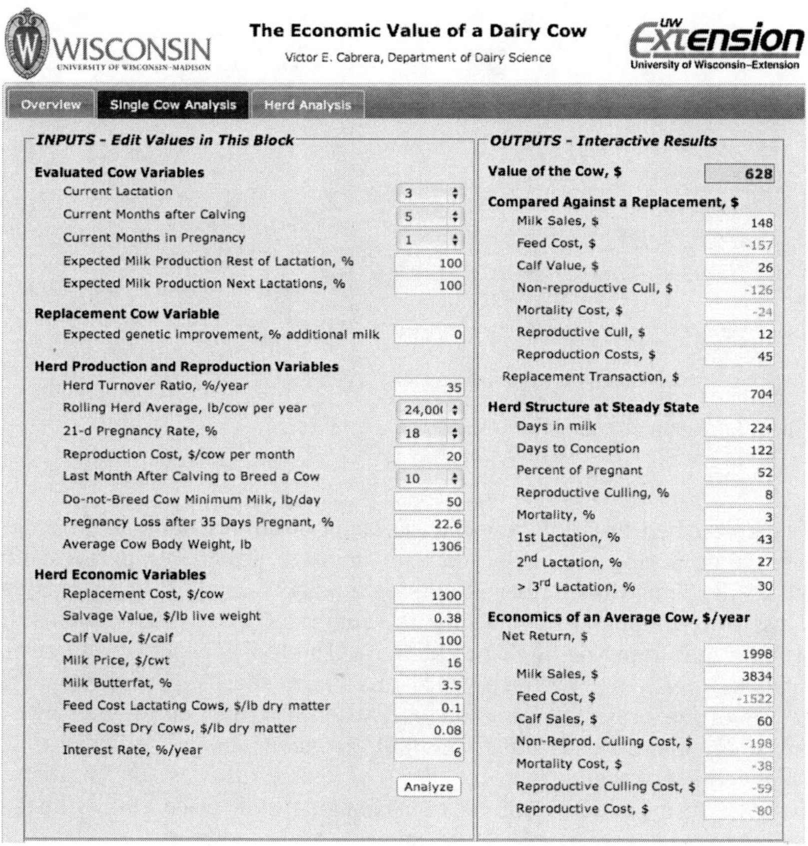

Figure 3. Screen snapshot of DSS *The Economic Value of a Dairy Cow.*

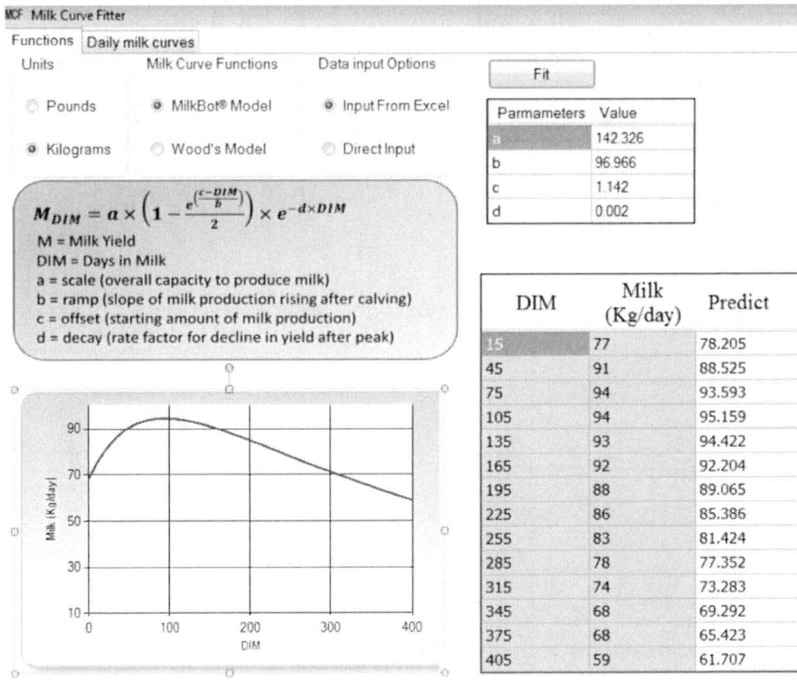

Figure 4. Screen snapshot of DSS *Milk Curve Fitter*.

As a result of benchmarking their herd's lactation curves, dairy farmers may contemplate a new set of decisions to improve productive performance such as switching the number of milking times per day [21] or re-consider the use of recombinant bovine somatotropin (rbST), a synthetic metabolic hormone that improves milk productivity. The tool "*Economic Analysis of Switching from 2X to 3X Milking*" performs a farm-specific partial budgeting analysis of the projected gain (or loss) when a farmer decides to milk 3 times a day instead of 2 times. The tool "*Economic Analysis of using rbST*" displays the economic gain (or loss) of using rbST as an interactive sensitivity analysis according to ever-changing milk price and estimated milk increase because of rbST under specific farm conditions.

Some dairy farmers are also interested in the possibility of either expand or modernize their farm facilities or increase their herd size. Therefore, they require support on important decisions that will drive the future of the dairy farm operation. The tool "*Decision Support System Program for Dairy Production and Expansion*" is a Spreadsheet application that allows dairy farmers to outline their current farm conditions regarding herd structure and market conditions, define a possible plan of expansion or modernization including required loans (for facilities and animals), and project the cash flow of the entire farm up to a period of 54 months in the future.

2.5. Price Risk Management And Financial Assessment (Dairymgt.Info → Tools→ Financial)

Unfavorable prices of milk and feed commodities together with increased price volatility create large uncertainty in the dairy farm business. Recent unprecedented uncertain times have prompted to re-visit farm's financial status and look for alternatives to stabilize net returns. It is critical to explore price risk management alternatives such as the relatively new revenue insurance program called Livestock Gross Margin for Dairy (LGM-Dairy) and to assess farm financial performance compared with peers [22].

In brief, the LGM-Dairy can protect the net margin (milk value less feed cost or milk income over feed cost) at a much lower cost than using comparable options in the future markets. The tool "*LGM-Analyzer*" (Figure 5) is an online, easy-to-use, suite of real-time, data intense, simulation, and optimization integrated modules to help on the decision of using LGM-Dairy. The *LGM-Analyzer* not only replicates the official premium calculation from the U.S. Department of Agriculture Risk Management Agency, but also is capable of perform historical sensitivity analysis as well as complex optimizations to minimize the premium cost at a level of target guaranteed income over feed cost. This suite of tools is also capable of comparing the LGM-Dairy with more traditional price risk management tools such as puts (Class III milk) and calls (corn and soybean meal) for feeds as bundled price options. The *LGM-Analyzer* connects live with the dairy and grain-based futures and market (through a structured query language) to determine the premium cost a particular farmer could expect according to a guarantee income over feed cost ("*Premium Estimator*"). Furthermore, a unique module ("*Least Cost Optimizer*") lets the user to minimize the LGM-Dairy premium cost at a defined level of income over feed cost insured. Other tools in the area of analysis of the LGM-Dairy include the "*LGM-Dairy Feed Equivalent*," a tool to covert feed diet ingredients to corn and soybean meal equivalents needed for a LGM-Dairy contract and the "*Net Guarantee Income over Feed Cost*," a tool to help dairy farmers determine the income over feed cost to break-even all other costs of production, which should be covered by using LGM-Dairy.

Also, performing a farm's financial benchmark assessment is critical in the process of measuring the financial health of a dairy farm. Moreover, this is usually required by lenders in order to consider loan applications. The "*Wisconsin Dairy Farm Benchmarking Tool*" is a database application that calculates 15 financial ratios including variables of liquidity, solvency, profitability, repayment capacity, and financial efficiency for a group of more than 500 Wisconsin dairy farms during a period of 10 years. The tool then compares each one of these ratios with those of a particular farm. Therefore, farmers can assess their financial health compared with their peers. Furthermore, the tool provides a *DuPont* analysis, in which a farm is compared against the population with respect to revenue and profit generated for every dollar invested. Another related tool, "*Working Capital Decision Support System*" assists dairy farmers in identifying cash flows, project expected incomes and expenses, and identifies cash excesses and shortfalls well in advance of their occurrence.

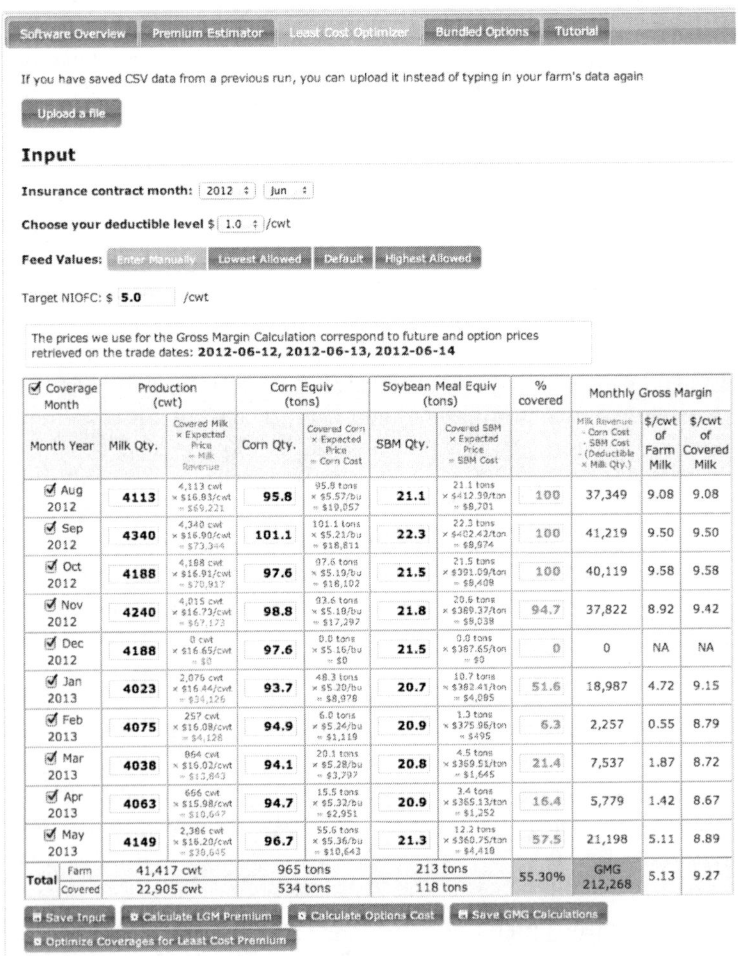

Figure 5. Screen snapshot of DSS *LGM-Analyzer*.

2.6. Environmental Stewardship (Dairymgt.Info → Tools→Environment)

The dairy farm business faces important challenges regarding increased environmental scrutiny. An increasingly important dairy farm management task is to maintain a farm nutrient balance and therefore avoid over-concentration of nutrients in or around the farm. Opportunities exist to better utilize nutrients in dairy farming and not only improve the balance of nutrients coming in and going out of the farm, but also decrease fertilizer expenses and therefore environmental concerns. Depending on the farm herd and crop characteristics, additional expenses might be required to comply with environmental regulations. In any case, an economic assessment along with the environmental requirements promotes better decision-making. A series of decision support tools deal with these sensitive aspects of dairy farming. The tool "*Dynamic Dairy Farm Model*" (Figure 6) is an integrated, whole-farm, simulation and optimization model that maximizes the net economic return while minimizing nitrogen leaching to surface and ground water sources. A simplified version of nutrient balance between nitrogen and phosphorus manure excretion for a fast assessment is the tool with name *"Dairy Nutrient Manager."* Also related, the "*Grazing-N*" is an application that balances nitrogen for dairy farms with grazing activities and the "*Seasonal Prediction of Manure Excretion,*" as its name says, helps dairy farmers project seasonally the amount of cow manure (and consequently nutrients in the manure) will be produced and will be needed to be recycled.

3. DECISION SUPPORT SYSTEMS DEVELOPMENT: CHALLENGES AND TRADE-OFFS

A number of methodologies and software applications were used to develop the decision support tools above described (Table 1). The goal always remained to provide solid, but user-friendly DSS tools. The methodology as well as the software application approach followed the tool development and the ultimate goal pursued and not vice versa. It was usual to combine and adapt methodologies within a particular tool development. Following is a succinct description of the most important methodologies used for the DairyMGT.info DSS tools and a discussion of the approaches used for the software applications.

Figure 6. Screen snapshot of DSS *Dynamic Dairy Farm Model.*

3.1. Methodologies Used For The Decision Support System Tools

3.1.1. Partial budgeting

Partial budgeting compares a current with an alternative technology by balancing the economics of 4 elements that are assessed before and after the adoption of the alternative technology: 1) additional returns (adds), 2) reduced costs (adds), 3) returns foregone (subtracts), and 4) additional costs (subtracts) [23]. Partial budgeting could be a robust methodology when a direct change is expected from the new technology without major interaction with other system components beyond the analyzed variables. Partial budgeting is the underline methodology to assess the break-even level of using feed additives, the economic benefit of milking 3 times a day, the economic evaluation of using rbST, the assessment of corn feeding strategies, and the assessment of income over feed cost by different diets under commodity price changes.

3.1.2. Cost Benefit

The cost benefit methodology is similar to partial budgeting but determines profitability of a new technology over longer periods of time and therefore requires the specification of a discount rate that is used to calculate a net present value [23]. The cost benefit is the underline methodology for calculating the value of adopting accelerated heifer liquid feeding programs and is as a supporting methodology to find out the economic value of sexed semen for dairy heifers, the value of reproductive programs in adult cows, and to assess the net present value of alternative scenarios of possible dairy farm expansion or modernization.

3.1.3. Decision Analysis

The decision analysis is appropriate when probabilistic distributions are important factors in determining the final outcomes [24] as it occurs when analyzing the value of using sexed semen on heifers, comparing the value of reproductive programs in adult cows, or projecting the replacement flow needed to maintain the herd size. In the first two cases, conditional probabilities were used to successively determine populations of pregnant, non-pregnant, and eligible to breed animals along with their respective expected monetary contributions. In the case of the replacement flow tool, transition probabilities are used to dynamically project the herd dynamics across time.

Table 1. Principal methodology and software application of DairyMGT.info decision support system tools.[1]Flash: Macromedia Flash. [2]Online tools use a combination of software including HTML, PHP, JavaScript, C, CSS, and MySQL. [3]Requires software installation in local machine.

Decision Support System Tool	Underline Methodology	Software Application
Feeding and Nutrition		
Corn Feeding Strategies	Partial Budgeting	Flash[1]
Income Over Feed Cost	Partial Budgeting	Flash
Income over Feed Supplement Cost	Linear Programming	Spreadsheet/ Online[2]

FeedVal 2012	Matrix Solution	Online
Dairy Extension Feed Cost Evaluator	Database Management	Online
Optigen® Evaluator	Matrix Solution	Online
Dairy Ration Feed Additive Break-Even Analysis	Partial Budgeting	Flash
Grouping Strategies for Feeding Lactating Dairy Cattle	Mathematical Simulation	Online
Reproductive Efficiency		
Exploring Pregnancy Timing Impact on Income over Feed Cost	Mathematical Simulation	Online
Economic Value of Sexed Semen for Dairy Heifers	Decision Analysis	Flash/Online
UW-DairyRepro$	Decision Analysis	Spreadsheet
Dairy Reproductive Economic Analysis	Markov Chains	Online
Heifer Management and Cow Replacement		
Heifer Break-Even	Enterprise Budgets	Online/Spreadsheet
Cost-Benefit of Accelerated Feeding Programs	Cost Benefit	Flash/Online
Heifer Replacement	Decision Analysis	Spreadsheet/Online
The Economic Value of a Dairy Cow	Markov Chains	Online/Spreadsheet
Production and Productivity		
LGM-Dairy Feed Equivalent	Matrix Solution	Online
LGM-Net Guarantee Income Over Feed Cost	Mathematical Simulation	Spreadsheet
Wisconsin Dairy Farm Benchmarking Tool	Database Management	Online/Spreadsheet
Working Capital Decision Support System	Enterprise Budgets	Spreadsheet
Environmental Stewardship		
Dynamic Dairy Farm Model	Markov Chains	Spreadsheet
Dairy Nutrient Manager	Mathematical Simulation	Spreadsheet
Grazing-N	Mathematical Simulation	Spreadsheet
Seasonal Prediction of Manure Excretion	Markov Chains	Spreadsheet

Milk Curve Fitter	Nonlinear Optimization	Installation[3]
Lactation Benchmark Curves for Wisconsin	Database Management	Flash/Spreadsheet
Economic Analysis of Switching from 2X to 3X Milking	Partial Budgeting	Flash
Economic Analysis of using rbST	Partial Budgeting	Flash
DSS Program for Dairy Production and Expansion	Markov Chains	Spreadsheet
Price Risk Management and Financial Assessment		
LGM-Analyzer	Mathematical Simulation	Online
LGM-Premium Estimator	Mathematical Simulation	Online
LGM-Least Cost Optimizer	Nonlinear Optimization	Online
LGM-Dairy Feed Equivalent	Matrix Solution	Online
LGM-Net Guarantee Income Over Feed Cost	Mathematical Simulation	Spreadsheet
Wisconsin Dairy Farm Benchmarking Tool	Database Management	Online/Spreadsheet
Working Capital Decision Support System	Enterprise Budgets	Spreadsheet
Environmental Stewardship		
Dynamic Dairy Farm Model	Markov Chains	Spreadsheet
Dairy Nutrient Manager	Mathematical Simulation	Spreadsheet
Grazing-N	Mathematical Simulation	Spreadsheet
Seasonal Prediction of Manure Excretion	Markov Chains	Spreadsheet

3.1.4. Enterprise Budgets

Enterprise budgets are a systematic way to list returns and costs and evaluate profits from inside a specific business enterprise [25] within the dairy farm. This methodology is used to calculatethe heifer break-even by contrasting heifers' rearing costs with potential benefits. This methodology is also used, in more detail, in the tool working capital to project the cash flow of a dairy farm enterprise.

3.1.5. Linear Programming

Linear programming is a mathematical optimization algorithm to maximize or minimize a goal (e.g., maximum profit or minimum costs) within a set of constraints represented as linear relationships [26]. Linear programming is at the core of the tool income over feed supplementation cost in determining the diet composition that results in the maximum net return within a set of constraints of available feed ingredients. Linear programming is also used recursively in the dynamic dairy farm model to maximize the farm net return while minimizing nitrogen leaching.

3.1.6. Markov Chains

Markov chains are a mathematical system that undergoes transitions from one state to the next within a finite space of states as random processes. In dairy farming, Markov chains are widely used for decision-making to predict herd demographics or to project cows' probabilistic life [2, 10, 12, 19-20]. Markov chains are also very useful to implement decision support tools, as these are less computationally demanding than alternative methods. Markov chains are therefore important part of the DairyMGT.info DSS tools and are the backbone structure of the tools: seasonal manure prediction, dynamic dairy farm model, reproductive economic analysis, and the economic value of a dairy cow. Markov chains are also important part of the tools dealing with expansion and modernization and the one comparing the value of different reproductive programs for adult cows.

3.1.7. Mathematical Simulation And Projection

Mathematical simulation and projection is a general description that encompass a group of diverse and integrated empirical techniques and algorithms that have as main goal to represent observed data as it happens in real-life situations when not a single method fits this condition to satisfaction. Mathematical simulation and projection is used in most of the DairyMGT tools. However, it is a core methodology in a group of them. For example, mathematical simulation is used in the grouping tool to calculate feed nutrient requirements for every single cow in a herd; in the timing of pregnancy tool to aggregate the overall milk production and feed consumption a cow will have depending on the time of pregnancy; and in all LGM related tools to generate thousands of replicates and calculate the statistics of net margins that will determine insurance premiums [27]. Also mathematical simulation and projection is important to predict cash flows within the expansion tool and to perform nutrient balances in tools such as dairy dynamic model, dairy nutrient manager, and grazing-N.

3.1.8. Nonlinear Optimization

Nonlinear optimization deals with finding an objective function of maximizing or minimizing a variable within a set of simultaneous constraints, where the objective function or some of the constraints have nonlinear relationships. Nonlinear optimization adds a set of complexity to the implementation of decision support tools because it is computational demanding. However, for some applications it is required. Since finding the global maxima for nonlinear problems it is not always possible, a compromise between finding a satisfactory answer and maintain the applications as user-friendly as possible is needed. Nonlinear optimization is used in the grouping, milk curve fitter, and LGM least cost optimizer tools. For the grouping tool, a nonlinear optimization algorithm groups lactating cows according to nutritional requirements with the objective function of finding the aggregated maximum income over feed cost through recursive iterations by allocating cows to size-defined groups. In the milk curve fitter tool, the user enters farm herd milk production and a nonlinear algorithm minimizes the residual difference between the farm observed data and the predicted data adjusted to a pre-defined milk lactation function such as Wood [28] or MilkBot [29]. The results are coefficients of the defined function that best represent farm-specific lactation curves. The LGM-least cost uses a nonlinear optimization to find out the minimum premium price to a defined target guarantee net income over feed cost according to future projected commodity prices and farm specific conditions, replicating the rules governing the insurance product. The result is the least cost premium for a determined level of coverage within the LGM-Dairy insurance structure [30-31].

3.1.9. Matrix Solution To Multiple Equations

Matrix or algebra simultaneous equation solution is helpful in the area of nutrition and feeding to replace feed ingredients and maintain same level nutritional of the diet and same level of feed intake. It is also useful to value feeds depending on their nutrients content. Each feed ingredient is defined in function of its nutrient contents and its market price. When the number of nutrients equals to the number of feed ingredients (same number of equations as unknowns) the result is an exact value for each nutrient and therefore the predicted value of a feed ingredient is equal to the input value as it is the case in the Optigen Evaluator tool [32]. Similar approach is used for the LGM-feed equivalent, which converts any feed ingredient into equivalents of corn and soybean meal, as it is required for LGM-Dairy insurance contracts. The tool FeedVal 2012 goes beyond and analyzes a set of user-defined matrix between 2 and 50 ingredients and between 2 and 13 nutrients to find out the difference between the feed ingredient market price and the estimated price based on the nutrient composition value of the ingredient.

3.1.10. Database Management And Analysis

Some tools require a database interface and some mechanism of querying the database to retrieve information and to perform analysis dynamically and efficiently. Databases are permanently being updated. Database tools are the lactation benchmark curves and the dairy farm ratio benchmarking. The user does not update these database applications directly, but a server manager. The user queries the database and is able to compare specific farm data with a set of filtered information within the databases. Other type of database application is the feed evaluator tool that registers users in the system and allows them to enter and save their data. The users update the database and the queries retrieve real-time information anytime. Users can then compare their own data against to a filtered group of other farms. A different concept is portrayed in all LGM related tools for which all the data (commodity prices of milk, corn, and soybean meal from the future markets) is retrieved real-time from the official sources anytime the user performs an analysis [29]. The calculation of either LGM premiums or least cost premiums changes depending not only on the user inputs, but also based upon the time of the query. The system saves historical information, so users can also do retrospective analyses.

3.1.11. External Simulation Models

Some tools require to be integrated with more complex, fully developed and established models. That is for example the case of the Dynamic Dairy Farm Model and the Grazing-N tools. In the first case, model requires assessments of crop production (corn, soybean, pastures, etc.), which are performed by using external crop simulation models from the family of Decision Support System for Agrotechnology Transfer [33]. The dynamic dairy farm model feeds the crop simulation model with data of soils, weather, and crop management schemes and the crop simulation models return predicted biomass produced, nutrient utilization, and nitrogen leaching from the soil. The Grazing-N application is integrated with the National Research Council model of nutrient requirement for dairy animals [34] according to a set of characteristics that include age, production, and live weight.

3.2. Software Applications

According to the type of application, the methodologies used in the tool, and, most importantly, the goal of the tool as a DSS, different software application approaches were used (Table 1). Most of the tools have been developed in different software applications with the objective of better meeting user styles and therefore capture larger audiences of users.

Spreadsheet applications are a very popular format among dairy farmers and consultants because of their familiarity with them, the possibility of using the same spreadsheet for further analyses, and the capacity of save and maintain a copy of it in a personal computer. Spreadsheet application was the elected method for a number of DairyMGT.info tools (Table 1). Most of the spreadsheet applications, however,required some type of Visual Basic code embedded into the application (macros).

Other group of tools uses Macromedia Flash as the software application. Macromedia Flash has the advantage of having a nice interactive visual interface connected with a calculator. From the point of view of the user, Flash tools are probably the easiest to use. They havethe additional advantage of becoming stand-alone applications and therefore of being used offline or embedded in Power Point presentations or Portable Document Format (PDF) files. One problem with Flash applications is, however, its limited computational functionality. Flash applications have only a set of limited mathematical functions without the possibility of using macros or combinethemwith code programming. Also Flash applications are not compatible with Apple smart phones and tablets. Current tools that are only Flash applications within the DairyMGT.info DSS tools will eventually be converted also to be online applications.

Other group of tools can be classified in the general category of online tools. These use an array of different software applications. What they all have in common is that these work in any web browser and eventually in any device and in any platform including smart phones, tablets, Apple, Linux, PC, etc. Calculations and analyses are normally performed in the DairyMGT.info web server, so the online tool is only an interface between the device of the user and the server. In general, online tools are very efficient and reliable tools that have the advantage to be always up-to-date: users always experience the latest version of the tool. Other important advantage is that complex processes and mathematical calculations can be managed using a combination of web code such as HTML (hyper text markup language), PHP (hypertext preprocessor), JavaScript (prototype-based scripting language), C (general-purpose language), CSS (style sheet language), MySQL (relational database management system), or others. Another advantage of online tools is that their design layout can be very efficient and solid once the tool is deployed. A drawback for developing online tools, however, is the need of expertise in web-based code writing. Nonetheless, online tools are very efficient and probably a trend to which many of the tools of DairyMGT.info will continue to gravitate.

4. ILLUSTRATION AND PRACTICAL DECISION-MAKING

4.1. Group Feeding

The value grouping feeding strategies was analyzed by applying the grouping tool to 30 dairy farms in Wisconsin. Test records were collected and adjusted to datasets consisting of cow identification, lactation, days after calving, milk production, and milk butterfat for each cow in each farm. The aim of this exercise was to demonstrate the value of grouping compared to no grouping without knowing studied farms' actual feeding strategies. Therefore, same procedure and assumptions were followed on each analyzed farm: 1) comparison of no grouping versus 3 same-size groups, 2) prices at $15.89/45.4 kg milk, $0.14337/0.454 kg CP, and $0.1174/4.19 mega joules (MJ) net energy, 3) average body weight of 500 kg for first lactation cows and 590 kg for cows in second and later lactations, 4) requirements of CP and net energy at the 83rd percentile level of the group (mean + 1 standard deviation), and 5) a cluster grouping criterion (grouping cows depending on their CP and net energy requirements for maintenance and milk production).

Evaluations clearly and consistently demonstrated that the income over feed cost (IOFC) in all analyzed farms was greater for the 3 feeding groups strategy than the no groping strategy (Table 2).

Table 2. Comparison of income over feed cost (IOFC) of no grouping versus 3 same-size feeding groups for Wisconsin dairy farms assessed by the tool: *Grouping Strategies for Feeding Lactating Dairy Cattle.*

	Number of lactating cows on analyzed farms (n = 30)	No grouping IOFC	3 same-size feeding groups IOFC	Additional IOFC of doing 3 same-size feeding groups
		----------$/cow per year----------		
Mean	788	2,311	2,707	396
Minimum	<200	697	1,059	161
Maximum	>1,000	2,967	3,285	580

The analysis indicated that farms could realize between $161 and $580/cow per year (mean = $396) of additional IOFC by switching from no grouping to 3 same-size feeding groups using the cluster criterion for grouping. These values represented an increase of between 7 and 52% of farm calculated IOFC. It was concluded then that grouping would have important economic implications in farm profitability and that further analysis should be done at farm-specific level and in a permanent basis by using the *Grouping Strategies for Feeding Lactating Dairy Cattle* DSS tool.

4.2. Sexed Semen

The *Economic Value of Sexed Semen for Dairy Heifers* tool was used for general conditions of Wisconsin dairy farms based on data of a sample of 309 dairy farms and 38 custom heifer growers, a survey performed by county extension agents [35]. At the time of the analysis, using the aggregated data of the 347 operations, the average economic benefit of using sexed semen, as calculated by the tool, was $30 per heifer. Results confirmed that most of these farmers were using optimally this new technology. They were using it for first and second service only, which was the same optimal strategy found by the tool [35]. A main conclusion of this analysis was that the sexed semen technology has an economic benefit, but it would be mostly recommended when the conception rate of the sexed semen is at least 80% of the conventional semen, the value of the heifer calf is high, and when the price of the sexed semen is twice or less than that of the conventional semen. Due that the conception rate of both the conventional and sexed semen and the market prices are important determinant parameters, a main recommendation was that the analysis should be performed on a farm-specific basis and on a permanent basis, for which the decision support tool plays an important role.

4.3. Dairy Reproductive Economic Analysis

Published data along with dairy farm records were collected and summarized to create a representative farm to assess the value of improving reproductive efficiency measured as improving the 21-day pregnancy rate using the tool *Dairy Reproductive Economic Analysis*. Data consisted of detailed information on transition probabilities arrays of replacement and abortion risks; definition of lactation curves, and several economic parameters. Then, the DSS was used multiple times to represent incremental gains in reproductive efficiency.

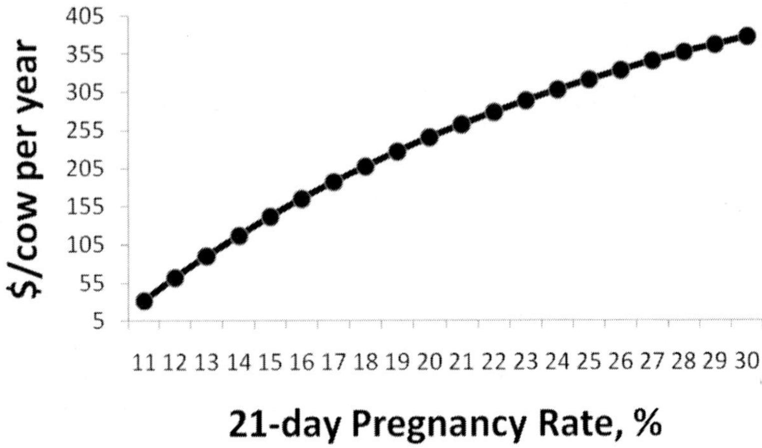

Figure 7. Projected net economic gain of improving 21-day pregnancy rate from a baseline of 10% assessed by the tool: *Dairy Reproductive Economic Analysis.*

Figure 7 portrays a marginally decreasing trend of economic gain with respect to 21-day pregnancy rate: the higher the original 21-d pregnancy rate, the lower the gain. Nonetheless the tool shows clearly that even at 30% 21-day pregnancy rate, an extremely (and unusual) good pregnancy rate, there is still an opportunity of additional gains because of improved reproductive efficiency. The tool, furthermore presents the main factors from which the additional value comes (in order): higher milk income over feed cost, lower culling costs, higher calf revenues, and lower reproductive costs. These results are being used in a large extension undertaking to promote improved reproductive efficiency in hundreds of dairy farms, but always with the final recommendation that specific farm data and information from current market conditions should be used with the DSS tool to have a more precise assessment.

4.4. Decision Support System for Expansion

Three hundred dairy farms completed a mailed questionnaire regarding their desires and needs of expansion or modernization [36]. Seventy eight percent of farms (26% of respondents) indicated that were planning to expand or modernize their installations and listed as the most important reason of doing that the expected increase on farm net return. Importantly, they acknowledge largely the uncertainty of the process of expansion as a large hindrance and therefore they asked for decision support tools that would allow them project systematically their options and analyze scenarios. More than 20 of these farmers were then contacted and offered to perform those projections by using the tool *Decision Support System Program for Dairy Production and Expansion*. The overall outcome was that all farmers visited agreed that the tool represented reasonably well their farm sand therefore they would trust its future projections. Further analyses were used to confirm or reject their pre-conceived evaluations and to assist farmers to make more informed decisions throughout the process of expansion or modernization. More than 10 farmers did some adjustments in their expansion or modernization process because of the tool and all of them indicated will continue using the DSS tool throughout their expansion or modernization operation.

4.5. The Economic Value of a Dairy Cow

Representative data from Wisconsin farms were collected from official sources, farm records, and market reports to become a baseline scenario [20] from which users could select modifications according to their own conditions. Results of these data contained in the tool *Economic Value of a Dairy Cow* indicated that the expected milk production of the cow was the single most important factor for replacement decisions. The impacts of increasing or decreasing up to 20% (120 to 80 in Table 3) the average milk production of a cow, a reasonable assumption, are portrayed in Table 3. It is evident that the milk production expectancy of following lactations is a much more important factor for pregnant cows whereas the impact of milk production expectancy of this lactation and future lactations are similarly important factors for non-pregnant cows.

Although these numbers are good indicators for farm decision-making, the need of using the tool with specific farm conditions and under current market condition could not be over emphasized.

This tool *Economic Value of a Dairy Cow* was also used to value the animal farm assets in a farm. The tool was first set with all parameters concerning to the specific farm and with economic variables representing the market conditions. Followed, the farmer created a list of all cows in the farm including their current state (lactation, month after calving, and pregnancy status) and, importantly, their projected milk production. Then, a cow value was calculated for every single animal in the herd. Finally, the calculated salvage value was added to the cow value. The farmer was then able to use these data for continued monetary support from a financial institution.

Table 3. Impact of expected milk production on the cow value of a 2-month pregnant, 8-month after calving cow and a non-pregnant, 7-month after calving cow assessed by the tool *Economic Value of a Dairy Cow*. Bolded values represent the cow with average production in the herd (100%). [1]Cow's expected milk production (% of the average cow) from the current state to the end of the present lactation. [2]Cow's expected milk production (% of the average cow) in all successive lactations.

Expected Milk Production (% of the average cow)		Cow Value of a 2-month pregnant, 8-month after calving cow, $			Cow Value of a non-pregnant, 7-month after calving cow, $		
Rest of Lactation[1]	Successive Lactations[2]	1st Lactation	2nd Lactation	3rd Lactation	1st Lactation	2nd Lactation	3rd Lactation
120	120	2,458	2,038	2,002	1,973	1,485	1,462
120	100	1,045	877	829	1,109	857	814
120	80	-380	-284	-345	244	230	165
100	120	1,891	1,499	1,477	1,184	796	809
100	**100**	**479**	**338**	**304**	**320**	**168**	**161**
100	80	-934	-823	-870	-545	-460	-487
80	120	1,325	961	952	395	106	157
80	100	-88	-200	-221	-469	-521	-491
80	80	-1,501	-1,361	-1,395	1,344	1,149	-1,139

4.6. The Lgm-Dairy Least Cost

During the months LGM-Dairy revenue insurance program was offered in year 2011, the average savings when using the LGM-Dairy Least Cost tool was 27.8% (Table 4). The tool was used during those months to assess the premium cost for a 200-cow farm producing 31 kg milk/cow per day. Based on experience and expertise with a number of dairy farmers and consultants in Wisconsin, the strategy was to insure a minimum income over feed cost of $5/46.4 kg milk during the effective insurance period that is 10 month s per contract (starting 2 months after the contract month).

Considering that the level of insurance protection is exactly the same whether to paying the regular premium or a least cost premium in Table 3, the savings are substantial. The main difference between regular and least cost premiums is the allocation of milk and feed being insured according to the covered months in the future. In the regular premium, the default situation is to assign the same level of milk quantity for protection every month. The least cost optimization, however, finds a better allocation that based on the underline simulated data determines a better plan that results in a much lower premium, but the same level of protection.

Table 4. Savings on premiums when insuring net margins using the LGM-Dairy insurance program during the year 2011 assessed by the tool *LGM-Dairy Least Cost* using default amounts of corn and soybean meal as feed insured and assuming a reasonable insurance deductible of $1/46.4 kg milk for a 200-cow dairy farm producing 31 kg milk/cow per day.

Month	Regular Premium, $	Least Cost Premium, $	Savings on Premium, %
January	4,384	3,389	22.7
February	4,904	3,429	30.1
March	5,209	3,863	25.8
October	4,019	2,685	33.2
November	4,216	3,064	27.3

4.7. Dynamic Dairy Farm Model

The *Dynamic Dairy Farm Model* was applied on a typical North Florida dairy farm of 400 cows with a production of 7,711 kg/cow per year having 62 ha of crop fields and pastures. A dual optimization including maximization of profit while relaxing N leaching indicated that the nitrogen leaching ranged between 4,800 to 5,000 kg/year whereas the profit would change between $70,000 and $70,600 (Figure 8) [2]. Furthermore, strategies to reduce nitrogen leaching would compromise profit. Depending on the farm goals and environmental regulations, the *Dynamic Dairy Farm Model* proved to be an effective tool to screen options and study whole farm management strategies. As in previous cases, farm specific conditions along with current market conditions need to carefully be defined before doing those assessments.

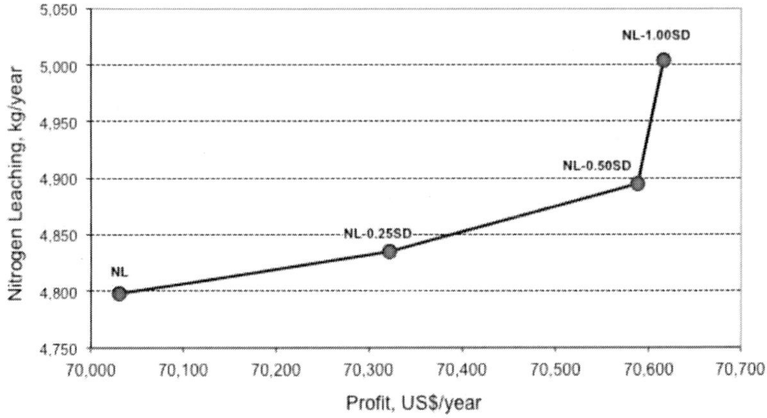

Figure 8. Dual optimization of profit maximization by relaxing nitrogen leaching assessed by using the tool *Dynamic Dairy Farm Model*. NL is average nitrogen leaching and SD is standard deviation.

5. EVALUATION OF DISSEMINATION AND ADOPTION: POTENTIAL IMPACT

Following is some evidence that indicates the DairyMGT.info Website has become the place-to-go for decision-making tools related to dairy farm management in Wisconsin and a trusted reference with increased visibility in other states and internationally. The DairyMGT website was officially launched at the end of 2009. A predecesor webpage existed since June 2008. between April 2012, and a rate of when email registration was required. According to *Google Analytics (http://www.google.com/analytics/)* the Wisconsin Dairy Management domain (DairyMGT.info or DairyMGT.uwex.edu) received 45,307 page views during the year period ending on April 30, 2012. Fifty nine percent were visitors from the U.S.A. and the rest from other 135 countries. From these, the most important countries were: India (5.5%), Australia (3.3%), Argentina (2.6%), Canada (1.9%), Mexico (1.8%), Kenya (1.6%), United Kingdom (1.5%), Italy (1.5%), Turkey (1.3%), Brazil (1.2%), Peru (1.2%), South Africa (1.0%), Pakistan (1.0%), and Spain (1.0%). Inside the U.S.A., visitors came from all states, but 63% of them were from Wisconsin. Other important states were: California (7.4%), Minnesota (3.1%), Illinois (2.8%), New York (2.6%), Iowa (1.6%), Texas (1.5%), Florida (1.3%), Pennsylvania (1.3%), Michigan (1.3%), and Washington (1.0%).

Table 5. Ranking of the most downloaded tools of DairyMGT.info Decision Support System tools during the period May 2011-April 2012.

Rank	Decision Support Tool	Downloads
1	The Wisconsin Dairy Farm Ratio Benchmarking Tool	1,280
2	LGM-Dairy Insurance Related Tools	1,279
3	Dairy Reproductive Economic Analysis	1,030
4	Corn Feeding Strategies	655
5	UW-DairyRepro$: A Reproductive Economic Analysis Tool	592
6	Optigen® Evaluator	482
7	Economic Analysis of Switching from 2X to 3X Milking	479
8	Lactation Benchmark Curves for Wisconsin	454
9	Grouping Strategies for Feeding Lactating Dairy Cattle	432
10	Heifer Break-Even	346
11	Milk Curve Fitter	313
12	The Economic Value of a Dairy Cow	312
13	Decision Support System Program for Dairy Production and Expansion	252
14	Economic Value of Sexed Semen Programs for Dairy Heifers	245
15	Dairy Ration Feed Additive Break-Even Analysis	240
16	Herd Structure Simulation	228
17	Milk Component Price Analysis	218
18	Heifer Replacement	215
19	Exploring Timing of Pregnancy Impact on Income Over Feed Cost	196
20	Dynamic Dairy Farm Model	156
21	Cost-Benefit of Accelerated Liquid Feeding Program for Dairy Calves	113
22	Dairy Nutrient Manager	98
23	Grazing-N: Application that Balances Nitrogen in Grazing Systems	97
24	Economic Evaluation of using rbST	90
25	Seasonal Prediction of Manure Excretion	48

During the same period of time, May 2011 to April 2012, 1,635 users of decision support tools elected to register their emails on the DairyMGT.info system. A thousand and fifty five did it during the months of 2011, a period in which email registration was optional. During Januaryto April 2012 a rate of 5 emails registrations a day was recorded. During the one year period May 2011 to April 2012 there were 9,336 downloads of the top 25 DSS tools as shown in Table 5.

A number of tools have been adjusted and translated to other languages to better represent conditions in other regions or in other countries following user inquiries and requests. This was the case for the tools: *Economic Value of Sexed Semen for Dairy Heifers, UW-DairyRepro$: A Reproductive Economic Analysis Tool, Value of a Springer,* and *Income Over Feed Supplement Cost* translated to Spanish and adjusted to Argentinian conditions. The *Economic Value of Sexed Semen for Dairy Heifers* tool was in addition translated to Chinese.

Another evidence of DairyMGT.info DSS demand is the world wide requests for talks regarding these tools. During the past 4 years (May 2008 to April 2012) 168 talks have been given regarding DairyMGT.info tools, a rate of 3.33 talks per month. These talks had a total attendance of about 6,500 people. One hundred and twelve of these talks were in Wisconsin (3,200 people); 25 in other states (1,700 people), and the rest, 31, in other countries such as Mexico, Chile, Peru, Argentina, Honduras, and Nicaragua (1,600 people).

Evidence of adoption together with functionality and benefits of the DSS tools can also be measured by comments and feedback reported by users and other stakeholders. Some anonymous test imonials about DairyMGT.info DSS Tools are listed below.

- "The Income over Feed Supplement Cost is a very useful tool that allows me to find out the best ingredients to buy and provide clear and practical advise in a number of clients I work with" – A dairy farm nutritionist.

- "I have used the tool 2X to 3X milking with a number of farms and consultants and it has always been well received. It does an excellent job of determining the economic impact of switching milking frequencies." – A county Extension agent.

- "The Optigen tool is a very simple application, yet it makes a quite powerful impact because it opens a realm of opportunities in the field." – A dairy industry service provider.

- "...the Sexed Semen evaluator brings very useful information and it is a tool that people can really use and apply within field situations. This is a very useful tool" – A veterinarian attending dairy farms.
- "I think that the information and spins of using the Income over Feed Cost database tool are great and powerful" – A dairy farm consultant.
- "The tools related to economic evaluations of reproductive programs in dairy cattle are going to be incredibly useful." – A dairy Extension specialist.
- "The State of Wisconsin has led the nation in number of contracts and milk insured under the LGM-Dairy program, which reflects, at least in part, the usage and practical application of the LGM-Dairy Analyzer tool of the UW-Madison" – An Extension specialist. "[The LGM-Dairy Analyzer] ...is having a direct and measurable impact [in our dairy industry]" – A University administrator.
- "We are defining reproductive strategies for our herd and we found invaluable the use of the [DairyMGT.info] management tools in our planning design. We specially appreciate the clarity of the applications and the simplicity of concepts that make these tools very practical and applicable." – A dairy farm manager of several dairy farms.
- "The [DairyMGT.info] decision support tools have really helped out our dairy farm in may aspects including financials, replacements, reproduction, and even nutrition." – A senior dairy science student and dairy farmer.
- "These [DairyMGT.info tools] are a collection of the most practical tools I have ever seen." – A well-established county Extension agent.

6. FUTURE DEVELOPMENTS: KEEP UP WITH TECHNOLOGY AND NEEDS

A number of emerging and evolving technologies are today available to dairy farmers more than ever. These include the use of smart phones, tablets and similar hardware devices; more efficient software resources; and improved data networks. There is no doubt the trend of fast technological improvement in the area of computer, software, and gadget development will continue even at a faster pace. Progressive farmers and an increasing proportion of Extension agents and dairy farm consultants are already using these technologies. New technologies bring challenges to keep information systems up-to-date, but at the same time bring great opportunities for improved DSS development.

One important advantage of smart phones and tablets is their portability along with connectivity. Nowadays farmers enjoy voice and, importantly, data network and therefore the capability to save and retrieve data eventually from anywhere at anytime. For example, a farmer can have complete information of a cow (e.g., age, lactation, pregnancy status, production history, today's production, genetic background, health incidence, etc.) at the time the cow is being registered through a smart phone system whether the cow is in a corral, in the milking parlor, or out in the field grazing. This gives the farmer the opportunity to make critic a land time-sensitive decisions right away. This could be one of the major benefits of smart phones and tablets applications. Decision support systems have to be integrated with these new technologies and need to take advantage of these important advantages.

One drawback, however, of smart phones and tablet applications is their restricted screen size and some hardware and software limitations. Applications need to be especially designed for smart phones and tablets. Normally, the information entered and retrieved would need to be summarized or would require additional layers of navigation. Extra design details could, though, lead to more compact, more intuitive, and overall more efficient DSS.

There is a trade-off of functionality and payback. The industry seems to favor both types: application for conventional computers and laptops in addition to those applications for smart phones and tablets. The decision-maker selects what type of tool to use for a particular situation. From the developmental standpoint, this is an additional challenge that requires additional work and expertise.

Important considerations regarding upcoming and developmental technologies are the increasing need for integration of DSS with information systems currently used in a farm. Most of the farmers are already using some type of software or information systems for operational management such as feeding, general record keeping, reproductive synchronization programs, identification, heat devices, or others. The DSS portrayed in this chapter and similar have the opportunity of becoming a bridge among these information systems. Decision support systems can use live information from farm records and provide predictions that go beyond the simple record keeping summaries. Farmer expertise combined with real-time DSS projections using farm record keeping systems is a powerful combination for efficient and effective decision-making in dairy farm management.

7. CONCLUSION

More than 30 computerized decision support system tools have been developed to assist dairy farmers in their continuous decision-making needs. All these tools are openly available at http://DairyMGT.info under the *Tools* section. Tools are grouped in major management areas of dairy farming such as feeding and nutrition, reproductive efficiency, heifer management and replacement, production and productivity, price risk management and financial assessment, and environmental stewardship. A number of methodologies and combinations of methodologies as well as different software applications were used to develop these decision support systems with the ultimate goal to always provide solid, but still user-friendly management tools for dairy practical farm decision-making. Methodologies included partial budgeting, cost benefit, decision analysis, enterprise budgets, linear programming, Markov chains, mathematical simulation and projection, nonlinear optimization, matrix solution, database management, and use of external simulation models. Software used to develop the tools included Macromedia Flash, HTML, PHP, JavaScript, C, CSS, MySQL, Spreadsheet applications, and executable programs. The DSS have proven to be effective decision-making tools for improved dairy farming operation. Large dissemination and impact of these DSS tools can be verified by having 9,336 downloads of these DSS tools during the one-year period between May 2011 and April 2012 and the request of 168 talks with 6,500 people in attendance across the world during the 4-year period between May 2008 and April 2012.

8. ACKNOWLEDGEMENTS

The development and maintenance of the DairyMGT.info tools has been possible by the partial support of several extra-mural grants: Agriculture and Food Research Initiative from the USDA National Institute of Food and Agriculture Competitive Grants No.: 2010-51300-20534, 2010-85122-20612, 2011-68004-30340 and several Hatch grants to V.E.C from the College of Agriculture and Life Sciences at the University of Wisconsin-Madison. Acknowledgement is extended to a number of people involved at different levels in the development of these tools; Collaborators: B.W. Gould, R.D. Shaver, M.A. Wattiaux, L. Armentano, J. Vanderlin, K. Bolton; Students: J.O. Giordano, J. Janowski, M. Valvekar, E. Demarchi, A. Kalantari; Programmers: A. Kalantari, N. Suryanarayana, K. Nathella, V. Vats, A. Gola.

REFERENCES

1. Corson. Rotz, Montes. F. Chianese, S. D. Hafner, C. U. Coiner, 2011 The Integrated Farm System Model: Reference Manual Version 3.5. USDA Agricultural Research Service. http://www.ars.usda.gov/SP2UserFiles/Place/19020000/ifsmreference.pdf (accessed 5 May 2012).
2. V. E. Cabrera, P. E. Hildebrand, J. W. Jones, D. Letson, A. de Vries, 2006 An Integrated North Florida Dairy Farm Model to Reduce Environmental Impacts under Seasonal Climate Variability Agriculture Ecosystems & Environment 113 82 97
3. H. Groenendaal, D. T. Galligan, H. A. Mulder, 2004 An Economic Spreadsheet Model to Determine Optimal Breeding and Replacement Decisions for Dairy Cattle. Journal of Dairy Science 87 2146 2157
4. C. Meadows, Schultz P. J. Rajala, G. S. Frazer, 2005 A Spreadsheet-Based Model Demonstrating the Nonuniform Economic Effects of Varying Reproductive Performance in Ohio Dairy Herds Journal of Dairy Science 88 3 1244 1254
5. J. O. Giordano, P. M. Fricke, M. C. Wiltbank, V. E. Cabrera, 2011 An Economic Decision-Making Support System for Selection of Reproductive Management Programs on Dairy Farms Journal of Dairy Science 94 6216 6232
6. V. E. Cabrera, N. E. Breuer, P. E. Hildebrand, 2008 Participatory Modeling in Dairy Farm Systems: A Method for Building Consensual Environmental Sustainability Using Seasonal Climate Forecasts Climatic Change 89 395 409
7. N. E. Breuer, V. E. Cabrera, K. T. Ingram, K. Broad, P. E. Hildebrand, 2008 AgClimate: A Case Study in Participatory Decision Support System Development Climate Change 87 385 403
8. V. E. Cabrera, N. E. Breuer, P. E. Hildebrand, D. Letson, 2005 The Dynamic North Florida Dairy Farm Model: A User-Friendly Computerized Tool for Increasing Profits while Minimizing Environmental Impacts Computer and Electronics in Agriculture 86 207 222
9. C. W. Fraisse, J. G. Bellow, N. E. Breuer, V. E. Cabrera, L. Hatch, G. Hogenboom, K. Ingram, J. W. O'Brien, J. J. Paz, D. Zierden, 2006 AgClimate: A Climate Forecast Information System for Agricultural Risk Management in the Southeastern USA Computer and Electronics in Agriculture 53 13 27
10. V. E. Cabrera, A. de Vries, P. E. Hildebrand, 2006 Manure Nitrogen Production in North Florida Dairy Farms: A Comparison of Three Models Journal of Dairy Science 89 1830 1841
11. V. E. Cabrera, S. Jagtap, P. E. Hildebrand, 2007 Strategies to Limit (Minimize) Nitrogen Leaching on Dairy Farms Driven by Seasonal Climate Forecasts Agriculture Ecosystems & Environment 122 479 489
12. V. E. Cabrera, 2010 A Large Markovian Linear Program for Replacement Policies to Optimize Dairy Herd Net Income for Diets and Nitrogen Excretion Journal of Dairy Science 93 394 406
13. T. J. Earleywine, 2001 Profitable Dietary Grain Concentrations and Grouping Strategies in Dairy Herds PhD Thesis Madison University of Wisconsin-Madison
14. N. J. Tessmann, H. D. Radloff, J. Kleinmans, T. R. Dhiman, L. D. Satter, 1991 Milk Production Response to Dietary Forage:Grain Ration Journal of Dairy Science 74 2696 2707
15. V. E. Cabrera, R. D. Shaver, M. A. Wattiaux, 2009 Optimizing Income over Feed Supplement Cost Four-State Dairy Nutrition and Management Conference Dubuque, IA
16. V. E. Cabrera, R. D. Shaver, P. Dyk, 2010 The Four-State Dairy Extension Feed Evaluator Four-State Dairy Nutrition and Management Conference Dubuque, IA
17. V. E. Cabrera, F. Contreras, R. D. Shaver, L. Armentano, 2012 Grouping Strategies for Feeding Lactating Dairy Cattle Four-State Dairy Nutrition and Management Conference Dubuque, IA

18. V. E. Cabrera, 2009 When to Use Gender Biased Semen: Economics Dairy Cattle Reproductive Council Annual Convention.Boise, ID and St. Paul, MN
19. J. O. Giordano, A. Kalantari, P. M. Fricke, M. C. Wiltbank, V. E. Cabrera, 2012 A Daily Herd Markov-Chain Model to Study the Reproductive and Economic Impact of Reproductive Programs Combining Timed Artificial Insemination and Estrous Detection Journal of Dairy Science: in press
20. V. E. Cabrera, 2012 A Simple Formulation of the Replacement Problem: A Practical Tool to Assess the Economic Value of a Cow, the Value of a New Pregnancy, and the Cost of a Pregnancy Loss Journal of Dairy Science: in press
21. R. A. Erdman, M. Varner, 1995 Fixed Yield Responses to Increased Milking Frequency Journal of Dairy Science 78 1199 1203
22. Risk Management Agency 2008 Livestock Gross Margin for Dairy Cattle Handbook Washington DC USDA RMAhttp://www.rma.usda.gov/handbooks /20000/2009/09lgm-dairy-handbook.pdf (accessed 14 May 2012).
23. R. B. M. Huirne, A. A. Dijkhuizen, 1997 Basic Methods of Economic Analysis In: Dijkhuizen AA, Morris RS (ed.) Animal Health Economics: Principles and Applications. Sidney University of Sidney 25 41
24. G. Gregory, 1988 Decision Analysis New York Plenum
25. M. F. Warren, 1986 Financial Management for Farmers London Hutchinson
26. V. E. Cabrera, P. E. Hildebrand, 2012 Linear Programming for Dairy Herd Simulation and Optimization: An Integrated Approach for Decision Making In: Zoltan AM (ed.) New Frontiers in Theory and Applications New York Nova 193 212
27. B. W. Gould, V. E. Cabrera, 2011 USDA's Livestock Gross Margin Insurance for Dairy: What is it and how it can be Used for Risk Management University of Wisconsin Department of Agricultural and Applied Economics http://future.aae.wisc.edu/lgm-dairy/lgmdairymanual.pdf (accessed 10 May 2012).
28. P. D. P. Wood, 1976 Algebraic Models of the Lactation Curves for Milk, Fat, and Protein Production, with estimates of seasonal variation Animal Production 22 35 40
29. J L. Ehrlich, 2011 Quantifying Shape of Lactation Curves, and Benchmark Curves for Common Dairy Breeds and Parities The Bovine Practitioner 45 1 88 94
30. M. Valvekar, V. E. Cabrera, B. W. Gould, 2010 Identifying Optimal Strategies for Guaranteeing Target Dairy Income Over Feed Cost Journal of Dairy Science 93 3350 3357
31. M. Valvekar, J. P. Chavas, B. W. Gould, V. E. Cabrera, 2011 Revenue Risk Management, Risk Aversion and the Use of LGM-Dairy Insurance Agricultural Systems 104 671 678
32. J. F. Inostroza, R. D. Shaver, V. E. Cabrera, J. M. Tricarico, 2010 Effect of Diets Containing a Controlled-Release Urea Product on Milk Yield, Composition and Component Yields in Commercial Wisconsin Dairy Herds and Economic Implications Professional Animal Scientist 26 175 180
33. G. Y. Tsuji, G. Hoogenboom, P. K. Thornton, 1998 Understanding Options for Agricultural Production London Kluwer Academic Pub
34. National Research Council 2001 Nutrient Requirement for Dairy Cattle Seventh Revised Edition Washington DC The National Academies Press
35. R. Sterry, D. Brusveen, V. E. Cabrera, K. Weigel, P. Fricke, 2009 Why They Use Sexed Semen Hoard's Dairyman Magazine 25 March 2009 25
36. V. E. Cabrera, J. M. Janowski, 2011 Wisconsin Dairy Business and Production Survey: Comparison between Farms Planning to Expand and Farms not Planning to Expand Journal of Extension: 3RIB1 http://www.joe.org/joe/2011june/rb1.php (accessed 10 May 2012).

CHAPTER 15

Health-Promoting Properties of Bioactive Peptides Derived from Milk Proteins in Infant Food: A Review

Vassilios Raikos [1] Theodore Dassios[2]

[1.]*Rowett Institute of Nutrition & HealthUniversity of AberdeenAberdeenUK*

[2.]*Addenbrooke's HospitalCambridge University Hospitals NHS Foundation TrustCambridgeUK*

ABSTRACT

Milk proteins have attracted extensive interest in terms of their bioavailability following ingestion. Enzymatic digestion of dairy products generates numerous peptides with various biological activities. Both human milk and infant formulas based on cow's milk are potential sources of bioactive peptides. This review aims to present current knowledge on the formation and fate of bioactive peptides from milk feeds intended for infants. Emphasis is placed on the source of the bioactive peptides with the nutritional impact of human milk and cow milk-based formulas on infant health being critically discussed from that perspective. Furthermore, the effect of processing and in vitro or in vivo digestion on the release and availability of peptides with bioactive sequences is evaluated. Considerable differences with respect to bioavailability and metabolic effects between the biologically active fragments generated following ingestion of human milk and infant formulas are documented. Peptides from milk protein of bovine origin could be a valuable supplement to human milk as multiple health-promoting properties are attributed to peptide fractions identified in standard cow milk-based infant formulas.

Keywords: Bioactive Infant formula Human milk Peptide

1 INTRODUCTION

Milk is very often described as the "ultimate" food which fulfills the nutritional needs of the mammal newborn and ensures safe development and growth during the first stages of its life (Mills et al. 2011). Individual milk components contribute to the high biological and nourishing value of this food. Milk proteins, for instance, are known to exert various actions that promote human health. These actions include stimulation of the immune system, digestion and absorption of nutritional elements, development of endogenous defense mechanisms against bacteria, fungi and viruses, prebiotic effects, and others (Lönnerdal 2003).

Recently, milk proteins have attracted extensive interest in terms of their bioavailability following ingestion. In vivo digestion of milk proteins within the gastrointestinal tract results in the release of peptides with new biological properties compared to the ones exhibited by the intact, precursor molecule (Chatterton et al. 2004). Many of these protein fragments, also known as bioactive peptides, consist of approximately 3–20 amino acids, which are encrypted within the parental protein and become activated following their release, induced by the action of gastric and pancreatic enzymes (Kekkonen 2009; Korhonen 2009). In addition, biologically active peptides may be formed during food processing. In vitro hydrolysis of milk proteins is a common method for producing hydrolyzed peptide fragments used for hypoallergenic infant formulas (Korhonen et al. 1998). The peptides generated either in vitro or in vivo differ markedly with respect to their biological activities depending on the digestion process and/or the parental protein from which they are released.

The identification of bioactive milk peptides was first reported in 1950 (Mellander 1950). Since then, numerous reports document the presence of peptides which exert some physiological effect on the human body in milk or milk-based products (Beaulieu et al. 2007; Foltz et al. 2007; Haque et al. 2009; Hayes et al. 2006; Kamau et al. 2010; LeBlanc et al. 2002; Meisel and Fitzgerald 2000; Nagpal et al. 2011). However, to the best of our knowledge, limited studies exist which focus on the health-promoting properties of milk peptides detected in infant feeding regimens. This paper reviews recent research on the formation and fate of bioactive peptides from milk feeds intended for infants. Furthermore, suggestions based on recently published research for improvement of the nutritional value of infant formulas through the inclusion of milk-derived peptides serving as functional ingredients are provided.

2 HUMAN MILK VERSUS INFANT FORMULA AS A POTENTIAL SOURCE OF BIOACTIVE PEPTIDES

Human milk is considered the reference standard for the manufacture of infant formulas. Thus, the latter are designed to simulate the content and the performance of the former in an attempt to mimic the effect of breast milk on metabolic and other physiologic functions. In addition to standard formulas, hydrolysates of cows' milk are also available for infant nutrition usually supplied as a hypoallergenic therapeutic or preventive formulation for allergenized or sensitized newborns (Caffarelli et al. 2002).

Qualitative and quantitative differences exist between human milk and infant formulas, and these include protein compositional dissimilarities. For instance, cows' milk-based infant formulas contain three to four times higher protein content compared to human milk (Rudloff and Kunz 1997). Furthermore, caseins dominate the protein fraction in bovine milk (~80%), whereas the principal class of protein in human milk is whey, particularly during the first days of lactation (Kunz and Lönnerdal 1989; Kunz and Lönnerdal 1990). Proteomic analysis indicates that human milk does not contain β-lactoglobulin and α_{s2}-casein, whereas lactoferrin and α-lactalbumin are the main protein components (Armaforte et al. 2010). The same study provides further evidence on qualitative similarities in terms of the protein profile of infant formula and bovine milk with α_{s1}-casein, α_{s2}-casein, β-casein, and β-lactoglobulin being the most abundant proteins. Other major differences between human and bovine milk proteins such as the different glycosylation degree of immunoglobulins, lactoferrin, and casein have been documented (Nwosu et al. 2012).

Considering human milk as the gold standard, several attempts have been made to "humanize" infant formulas. Commercially available products include whey-enriched infants formulas based on cow's milk. In this case, formulations include a casein-to-whey protein ratio of approximately 40:60, which is somewhat closer to the one encountered in human milk. Whey-dominant formulas which resemble the human milk compositional standard seem to a have the same impact on growth of premature infants as compared to casein-dominant formulations (Bernbaum et al. 1989).

However, other advantages particularly for low-birth weight and preterm infants may be related to whey-dominant formulas, which is an expected finding as milk is a biological fluid that has evolved to meet the nutritional demands of the human species (Rudloff and Kunz 1997). Other attempts to narrow the gap between the protein content of human milk and infant formulas involve the fortification of infant formulas with purified proteins of bovine origin, which are abundant in human milk such as α-lactalbumin, lactoferrin, and others (Chatterton et al. 2004).

Despite the efforts to replicate the protein compositional recipe of human milk, infant formulas and protein hydrolysate products display qualitative and quantitative differences in terms of protein content that may affect physiological process in the newborn. In general, formulas provide a different profile of amino acids in addition to more protein content. Differences in the parental (intact) proteins result most likely in the generation of peptides with different biological activities following the process of digestion. It has been documented that hydrolysis products of the native human α-caseins generated by the proteolytic activity of plasmin cannot be detected in commercial formulations intended for infants (Armaforte et al. 2010). This is because bovine plasmin preferentially acts upon β-casein (Fox and Kelly 2005). As a result, considerable differences are expected in terms of bioavailability and metabolic effects between the biologically active fragments generated following ingestion of human milk, infant formulas, and hydrolyzed products, which may affect infant growth.

3 ASPECTS OF IN VIVO AND IN VITRO GENERATION OF BIOACTIVE PEPTIDES

There are various ways for releasing bioactive peptides. First, milk proteins may be digested in vitro during food processing by maturation with selected enzymes. In this case, where milk is fermented with a proteolytic culture, the process is carefully controlled (enzyme specificity, temperature, incubation time, etc.), and thus the release of bioactive peptides may be predicted (Exl 2001). Common enzymes in the commercial production of bioactive peptides via in vitro digestion include pepsin, trypsin, and chymotrypsin (Korhonen et al. 1998). Bioactive peptides may also be generated following ingestion in vivo by either digestive enzymes or by microbial enzymes (Kekkonen 2009). Thus milk proteins during their passage through the digestive system may be the target of enzymes produced either by the human body or by microorganisms which normally inhabit the gastrointestinal tract. A wide range of enzymes with different cleavage sites may act throughout the whole gastrointestinal tract and under conditions which may differ depending on various physiological factors (Antalis et al. 2007). In this case, the process is far less easy to control and to monitor. This is the reason why in vitro digestion, also known as simulated gastrointestinal digestion (SGID), is often employed to investigate the stability of milk proteins and the fate of bioactive peptides. SGID takes into account key factors affecting the digestion process such as the rates of gastric emptying, which are used to determine the period of incubation with gastric proteases, and changes in gastric pH value which determine to a large extent the enzyme activity (De Noni 2008).

Nevertheless, SGID may be regarded as an oversimplified method to determine the fate of precursor milk proteins. In reality, things are more complicated as physiological processes in vivo are not static, and the environment within the gastrointestinal tract of an infant is subject to changes with time or cause. Gastric pH, for instance, is known to be high in early neonatal life and has been determined to be 6.0 in term newborn infants with the tendency to decrease sharply during the first 24 h of life (Widstrom et al. 1988). According to Chatterton et al. (2004), proteins in human milk and infant formula showed a pH-dependent extent of hydrolysis when incubated with gastric juice in a pH range between 6.5 and 2.0. This is somewhat expected as gastric enzymes show optimum activity at specific pH values. Furthermore, the concentration of milk proteases (e.g., plasmin) present in breast milk is increased for premature infants, which results in extensive hydrolysis of milk proteins to facilitate the digestion process (Armaforte et al. 2010). This could in turn affect the biological activity of the digests, as peptides have to be absorbed from the intestine and reach their target cells in adequate concentrations. As a result, one may conclude that the digestion process of milk proteins in vivo is a complex process and depends on various factors which cannot be adequately simulated in vitro. In vivo studies are more reliable in order to draw safe conclusions with respect to the formation of bioactive peptides following the ingestion of human milk and infant formulas. The formation of bioactive peptides in vivo following the digestion of milk proteins has been documented in the past both in human (Boutrou et al. 2013; Foltz et al. 2007) and animal (Bouzerzour et al. 2012) models. According to these studies, although evidence exists to support their presence in the human and piglet gastrointestinal tract, further research is required to investigate the physiologic conditions which would allow the bioactive peptides to exert their bioactivity.

4 BIOACTIVE PEPTIDES IN HUMAN MILK AND INFANT FORMULAS

Peptides released from in vivo or in vitro digestion of milk proteins are known to affect major systems such as the cardiovascular, nervous, digestive, and immune system. According to Hayes et al. (2007), bioactive peptides from milk having a physiological impact on human body can be categorized as antihypertensive, antithrombotic, opioid, casein phosphopeptides, antimicrobial, cytomodulatory, immunomodulatory, and miscellaneous. Recent findings indicate that milk peptides can have additional physiological effects such as antioxidant activity or may even possess functional significance which is not yet clearly understood. Therefore, the consideration of milk peptides as dietary supplements or food additives could be promising for the food industry in the future. Table 1 summarizes the bioactive peptides identified in both human and bovine milk with potential beneficial effect on human health.

4.1 Effects on the cardiovascular system

The release of angiotensin converting enzymes (ACE)-inhibitory peptides by SGID of infant formulas has been documented by Hernández-Ledesma et al. (2004). According to this study, enzymatically hydrolyzed casein and whey formulas exhibited higher ACE-inhibitory activities than non-hydrolyzed milk protein formulations. It was hypothesized that low molecular weight peptides (<3,000 Da) were mainly responsible for this activity with several potent peptides being identified (β-Lg: RVY and LVR; β-casein: YPFPGPI and HLPLP). The size of the hydrolyzed fragments could be critical in terms of the antihypertensive activity as di- and tripeptides are easily absorbed in the intestine and may reach the blood and other target sites (Hara et al. 1984). Furthermore, SGID had a positive impact on the unhydrolyzed samples and did not affect the hydrolyzed whey-based formulas with respect to ACE inhibition. This suggests that whey hydrolysates survive physiological digestion. Similar findings were reported by other researchers (Martin et al. 2008) who compared the ACE-inhibitory activity of hypoallergenic formulas and breast milk. The dipeptide Ile-Trp (IW), present at positions 59–60 in bovine α-lactalbumin from hypoallergenic formulas, was identified as the most potent ACE inhibitor in this study. According to the authors, conventional milk-based formulas and breast milk samples did not exhibit ACE inhibition. In agreement with the previous studies, SGID did not affect the overall inhibitory effect of the hydrolysate formula, indicating that the peptides survive gastrointestinal hydrolysis. The presence of peptides with potent ACE-inhibitory activity was also confirmed in more recent studies with hypoallergenic formulas containing partially hydrolyzed whey, highly hydrolyzed whey and casein, and highly hydrolyzed casein bovine proteins (Català-Clariana et al. 2010). In this case, five bioactive peptides namely LKP, IPY, ALPM, PGPIHN, and VAGTWY were detected by means of capillary electrophoresis time-of-flight mass spectrometry.

Table 1. Biofunctional peptides derived from human and bovine milk proteins with effects on human health

Function	Precursor protein	Fragment	Peptide sequence	Origin	Reference
Antihypertensive	β-lactoglobulin	f (40–42)	RVY	Bovine	(Hernández-Ledesma et al. 2004)
Antihypertensive	β-lactoglobulin	f 122–124)	LVR	Bovine	(Hernández-Ledesma et al. 2004)
Antihypertensive	β-casein	f (60–66)	YPFPGPI	Bovine	(Hernández-Ledesma et al. 2004
Antihypertensive	β-casein	f (134–138)	HLPLP	Bovine	(Hernández-Ledesma et al. 2004)
Antihypertensive	α-lactalbumin	f (59–60)	IW	Bovine	(Martin et al. 2008)
Antihypertensive	β-lactoglobulin	f (46–48)	LKP	Bovine	(Català-Clariana et al. 2010)
Antihypertensive	β-lactoglobulin	f (142–145)	ALPM	Bovine	(Català-Clariana et al. 2010)
Antihypertensive	β-lactoglobulin	f (15–20)	VAGTWY	Bovine	(Català-Clariana et al. 2010)
Antihypertensive	αs2-casein	f (201–203)	IPY	Bovine	(Català-Clariana et al. 2010)
Antihypertensive	β-casein	f (63–68)	PGPIHNI	Bovine	(Català-Clariana et al. 2010)
Antihypertensive	β-casein	f (125–129)	HLPLP	Human	(Hernández-Ledesma et al. 2007)
Antihypertensive	β-casein	f (80–90)	TPVVVPPFLQP	Human	(Bouzerzour et al. 2012)
Antioxidant	Unknown	N/A	YGYTGA	Human	(Tsopmo et al. 2011)
Antioxidant	Unknown	N/A	ISELGW	Human	(Tsopmo et al. 2011)
Antioxidant	β-casein	f (154–160)	WSVPQPK	Human	(Hernández-Ledesma et al. 2007)
Antithrombotic	κ-casein	f (114–124)	IAIPPKKIQDK	Human	(Chabance et al. 1995)
Antithrombotic	κ-casein	f (106–116)	MAIPPKKNQDK	Bovine	(Chabance et al. 1995)
Opioid	β-casein	f (60–66)	YPFPGPI	Bovine	(De Noni, 2008; De Noni and Cattaneo, 2010)
Opioid	β-casein	f (60–64)	YPFPG	Bovine	(Jarmolowska et al. 2007a)
Opioid	κ-casein	f (25–34)	YIPIQYVLSR	Bovine	(Jarmolowska et al. 2007a)
Opioid	β-casein	f (51–55)	YPFVE	Human	(Jarmolowska et al. 2007b)
Opioid	β-casein	f (51–57)	YPFVEPI	Human	(Jarmolowska et al. 2007b)
Immunomodulatory	β-casein	f (60–66)	YPFPGPI	Bovine	(Bouzerzour et al. 2012)
Mineral binding	αs1-casein	–	SpSpSpEE (cluster sequence)	Bovine	(Miquel et al. 2006; Miquel et al. 2005)
Mineral binding	αs2-casein	–	SpSpSpEE (cluster sequence)	Bovine	(Miquel et al. 2006; Miquel et al. 2005)
Mineral binding	β-casein	–	SpSpSpEE (cluster sequence)	Bovine	(Miquel et al. 2005)
Antimicrobial	κ-casein	f (106–169)	–	Bovine	(Brück et al. 2003)

In other studies, breast milk and bovine milk protein-based infant formulas were digested with pepsin and pancreatin and the contribution of the fragments generated to ACE-inhibitory activity was determined (Hernández-Ledesma et al. 2007). This study revealed that the peptide fraction obtained from human breast milk showed limited ACE-inhibition compared with the one exhibited by the corresponding fraction obtained from infant formulas. According to the authors, this is due to the different peptide fragments obtained from human milk samples and infant formulas following hydrolysis with the specific enzymes. Interestingly, only one peptide (HLPLPL) was common between the digests of human and bovine origin. Nevertheless, a potent ACE inhibitor was identified in human milk and corresponded to the β-casein fragment f(125–129) with the sequence HLPLP. In this case, it was hypothesized that the presence of the amino acid proline at the C-terminal end could be crucial in terms of the activity observed (Cheung et al. 1980). Finally, the peptide f 80-90 (TPVVVPPFLQP) from β-casein which is known to possess antihypertensive activities was detected in animal models fed with infant formulas (Bouzerzour et al. 2012). In this study, piglets were used as a model to study aspects of protein digestion and absorption in human infants.

The production of reactive oxygen species (ROS) is part of normal cell metabolism. In some cases, such as in preterm infants, high levels of ROS are produced that disturb the redox balance and result in a state formerly known as oxidative stress (Buonocore et al. 2002). Oxidative stress induced by free radicals is linked to the pathogenesis of various diseases including heart disease (Aruoma 1998). The antioxidant potential of human milk has been documented in the past (Friel et al. 2002). According to Samaranayaka and Li-Chan (2011), peptides and proteins are implicated in the overall antioxidative status of cells and contribute towards maintaining the health of biological tissue. Recent findings indicate that peptides released from human milk after digestion with pepsin and pancreatin could be beneficial with respect to antioxidant activity (Tsopmo et al. 2011). In this study, the peptides YGYTGA and ISELGW showed high radical scavenging activity, and this property was attributed to the presence of the amino acid tryptophan which can break radical chain reactions thanks to the donation of the hydrogen attached to the nitrogen of its indole ring. The same hypothesis was made by other researchers who reported moderate antioxidant activity by peptides derived from digested human samples with pepsin and pancreatin (Hernández-Ledesma et al. 2007). In this case, the peptide with the highest antioxidant activity was WSVPQPK, which also contains tryptophan and, thus, may serve as hydrogen donor and/or free radical scavenger.

Peptides of human (IAIPPKKIQDK) and bovine (MAIPPKKNQDK) origin with antithrombotic effects were detected at physiologically active levels in the plasma of newborn infants after breastfeeding or ingestion of cow milk-based formula (Chabance et al. 1995). These peptides, known as κ-caseinoglycopeptides, exert their biological function by reducing platelet aggregation and enhancing fibrinolysis, which results in the prevention of thrombosis development.

4.2 **Effects on the nervous system**

Opioid peptides are small molecules which are synthesized in vivo and may function both as hormones and neurotransmitters. These peptides depending on their bioactivity are classified as opioid agonists or antagonists. β-casomorphins derived from β-casein, and particularly β-casomorphin-5 (BCM5) and β-casomorphin-7 (BCM7), are the best-known opioid peptides (Schlimme et al. 1989). The release of BCM7 from commercial milk-based infant formulas following SGID was confirmed (De Noni 2008). The peptides identified were released from genetic variants A1 and B of β-casein. According to the European Food Safety Authority, there is no established relationship between the dietary intake of BCM7 and non-communicable diseases (EFSA European Food Safety Authority 2009). Although no safe conclusions can be drawn at the moment regarding the minimum amount of BCM7 capable to elicit pharmacological effects in vivo or ex vivo, further studies confirm its release by the action of gastrointestinal proteases during in vitro digestion in commercial infant formulas at concentrations between 0.04 and 0.21 mg.l (De Noni and Cattaneo 2010). The presence of four opioid peptides in the commercially available Humana formula (Humana GmbH, Germany) for newborns after digestion with pepsin and trypsin has been documented (Jarmolowska et al. 2007a). Both opioid agonists (BCM5) and antagonists (casoxin C, casoxin 6, and lactoferroxin A) deriving from β-, κ-casein, and lactoferrin were identified. All four opioid peptides were detected in the formulas prior to enzymatic hydrolysis at concentrations between 0.075 to 0.39 μg.ml with their concentration to rise sharply after enzyme hydrolysis. The presence of BCM5 and BCM7 in human milk has also been confirmed with the content of the opioid peptides being dependent on the lactation stage (Jarmolowska et al. 2007b).

BCM7 and related compounds have been linked in the past with the development of human diseases such as juvenile diabetes, heart disease, autism, and schizophrenia (McLachlan, 2001). Although evidence regarding the abundance of such peptides in infant formulas is inconclusive, the formation of BCMs in infant feeds following SGID has been demonstrated. The in vivo effects of these peptides on the human physiology have not yet been fully described.

4.3 Other health-promoting effects

It has been demonstrated that protein peptides exert biological action on the metabolism of minerals. Minerals play a multifunctional role in human health. They participate in bone structure and teeth and are involved in several metabolic processes such as energy production, nutrient absorption, protein building, and blood formation (Soetan et al. 2010). Casein phosphopeptides (CPPs) deriving from the digestion of milk proteins may act as mineral carriers and affect their bioavailability in the human body (Meisel and Fitzgerald 2003). Recent studies document the presence of bioactive peptides with mineral carrier properties in infant formulas (Miquel et al. 2006). The peptides originating from α_{s1} and α_{s2}-casein survived SGID with pepsin and pancreatin, and their identified molecular masses were in the range of 1,400 to 9,600 Da, which makes them good candidates for intestinal absorption. According to the authors, calcium could be bound preferentially to CPPs with the cluster sequence SpSpSpEE and iron and zinc to CPPs with the phosphorylated cluster and phosphoserine residues. The release of CPPs with metal-chelating properties following SGID of infant formulas was also reported in previous studies (Miquel et al. 2005). Many of these CPPs were only detected in probiotic formulas indicating that they are formed by the action of bifidobacteria.

Dietary protein is known to play a major role in the human body's defense mechanism against infection by either stimulating the immune system (immunomodulatory) or contributing to the inhibition of microbial growth (antimicrobial). Supplementation of infant formulas with α-lactalbumin and glycomacropeptide (GMP), a derivative of κ-casein, which were digested with pepsin and pancreatin, resulted in decreased counts of *Bacteroides* (Brück et al. 2003). This was attributed to the ability of the supplemented formula to mimic the fermentative qualities of breast milk and induce the growth of protective microflora in the gastrointestinal tract. Furthermore, the bacteriostatic and bactericidal effects of α-lactalbumin and GMP have been demonstrated when a significant reduction in enteropathogenic bacteria including enteropathogenic *Escherichia coli* and *Salmonella* was observed following supplementation of the infant formulas with the specific milk substrates (Brück et al. 2003). The release of bioactive peptides with immunomodulatory activities from infant formulas is also documented by more recent studies. In this case, the peptide f 60-66 (YPFPGPI) from β-casein was detected following in vivo digestion by piglets which were fed with infant formulas (Bouzerzour et al. 2012). According to the findings of this study, whey proteins were more resistant to enzymatic digestion in the stomach of piglets compared to caseins with the latter being hydrolyzed to a large extent just 30 min after meal ingestion.

5 CONCLUSIONS

Extensive research has underlined the important contribution of breastfeeding in promoting general health, growth, and development of the human species and the significant reduction in the risk of acute and chronic diseases (Gartner et al. 2005). Proteins contribute to the nutritional value of milk and also have other biological roles as intact molecules, as many of them are resistant to the effects of digestion (Chatterton et al. 2004). On the other hand, digestion of milk proteins in the gastrointestinal tract results in the release of bioactive agents which may affect several physiological processes. Despite the fact that many of these bioactive peptides have been identified as integral parts of milk proteins, their contribution in human nutrition has not been sufficiently described in the literature. Furthermore, limited information is available with respect to their passage through the human, and most importantly the infant's digestive tract. Up-to-date, bioactive peptides have been identified in both human milk and cow milk-based infant formulas, which differ in their amino acid profiles and properties. Modern methods of analysis enable fast and reliable separation and identification of bioactive peptides in milk-derived products (Català-Clariana et al. 2013). The peptides identified following SGID have antihypertensive, antioxidant, antithrombotic, opioid, mineral carrier, and immunogenic effects on human physiology. Further research in this field is required to establish the beneficial role of human or bovine milk from that perspective. It seems likely though that peptides from milk protein of bovine origin could be a valuable supplement to human milk. The manufacture of "chimeric" infant formulas with multiple health-promoting effects could be used as a model example in the future for the manufacture of ultimately functional foods.

REFERENCES

1. Antalis TM, Shea-Donohue T, Vogel SN, Sears C, Fasano A (2007) Mechanisms of disease: protease functions in intestinal mucosal pathobiology. Nat Clin Pract Gastroenetrol Heapatol 4:393–402
2. Armaforte E, Curran E, Huppertz T, Ryan CA, Caboni MF, O'Connor PM, Ross RP, Hirtz C, Sommerer N, Chevalier F, Kelly AL (2010) Proteins and proteolysis in

preterm and term human milk and possible implications for infant formulae. Int Dairy J 20:715–723

3. Aruoma OI (1998) Free radicals, oxidative stress, and antioxidants in health and disease. J Am Oil Chem Soc 75:199–212
4. Beaulieu J, Dubuc R, Beaudet N, Dupont C, Lemieux P (2007) Immunomodulation by a malleable matrix composed of fermented whey proteins and lactic acid bacteria. J Med Food 10:67–72
5. Bernbaum JC, Sasanow SR, Churella HR, Daft A (1989) Growth and metabolic response of premature infants fed whey- or casein-dominant formulas after hospital discharge. J Pediatr 115:652–656
6. Boutrou R, Gaudichon C, Dupont D, Jardin J, Airinei G, Marsset-Baglieri M, Benamouzig R, Tomé D, Leonil J (2013) Sequential release of milk protein-derived bioactive peptides in the jejunum in healthy humans. Am J Clin Nutr 97:1314–1323
7. Bouzerzour K, Morgan F, Cuinet I, Bonhomme C, Jardin J, Le Huëou-Luron I, Dupont D (2012) In vivo digestion of infant formula in piglets: protein digestion kinetics and release of bioactive peptides. Brit J Nutr 108:2105–2114
8. Brück WM, Graverholt G, Gibson GR (2003) A two-stage continuous culture system to study the effect of supplemental α-lactalbumin and glycomacropeptide on mixed cultures of human gut bacteria challenged with enteropathogenic Escherichia coli and Salmonella serotype Typhimurium. J Appl Microbiol 95:44–53
9. Buonocore G, Perrone S, Longini M, Vezzosi P, Marzocchi B, Paffetti P, Bracci R (2002) Oxidative stress in preterm neonates at birth and on the seventh day of life. Pediatr Res 52:46–49
10. Caffarelli C, Plebani A, Poiesi C, Petroccione T, Spattini A, Cavagni G (2002) Determination of allergenicity to three cow's milk hydrolysates and an amino acid-derived formula in children with cow's milk allergy. Clin Exp Allergy 32:74–79
11. Català-Clariana S, Benavente F, Giménez E, Barbosa J, Sanz-Nebot V (2010) Identification of bioactive peptides in hypoallergenic infant milk formulas by capillary electrophoresis–mass spectrometry. Anal Chim Acta 683:119–125
12. Català-Clariana S, Benavente F, Giménez E, Barbosa J, Sanz-Nebot V (2013) Identification of bioactive peptides in hypoallergenic infant milk formulas by CE-TOF-MS assisted by semiempirical model of electromigration behavior. Electrophoresis 34:1886–1894
13. Chabance B, Jollès P, Izquierdo C, Mazoyer E, Francoual C, Drouet L, Fiat A-M (1995) Characterization of an antithrombotic peptide from κ-casein in newborn plasma after milk ingestion. Brit J Nutr 73:583–590
14. Chatterton DEW, Rasmussen JT, Heegaard CW, Sørensen ES, Petersen TE (2004) In vitro digestion of novel milk protein ingredients for use in infant formulas: Research on biological functions. Trends Food Sci Technol 15:373–383
15. Cheung H-S, Wang F-L, Ondetti MA, Sabo EF, Cushman DW (1980) Binding of peptide substrates and inhibitors of angiotensin-converting enzyme. Importance of the COOH-terminal dipeptide sequence. J Biol Chem 255:401–407
16. De Noni I (2008) Release of β-casomorphins 5 and 7 during simulated gastrointestinal digestion of bovine β-casein variants and milk-based infant formulas. Food Chem 110:897–903
17. De Noni I, Cattaneo S (2010) Occurrence of β-casomorphins 5 and 7 in commercial dairy products and in their digests following in vitro-simulated gastrointestinal digestion. Food Chem 119:560–566
18. EFSA European Food Safety Authority (2009) Review of the potential health impact of β-casomorphins and related peptides. Sci Rep 231:1–107
19. Exl BM (2001) A review of recent developments in the use of moderately hydrolyzed whey formulae in infant nutrition. Nutr Res 21:355–379
20. Foltz M, Meynen EE, Bianco V, van Platerink C, Koning TM, Kloek J (2007) Angiotensin converting enzyme inhibitory peptides from a lactotripeptide-enriched milk beverage are absorbed intact into the circulation. J Nutr 137:953–958

21. Fox PF, Kelly AL (2005) Indigenous enzymes in milk: overview and historical aspects—part 1. Int Dairy J 16:500–516
22. Friel JK, Martin SM, Langdon M, Herzberg GR, Buettner GR (2002) Milk from mothers of both premature and full-term infants provides better antioxidant protection than does infant formula. Pediatr Res 51:612–618
23. Gartner LM, Morton J, Lawrence RA, Naylor AJ, O'Hare D, Schanler RJ, Eidelman AI (2005) Breastfeeding and the use of human milk. Pediatrics 115:496–506
24. Haque E, Chand R, Kapila S (2009) Biofunctional properties of bioactive peptides of milk origin. Food Rev Int 25:28–43
25. Hara H, Funabiki R, Iwata M, Yamazaki K (1984) Portal absorption of small peptides in rats under unrestrained conditions. J Nutr 114:1122–1129
26. Hayes M, Ross RP, Fitzgerald GF, Hill C, Stanton C (2006) Casein-derived antimicrobial peptides generated by Lactobacillus acidophilus DPC6026. Appl Environ Microbiol 72:2260–2264
27. Hayes M, Stanton C, Fitzgerald GF, Ross RP (2007) Putting microbes to work: dairy fermentation, cell factories, and bioactive peptides. Part II: bioactive peptide functions. Biotech J 2:435–449
28. Hernández-Ledesma B, Amigo L, Ramos M, Recio I (2004) Release of angiotensin converting enzyme-inhibitory peptides by simulated gastrointestinal digestion of infant formulas. Int Dairy J 14:889–898
29. Hernández-Ledesma B, Quirós A, Amigo L, Recio I (2007) Identification of bioactive peptides after digestion of human milk and infant formula with pepsin and pancreatin. Int Dairy J 17:42–49
30. Jarmolowska B, Sidor K, Iwan M, Bielikowicz K, Kaczmarski M, Kostyra E, Kostyra H (2007a) Changes of ß-casomorphin content in human milk during lactation. Peptides 28:1982–1986
31. Jarmolowska B, Szlapka-Sienkiewicz E, Kostyra E, Kostyra H, Mierzejewska D, Darmochwal-Marcinkiewicz K (2007b) Opioid activity of Humana formula for newborns. J Sci Food Agric 87:2247–2250
32. Kamau SM, Lu RR, Chen W, Liu XM, Tian FW, Shen Y, Gao T (2010) Functional significance of bioactive peptides derived from milk proteins. Food Rev Int 26:386–401
33. Kekkonen R (2009) Bioactive milk protein and peptide functionality. In: Corredig M (ed) Dairy Derived Ingredients. Woodhead Publishing Limited and CRC Press LLC
34. Korhonen H (2009) Milk-derived bioactive peptides: from science to applications. J Funct Foods 1(2):177–187
35. Korhonen H, Pihlanto-Leppälä A, Rantamäki P, Tupasela T (1998) Impact of processing on bioactive proteins and peptides. Trends Food Sci Technol 9(8):307–319
36. Kunz C, Lönnerdal B (1989) Human milk protein: separation of whey proteins and their analysis by polyacrylamide gradient gel electrophoresis, fast protein liquid chromatography (FPLC), gel filtration, and anion-exchange chromatography. Am J Clin Nutr 49:464–470
37. Kunz C, Lönnerdal B (1990) Human-milk proteins: analysis of casein and casein subunits by anion-exchange chromatography, gel electrophoresis, and specific staining methods. Am J Clin Nutr 51:37–46
38. LeBlanc JG, Matar C, Valdez JC, LeBlanc J, Perdigon G (2002) Immunomodulating effects of peptidic fractions issued from milk fermented with Lactobacillus helveticus. J Dairy Sci 85:2733–2742
39. Lönnerdal B (2003) Nutritional and physiologic significance of human milk proteins. Am J Clin Nutr 77:1537S–1543S
40. Martin M, Wellner A, Ossowski I, Henle T (2008) Identification and quantification of inhibitors for angiotensin-converting enzyme in hypoallergenic infant milk formulas. J Agric Food Chem 56:6333–6338
41. McLachlan CNS (2001) ß-casein A1, ischaemic heart disease mortality, and other illnesses. Med Hypotheses 56:262–272

42. Meisel H, Fitzgerald RJ (2000) Opioid peptides encrypted in intact milk protein sequences. Brit J Nutr 84:S27–S31
43. Meisel H, Fitzgerald RJ (2003) Biofunctional peptides from milk proteins: mineral binding and cytomodulatory effects. Curr Pharm Des 9:1289–1295
44. Mellander O (1950) The physiological importance of the casein phosphopeptide calcium salts. II. Peroral calcium dosage of infants. Acta Soc Med Ups 55:247–255
45. Mills S, Ross RP, Hill C, Fitzgerald GF, Stanton C (2011) Milk intelligence: mining milk for bioactive substances associated with human health. Int Dairy J 21:377–401
46. Miquel E, Alegría A, Barberá R, Farré R (2006) Casein phosphopeptides released by simulated gastrointestinal digestion of infant formulas and their potential role in mineral binding. Int Dairy J 16:992–1000
47. Miquel E, Gómez JA, Alegría A, Barberá R, Farré R, Recio I (2005) Identification of casein phosphopeptides released after simulated digestion of milk-based infant formulas. J Agric Food Chem 53:3426–3433
48. Nagpal R, Behare P, Rana R, Kumar A, Kumar M, Arora S, Morotta F, Jain S, Yadav H (2011) Bioactive peptides derived from milk proteins and their health beneficial potentials: an update. Food Funct 2:18–27
49. Nwosu CC, Aldredge DL, Lee H, Lerno LA, Zivkovir AM, German JB, Lebrilla CB (2012) Comparison of the human and bovine milk N-Glycome via high-performance microfluidic chip liquid chromatography and tandem mass spectrometry. J Prot Res 11:2912–2924
50. Rudloff S, Kunz C (1997) Protein and nonprotein nitrogen components in human milk, bovine milk, and infant formula: quantitative and qualitative aspects in infant nutrition. J Pediatr Gastroenterol Nutr 24:328–344
51. Samaranayaka AGP, Li-Chan ECY (2011) Food-derived peptidic antioxidants: a review of their production, assessment, and potential applications. J Funct Foods 3(4):229–254
52. Schlimme E, Meisel H, Frister H (1989) Bioactive sequences in milk proteins. In Barth C, Schlimme E. (eds.) Milk proteins: Nutritional, clinical, functional and technological aspects. Steinkoppf, Darmsdadt, Springer
53. Soetan KO, Olaiya CO, Oyewole OE (2010) The importance of mineral elements for humans, domestic animals, and plants—a review. Afr J Food Sci 4:200–222
54. Tsopmo A, Romanowski A, Banda L, Lavoie JD, Jenssen H, Friel JK (2011) Novel antioxidative peptides from enzymatic digestion of human milk. Food Chem 126:1138–1143
55. Widstrom AM, Christensson K, Ransjo-Arvidson AB, Mathiesen AS, Winberg J, Uvnas-Moberg K (1988) Gastric aspirates of newborn infants: pH, volume, and levels of gastrin- and somatostatin-like immunoreactivity. Acta Pediatr Scand 77:502–508

CHAPTER 16

Conventional and Innovative Processing of Milk for Yogurt Manufacture; Development of Texture and Flavor: A Review

Panagiotis Sfakianakis and Constatnina Tzia *

Laboratory of Food Chemistry and Technology, School of Chemical Engineering, National Technical University of Athens, 5 Iroon Polytechniou St., Polytechnioupoli, 15780, Zografou, Greece

ABSTRACT

Milk and yogurt are important elements of the human diet, due to their high nutritional value and their appealing sensory properties. During milk processing (homogenization, pasteurization) and further yogurt manufacture (fermentation) physicochemical changes occur that affect the flavor and texture of these products while the development of standardized processes contributes to the development of desirable textural and flavor characteristics. The processes that take place during milk processing and yogurt manufacture with conventional industrial methods, as well as with innovative methods currently proposed (ultra-high pressure, ultrasound, microfluidization, pulsed electric fields), and their effect on the texture and flavor of the final conventional or probiotic/prebiotic products will be presented in this review.

Keywords: milk; yogurt; yogurt manufacture; thermal treatment; homogenization; pressure; ultra high pressure; ultrasound; microfluidization; pulsed electric fields; prebiotic; probiotic

1. INTRODUCTION—HISTORY OF YOGURT

Milk and dairy products have been consumed since the domestication of mammals; yogurt and similar fermented milk products in particular are thought to originate from the Middle East. The original production of fermented milk products derived from the need to prolong the shelf life of milk instead of being disposed [1]. Yogurt manufacture was initially based on knowledge and empirical processes without standard procedures or investigation of the steps that occur during the entire process. Only after the late 20th century, when yogurt became a profitable commercial good, its manufacture became industrialized and the processes were standardized. During the last 20 years, interest in yogurt manufacture has increased tremendously for scientific and commercial reasons. Scientific findings have suggested new dairy products that benefit human health (probiotic cultures, fortification with bioactive compounds) as well as with improved sensory, especially textural characteristics. Thus, consumer demand for yogurt and similar fermented dairy products has increased.

Yogurt is defined as the product being manufactured from milk—with or without the addition of some natural derivative of milk, such as skim milk powder, whey concentrates, caseinates or cream—with a gel structure that results from the coagulation of the milk proteins, due to the lactic acid secreted by defined species of bacteria cultures. Furthermore, these bacteria must be "viable and abundant" at the time of consumption [2]. The above definition is part of the food legislation of many countries, ensuring that the essential characteristics of yogurt will be preserved, as well as that its traditional "concept" will not be compromised. The most common types of yogurt commercially available are set type yogurt and strained yogurt; though lately frozen yogurt and drinking yogurt have become quite popular as well. Set type yogurt is fermented in retail containers and no further stirring or water removal takes pace after the fermentation process. Strained (or stirred, or Greek style) yogurt is fermented in tanks under continuous mild stirring and after the completion of fermentation a portion of the whey is removed. Due to the manufacturing process, the two types develop a different texture; set type yogurt develops a continuous gel texture, whereas strained yogurt displays a viscous, creamy smooth texture [1].

2. STANDARDIZED YOGURT MANUFACTURING PROCESS

Yogurt manufacture begins with the milking of the mammal, includes several processes, ending with the packaging of the final product, yogurt. The aim of this work is to present the processes that take place in the dairy industry; no further details about the milking and transportation of the milk will be included, even though these critical stages affect the quality and safety of the final product. Yogurt is mainly produced from bovine milk, although milk from other mammals is utilized for yogurt production as well. Yogurt derived from the milk of species other than bovine tends to vary in several sensory and physicochemical characteristics, due to differences per milk composition. For instance, yogurt derived from milk with high fat content (e.g., sheep, goat, and buffalo) has a more creamy texture compared to that derived from milk with lower fat content (e.g., bovine, mare, and ass). Therefore, the species of the milk-producing mammal significantly influence the characteristics of the produced yogurt [3].

2.1. Initial Treatment of Milk

Raw milk undergoes, in the dairy industry, centrifugal clarification to remove somatic cells and any other solid impurities [3]. Afterwards, a mild heating process, known as thermalization, is performed at temperature range 60–69 °C for 20–30 s, aiming at the killing of many vegetative microorganisms and the partial inactivation of some enzymes. This process causes almost no other irreversible change in the milk [4]. After thermalization, milk is cooled <5 °C or inoculated with lactic acid bacteria or other microfloras to control the growth of the psychrotrophic bacteria [3].

2.2. Standardization of Milk Components—Fat and SNF (Solid Non-Fat) Content

The standardization of milk refers to the standardization of fat and solid-non-fat content (SNF). Bovine milk fat content varies from 3.2%–4.2% w/w. The fat content of the milk is adjusted to range from <0.5%, for skim milk, to 1.5%–2%, for semi-fat milk, to 3.5% for full fat milk. As far as yogurt is concerned, the fat content ranges from 0.1%–10% according to consumer demands. In practice, to achieve the designed fat level, either the addition of skim milk or milk fat or the separation of fat from milk via centrifuge and mixing milk fat with skimmed milk is carried out

[3]. The standardization process is of paramount importance, because the fat content of the milk influences the yogurt characteristics; increasing the fat content of milk results in an increase in the consistency and viscosity of yogurt [5,6]. Also, the milk fat content affects the maximum rate of pH decrease and pH lag phase during yogurt fermentation [7].

The term of standardization is also applied to the SNF content of the milk. The SNF components of milk mainly consist of lactose, protein and minerals; SNF content of milk varies from 11% to 14% of the total weight of the milk while the SNF of yogurt ranges from 9% to 16%. The SNF content of milk used for yogurt manufacture is altered, in some cases, by producers in order to attain the desired characteristics of the coagulum; the higher the SNF level, the higher the resulting yogurt's viscosity and firmness. The addition of native milk components is permitted to yogurt and fermented milk products in some countries. It is quite common in yogurt manufacturing to fortify the milk mixture with milk powder (skimmed or full fat), whey protein concentrates or casein powder, to achieve the desired SNF content and subsequently an increase in firmness and cohesiveness [5]. It must be noted that the fat and SNF content of milk has an impact on the fermentation process. In particular, the interaction of milk SNF content and fermentation temperature has a significant effect on the duration of the fermentation process; an increase of SNF increases the duration of the fermentation process [8].

2.3. Homogenization

Milk is a typical oil in water (o/w) emulsion with milk fat globules (MFG) acting as the oil droplets and the milk fat globules membrane as the emulsifier. However, because of the reaction of agglutinins and interfacial tension, the fat globules tend to collide, either by sharing the membrane or because of the Laplace principle, according to which the pressure is greater inside small globules than inside large globules and, hence, there is a tendency for large fat globules to grow at the expense of the smallest. This phenomenon, in addition to Brownian motion, forces the milk fat to rise to the surface of the milk and thus creates the undesirable effect of separation [9,10]. In order to prevent this effect, standardized milk undergoes homogenization. The basic principle of milk homogenization is to subject MFG to severe conditions in order to disrupt the membrane surrounding them and then maintain the new globules in dispersion while a new membrane is formed at the fat serum interface. The severe conditions that cause milk homogenization can be achieved by the application of pressure, high velocity flow of the milk, or high frequency vibrations (>10 kHz). The shear stress and temperature gradient developed under these conditions lead to a cavitation phenomenon that contributes to the homogenization process. Homogenization is carried out by the application of pressure. In particular, the pressure commonly applied in the dairy industry is 10–20 MPa [11,12]. The main homogenization effects are a reduction of the diameter of the MFG from 2–10 μm to 0.1–1 μm and altering the composition of the MFG membrane. According to Cano-Ruiz and Richter [13], the membrane absorbs protein molecules, mostly caseins, from the milk serum to become sufficient to emulsify the new-formed globules, since the fat surface area increases due to homogenization. Aguilera and Kessler [14] showed that a reduction of MFG size and the alterations in the MFG membrane caused by homogenization contribute to milk emulsion stability. Furthermore, homogenization affects the characteristics of acidified milk gels, like yogurt. According to Cho, *et al.* [15], the smaller MFG facilitate the incorporation of fat into the protein network [5], while their increased surface area favors the interactions between fat and milk proteins, casein and denatured whey, during acidification and subsequent gel formation [5,15].

2.4. Heat Treatment

Heat treatment of milk is carried out to ensure the safety of the product, whether it is milk itself or any other dairy product, and to exploit several effects that increased temperature has on certain milk components facilitating further processes for dairy products manufacture [16]. Heat treatment of milk reduces the number of pathogenic microorganisms to safe limits for the consumer's health. Various heat treatments can be applied, which are classified based on the duration and the temperature (Table 1). The most common are known as thermalization (referred in Section 2.1), low and high pasteurization, sterilization and UHT (Ultra Heat Treatment) [3,4,17]. Low pasteurization refers to heat treatment of milk at 63–65 °C for 20 min or at 72–75 °C for 15–20 s (HTST, High Temperature Short Time). During this process, most pathogens, vegetative bacteria, yeast and molds are killed. Additionally, with low temperature pasteurization, several enzymes become inactive, while the flavor of milk is hardly altered. Furthermore, little or no serum proteins are denatured, and cold agglutination and bacteriostatic properties remain virtually intact [2,4]. A more intense heat treatment is high temperature pasteurization that requires a temperature of 85 °C for 20–30 min or 90–95 °C for 5 min. During high temperature pasteurization most vegetative microorganisms are killed, except from spores; most enzymes are deactivated (except milk proteinase, plasmin in particular, some bacterial proteinases and lipases); most whey proteins are denatured, and a distinct "cooked" flavor is developed due to the formation, mostly, of ketones [4,18]; no further irreversible changes occur. Sterilization results in extermination of all microbial content of milk, including bacterial spores, and it is achieved at 110 °C for 30 min or at 130 °C for 40 s. In addition, sterilization causes inactivation of most milk enzymes (except several bacterial lipases), darkening of the milk color due to the Maillard reaction, evaporation of most flavor volatiles, thus weakening the flavor of the milk, and considerable damage to all milk proteins, even caseins. Finally, UHT is carried out at 145 °C for 1–2 s and achieves equal bacterial eradication as from sterilization, minimal flavor deterioration and causes denaturation of several whey proteins (β-lactoglobulin, serum albumin, and some immunoglobulins). UHT treatment and high pasteurization produces many volatiles in milk, such as: 2-pentanone, 2-heptanone, 2-nonanone, 2-undecanone, 2,6-dimethylpyrazine, 2-ethylpyrazine, 2-ethyl-3-methylpyrazine, methional, pentanoic acid, benzothiazole vanillin, hexanal, benzothiazole, decalactone, H_2S, methanethiol, dimethylsulphide and carboxylsulphide. These sulfur containing molecules are responsible for the "cooked" off flavor developed during UHT and high temperature pasteurization [18]. It should be mentioned that the most commonly used heat treatment in the yogurt manufacturing process is the high temperature pasteurization at 85 °C for 20 min [3,4].

Pathogens that can grow in milk, due to bad hygiene practices or hardware failure during the stages of processing, include *Mycobacterium tuberculosis*, *Coxiella burnetii*, *Staphylococcusaureus*, *Salmonella* species, *Listeria monocytogenes*, and *Campylobacter jejuni*. These microorganisms are killed by even mild heat treatment ensuring that processed milk is safe for consumption. The claim that milk is safe after a mild heat treatment might sound frivolous, but most high-heat resistant pathogens either do not occur in milk (e.g., *Bacillus anthracis*) or are outnumbered by other native microorganisms (e.g., *Clostridium perfringens*), or cause spoilage before their quantity is enough to cause health issues (e.g., *Bacillus cereus*).

In addition to the reduction or complete extermination of microbiological load, heat treatment causes release of CO_2 and O_2, an increase in the amount of insoluble colloidal calcium phosphate, a decrease in calcium cations, and forces lactose isomerization, degradation and Maillard reaction, thus affecting the pH of the milk and flavor. Finally for yogurt, the most important changes during heat treatment of milk concern the milk proteins; the reactions of milk proteins, during heat treatment, have a serious impact on the yogurt curd formation and will be described more thoroughly [4].

The casein molecules in milk are in the form of micelles or aggregates of submicelles which are formed from α_{s1}-, α_{s2}- and β-caseins stabilized by κ-casein molecules held together by calcium and calcium phosphate. This structure is stable and requires a high amount of energy to be disrupted. On the other hand, whey proteins in solution have a globular shape. Whey proteins, due to their structure, are fairly stable and do not interact with fatty molecules, calcium ions or caseins in their native state. However, in the case of whey proteins (β-lactoglobulin, serum albumin) over 80 °C, their peptide chains unfold, thus denaturating irreversibly. This deformation of the peptide chains exposes their thiol groups and enables them to interact with other molecules forming S–S bonds. Depending on the pH of the environment and the proximity of molecules available, whey proteins can form bonds with other whey proteins and caseins (κ- and α_{s1}- mostly) and also they can be incorporated at the MFG membrane. The denatured whey proteins, especially at pH values lower than 6.5, have the tendency to associate with casein micelles [4,11,16]. All the above phenomena are of paramount importance and are exploited during yogurt manufacture. Yogurt curd formation is based on the isoelectric precipitation of casein. However, whey proteins can be involved; if the thiol groups of the whey proteins are exposed, an interaction between casein and whey protein molecules occur, and the formation of casein-whey bonds are facilitated. Thus, whey proteins are incorporated into the curd matrix, strengthening the latter and resulting in a more firm yogurt. Therefore, the heat induced denaturation of whey protein favors the yogurt formation with high firmness and viscosity values [5,19]. Table 1 summarizes the thermal treatments utilized in dairy processing and the respective effects on milk itself and yogurt.

2.5. Fermentation Process

The fermentation process is the most important stage of yogurt manufacture. During this stage, the yogurt curd is formed, and its textural characteristics and distinct flavor are developed [3,5]. The key factor of the fermentation process is the starter culture that acts through biochemical reactions and inductively causes the formation of the curd and the development of flavor components [5]. For a fermented dairy product to be labeled as "yogurt", it should contain the two live bacterial strains of *Streptococcus salivarius* subsp. *thermophilus* and *Lactobacillus delbrueckii* subsp. *bulgaricus* in abundance. However, yogurt starter cultures may include other microorganisms as well, like *Lactobacillus acidophilus, Lactobacillus casei, Lactobacillus lactis, Lactobacillus jugurti, Lactobacillus helveticus, Bifidobacterium longum, Bifidobacterium bifidus* and *Bifidobacterium infantis. Streptococcus*

Table 1. Impact of different thermal treatment techniques on milk and yogurt properties affecting flavor and texture.

Milk Treatment	Treatment Description	Effect on Milk	Effect on Yogurt
Thermalisation	Heating at 60–69 °C, for 20–30 s	Death of non-heat resistance bacteria. Inactivation of several enzymes [4].	No significant effect. Characteristics affected by further processing [4].
Low Pasteurization	Heating at 63–65 °C for 20 min/at 72–75 °C for 15–20 s (HTST)	Death of most pathogens, vegetative bacteria, yeast and molds. Several enzymes denatured, denaturation of several whey proteins [4].	Slight increase in viscosity and firmness [1].
High Pasteurization	Heating at 85 °C for 20–30 min/at 90–95 °C for 5 min	Death of most vegetative microorganisms, except spores. Deactivation of most enzymes. Denaturation of most whey proteins. Development of "cooked" flavor [4,18].	Large increase in viscosity and firmness [1].
Sterilization	Heating at 110 °C for 30 min/at 130 °C for 40 s	Extermination of all microorganisms. Deactivation of most enzymes. Denaturation of whey proteins and aggregation of caseins (casein micelles) and MFG. Weakening of flavor intensity. Color darkening [4,18].	Incorporation of whey proteins into casein matrix. Very large increase in viscosity and firmness [1,4].
Ultra Heat Treatment (UHT)	Heating at 145 °C for 1–2 s	Extermination of all microorganisms. Mild flavor deterioration. Denaturation of whey proteins (β-lactoglobulin, serum albumin, several immunoglobulins) Development of off-flavors. Color darkening [4,18].	Medium increase in viscosity and firmness [1].

thermophilus subsp. *thermophilus* (ST) is the only species in the streptococcus genus that is used in dairy starter cultures. ST is Gram positive and usually considered thermophilic, however, as the optimum temperature for its growth is 35–53 °C; therefore, ST can be considered as "thermotolerant". Its cells are spherical in shape, forming chains, during the early stage of their lives and as they mature develop a more rod-like morphology and favor colonial growth. *Lactobacillus delbrueckii* subsp. *bulgaricus* (LB) is rod-shaped, Gram-positive, anaerobic bacteria and its optimum growth temperature is 40–44 °C. LB can produce very high amounts of lactic acid by metabolizing lactose [5,20]. These two species display synergy in the milk environment, metabolizing lactose into lactic acid and causing reduction of milk pH. The synergism between ST and LB is based on their individual characteristics, and as a result higher lactose metabolism and lactic acid production is attained compared to each one acting individually. ST is more "aerotolerant" than LB, lacks good proteolytic ability in comparison to LB, but possesses greater peptidase activity. When grown together in milk, ST grows vigorously at first, whereas LB grows slowly. ST, because of its great proteolytic activity, creates an abundance of peptides to stimulate the growth of LB. During the early stage of fermentation, milk lactose is transported through the cell membrane of ST with the help of the enzyme galactoside permease located in the membrane. The lactose in the cell is then hydrolyzed by lactase or β-galactosidase enzyme. ST produces significant levels of lactase, which catalyzes the hydrolysis of lactose to glucose and galactose. Glucose is converted to pyruvate which is metabolized to lactic acid by the enzyme lactic dehydrogenase. Lowered oxygen tension and formate (byproduct of ST metabolism) in turn stimulate LB growth, which is further aided by the amino acids released by the active peptidases secreted by ST. Through coordinated tandem activities, both bacteria accelerate the entire fermentation, which none of them would be able to achieve individually. When the pH of the yogurt approaches 5.0, activity of ST subsides and LB gradually dominates the overall fermentation process until the target value of pH is reached and the fermentation process ceases. Normally, the fermentation period is terminated by lowering the temperature to 4 °C. At this temperature, the culture is still alive, but its activity is drastically limited to allow controlled flavor during storage and distribution [5,20].

The growth of the symbiotic culture induces changes in the native components of the milk that are responsible for the physicochemical and sensory characteristics of yogurt. During fermentation, lactose, milk proteins and microbial content, as well as several carbon compounds, suffer major changes, whereas minor changes occur for vitamins and minerals. Lactose is reduced by 30% and produces double the molar amount of lactic acid. Proteins (caseins and whey) aggregate, increasing the consistency of yogurt. Due to proteolysis caused by starter culture, amino acids (mainly proline and glycine) are released into the yogurt, even during storage at 4 °C [20]. During incubation, the starter culture growth results in an increase in the system's microbial content from 10^8 to 10^{10} CFU g^{-1}. The carbonyl compounds formed during fermentation are mostly lactic acid, acetaldehyde, dimethyl sulfide, 2,3-butanedione, 2,3-pentanedione, 2-methylthiophene, 3-methyl-2-butenal, 1-octen-3-one, dimethyl trisulfide, 1-nonen-3-one, acetic acid, methional, (*cis,cis*)-nonenal, 2-methyl tetrahydrothiophen-3-one, 2-phenyacetaldehyde, 3-methylbutyric acid, caproic acid, guaiacol and benzothiozole. These compounds contribute to the distinctive flavor of yogurt. As far as the lipids are concerned, several free fatty acid molecules are liberated due to lipase activity, mostly stearic and oleic acid. The only change in vitamin content is an increase in Vitamin B, throughout the fermentation process and storage. Finally, the quantity of minerals in yogurt remain the same as in milk, the only change is that, due to pH lowering, these minerals are in ionic rather than colloidal form [5,18].

The concentration of lactic acid in milk during fermentation increases, pH decreases, therefore the carboxyl groups dissociate, serine phosphate is ionized, and the negative charge between casein micelles is increased. However, the presence of calcium phosphate neutralizes this negative charge, keeping electrostatic repulsion down to a level where attractive forces between the protein molecules are dominant. Due to these attractive forces, the casein micelles aggregate and eventually coagulate into a network of small chains; this is responsible for the increase of viscosity and formation of the yogurt coagulum [3,19,21,22].

The milk fermentation process of yogurt can be described adequately by the evolution of pH and viscosity with respect to time; the model that expresses the evolution of pH during fermentation time is the modified Gompertz models of de Brabandere and de Baerdemaeker (1999) (Equation (1)) [23]:

$$pH = pH_0 + (pH_0 - pH_\infty) - \left\{-\exp\left[\frac{e \cdot \mu_{pH}}{(pH_0 - pH_\infty)} \cdot (\lambda_{pH} - t) + 1\right]\right\} \quad (1)$$

pH_0, pH_∞ = initial and end values of pH, respectively; μ_{pH} (min^{-1}) = maximum rate of pH decrease; λ_{pH} (min) = duration of pH lag phase. Furthermore, the model that describes the evolution of viscosity during fermentation is the modified Gomperz model of Soukoulis, *et al.* (2007) (Equation (2)) [7]:

$$\mu_\alpha = \mu_{\alpha 0} + (\mu_{\alpha 0} - \mu_{\alpha\infty}) - \left\{-\exp\left[\frac{e \cdot \mu_v}{(\mu_{\alpha 0} - \mu_{\alpha\infty})} \cdot (\lambda_v - t) + 1\right]\right\} \quad (2)$$

$\mu_{\alpha 0}$, $\mu_{\alpha\infty}$ (Pa·s) = initial and end values of viscosity respectively; μ_v (min^{-1}) = maximum rate of viscosity decrease; λ_v(min) = duration of viscosity lag phase.

2.6. Cooling

After the pH of yogurt reaches the value of 4.7–4.3, the yogurt is cooled to around 5 °C. This inhibits the growth and metabolic reaction of the starter culture and prevents the rise in acidity. Cooling of yogurt can be in one or two phases. One-phase cooling involves the rapid decrease of the coagulum temperature to less than 10 °C, where the fermentation process is inhibited leading to yogurt with low viscosity. Two-phase cooling is initiated by rapidly decreasing the temperature to less than 20 °C and then gradually reaching the storage temperature of 5 °C leading to yogurt with an increased viscosity and limited syneresis. This is quite common in the yogurt manufacture process, especially when fruits are to be added [3,24].

3. INNOVATIVE METHODS FOR MILK AND YOGURT PROCESSING

Among the processes involved in milk and yogurt processing, the most important are homogenization, pasteurization and fermentation. Apart from conventional processes previously mentioned, new trends in milk processing that involve the utilization of ultra high pressure, ultrasound, pulsed electric field and microfluidization are presented in this section. Last, but not least, the probiotic bacteria and prebiotics are discussed that have beneficial effects on human health and have extensively been utilized in the dairy industry. Probiotics and prebiotics are worth mentioning, as dairy products are the most common medium for delivering them to the human intestine. In Table 2, the effects of conventional or innovative methods of homogenization on milk and yogurt properties are presented that generate the flavor and texture characteristics.

3.1. Ultra High Pressure Milk Treatment and Effect on Yogurt Characteristics

Ultra high pressure (UHP) involves the application of pressures from 100 to 1000 MPa. UHP utilization in food products was initiated during the early 1980s and is a non thermal pasteurization method. Studies have shown that milk treatment with pressures of 400–600 MPa for 10 min at 25 °C can achieve a similar result to low temperature pasteurization in terms of pathogenic and spoilage microorganisms inactivation [25]. Similar studies on UHP treatment of milk from Johnston, *et al.* [26] and Law, *et al.* [27] described the disintegration of the casein micelles into smaller particles and the simultaneous increase in the amount of caseins and calcium phosphate in the serum phase. Also, it was found by the same researchers that after UHP treatment, especially with pressure higher than 500 MPa, that the denaturation of several whey proteins occurs, in particular β-lactoglobulin, several immunoglobulins and α-lactalbumin [25,28]. However, Gervilla, *et al.* [29] observed a strange effect of UHP on the size and distribution of MFG. UHP up to 500 MPa at 25 and 50 °C reduced the diameter of MFG in the range of 1–2 μm, but at 4 °C the MFG displayed no tendency to shrink. This could be attributed to the fact that the MFG membrane remained unchanged.

The application of UHP in milk acid gel formation improves the texture and firmness, reduces syneresis and increases the water holding capacity in comparison to conventional yogurts [25]. The combination of UHP and thermal treatment is reported by Ferragut, *et al.* [30] to increase yogurt viscosity and lower gelation times compared to UHP treated samples.

3.2. Ultrasound Milk Treatment and Yogurt Characteristics

Ultrasound (US) is a sound wave with a frequency higher than the upper limit of human hearing, typically higher than 20 kHz. US has been utilized in the food industry since the late 1960s, for cleaning, monitoring and food component characterization. High intensity US (power level higher than 10 W), when propagated through a solution, generate immense pressure, temperature and shear gradients and thus cause cavitation [31,32]. Therefore, US is considered as an alternative method for reducing MFG size and can be effectively applied to homogenize milk. Wu, *et al.* [33] and Nguyen and Anema [34] showed that the application of US in milk reduces the MFG diameter to between 0.1 and 0.6 μm. In addition, US treatment has been referred by Krešić, *et al.* [35] and Chandrapala, *et al.* [36] to cause alterations of the MFG membrane composition and structure leading to an efficient homogenization effect compared to conventional methods. Additionally, the effect of US on milk proteins has been studied by Madadlou, *et al.* [37] and by Chandrapala, *et al.* [36]. US treatment has been shown to cause alteration in the secondary structure of the milk proteins, aggregation of protein particles as well as denaturation [36,37]. Riener, *et al.* [38] combined US with heat treatment (thermosonication) of milk and achieved a similar effect on the MFG as obtained with US treatment without heat, leading to reduction in size and changes of the membrane allowing interaction with casein micelles. Specifically, thermosonication treatment leads to an average diameter of 0.6 μm of MFG and a MFG membrane richer in casein molecules than the native [38]. Furthermore, high amplitude US has been reported to reduce the microbial content of milk [32,39]. Finally, high intensity US treatment causes the emission of volatiles from milk and formation of off flavors. Based on the study conducted by Riener, *et al.* [40], when milk is ultrasonicated, benzene, toluene, 1,3-butadiene, 5-methyl-1,3-cyclopentadiene, 1-hexene, 1-octene, 1-nonene, p-xylene, *n*-hexanal, *n*-heptanal, 2-butanone, acetone, dimethylsulfide and chloroform are emitted. The aldehydes can be produced from the breakdown of hydroperoxides generated by photo-oxidation induced by US, whereas the series of C6–C9 1-alkenes could arise from pyrolytic cleavage of fatty acid chains. The benzene formation may be attributed to cleavage of side chains of amino acids such as phenylalanine. These volatiles cause a rubbery and burned aroma [40].

The implementation of US treatment on the production of fermented dairy products has been studied with promising results. Milk gels and yogurt produced from milk treated by high intensity US have shown improved physical properties and high value of texture characteristics (firmness, cohesiveness). Increased amplitude level US treatment (20 kHz, 50–500 W, 1–10 min) significantly improved the water holding capacity of yogurt and increased viscosity and reduced syneresis; moreover, higher US intensity and a higher US exposure time of the milk resulted in increased yogurt viscosity [33]. Increased yogurt viscosity was reported by Riener, *et al.* [38] even from skim milk treated with US (22 kHz, 50 W, 0–30 min), due to high thermal denaturation of whey proteins. Milk thermosonication prior to fermentation, (25 kHz, 400 W, 45 or 75 °C for 10 min) resulted in the formation of yogurt with greater viscosity and higher water holding capacities compared to conventionally treated milk. The same treatment altered the microstructure of yogurt resulting in a honeycomb like network and exhibiting a more porous nature, whose average structural size was smaller (~2 μm) compared to conventionally heated yogurt [40]. A study by Vercet, *et al.* [41] combined thermosonication treatment of milk (40 °C, 20 kHz for 12 s) with moderate pressure (2 kg × cm^{-2}) and showed that the apparent viscosity, yield stress, and viscoelastic properties of yogurt were increased in addition to its structure being strengthened. The increased viscosity and texture of yogurt derived from ultrasonicated milk can be attributed to the denaturation of whey proteins and association of the latter with caseins. Denatured whey proteins are more susceptible to association with casein and casein micelles. Additionally, during acidification the denatured whey proteins, associated or not with casein micelles, aggregate due to the reduction of the repulsive charge. Therefore, denatured whey proteins associated with casein micelles could act as bridging material between casein micelles and as a result the bonds that form in the yogurt matrix are formed more easily, resulting in stronger yogurt coagulum [42].

3.3. Application of Microfluidization in Milk and Yogurt Manufacture

The Microfluidizer® is an apparatus that causes homogenization via shear, turbulence and cavitation. Initially, it accelerates the fluid and separates it into two microstreams that intersects in a chamber and collide. The impact causes intense turbulence and cavitation and thus the homogenization effect is achieved [43,44]. In the case of milk studied by Ciron, *et al.* [43], the microfluidization treatment reduced the diameter of the MFG to less than 2 μm [45]. The application of microfluidization in yogurt manufacture has often been used. A comparison between yogurts derived from microfluidized milk of 0% and 1.5% fat content with conventionally homogenized milk showed that non-fat yogurt from microfluidized milk displayed increased syneresis, and reduced viscosity and firmness compared to conventionally manufactured yogurt, whereas low fat yogurts from microfluidized milk had similar texture characteristics with those from conventionally homogenized milk. Microfluidization of low-fat milk resulted in yogurt with modified microstructure, giving more interconnectivity in the protein networks with embedded fat globules, but with similar texture profiles and water retention compared to yogurt prepared from conventionally homogenized milk [43]. However, this technique requires more study to assess the efficiency of yogurt manufacture [43,46].

3.4. Pulsed Electric Field Application in Milk and Yogurt Manufacture

Pulsed electric field (PEF) treatment emits intense electric pulses through a continuous medium, to inactivate microorganisms with the best results achieved in fluids. PEF has been applied in dairy systems combined with probiotic cultures [47]. The intensity of fields range between 15 and 50 kV/cm, and the treatment lasts only a few seconds [48]. The PEF principle is to destabilize the microbial cells with a high-pressure pulse. Subsequently, electroporation to the cellular membrane makes it more permeable; therefore, the cells rupture and expel their contents [49]. Additionally, Lin, *et al.* [50] combined PEF, UHP and thermal treatment in milk and noticed even more of a decrease in microbiological content. The efficiency of PEF depends on the intensity of the electric field, and the number and duration of pulses [51]. Despite its potential, PEF application requires a high tolerance to elevated electric fields, low electric conductivity and absence of bubbles [51,52].

3.5. Probiotic and Dairy Products

Modern nutritional probiotic products are classified as those that contain probiotic microorganisms. Probiotic microorganisms are defined as live microbes, which when ingested benefit the health of the host through their effect on the intestinal microflora. In addition, probiotic cultures must be able to survive throughout the intestinal tract, resist acidic conditions during gastric passage and bile digestion. To have their beneficial effect on the health of the host, probiotic strains must, at least temporarily, establish themselves among the natural microflora of the intestine. The initial mention about microorganisms that benefit the health of host being, while delivered via food consumption is attributed

Table 2. Impact of different homogenization techniques on milk and yogurt properties affecting flavor and texture.

Milk Treatment	Treatment Description	Effect on Milk	Effect on Yogurt
Conventional with Pressure	10–20 MPa	Decrease of MFG size. Stability of milk as an emulsion. Whey proteins absorption to the MFG membrane.	Facilitation of curd formation. Whey protein incorporation into the casein matrix. Slight increase in viscosity and firmness.
Ultra High Pressure (UHP)	100–1000 MPa	Inactivation of spoilage and pathogenic microorganism. Casein micelles disruption. Denaturation of several whey proteins. MFG size decrease with a tendency for collision and re-aggregation.	Higher value of texture characteristics. Higher viscosity. Lower syneresis. Increased water holding capacity.
High Intensity Ultrasonication	Higher than 20 kHz, amplitude higher than 100 W	MFG size reduction. Stability of milk as an emulsion. Interaction of whey proteins with casein micelles and MFG. Reduction of microbial content. Development of off-flavor volatiles.	Higher value of texture characteristics. Higher viscosity. Lower syneresis. Increased water holding capacity.
Microfluidization	Separation of milk into two steams, moving at high velocity with subsequent collision.	MFG size reduction.	Non-fat yogurt: increased syneresis and lower viscosity. Low fat yogurt: similar texture characteristics as for conventionally manufactured yogurt.
Pulsed Electric Field (PEF)	Application of electric pulses through milk. Intensity: 1–50 kV/cm for a few seconds.	Microbial content reduction.	Similar texture and water holding capacity as for conventionally manufactured yogurt.

to Élie Metchnikoff in the early years of the 20th century [53]. The terminology "probiotic" is attributed to Lilley and Stillwell [54] in order to differentiate them from antibiotics. These authors defined probiotics as a substance produced by one microorganism stimulating the growth of another microorganism [54]. Later on, in 1971, Sperti gave this name to tissue extracts promoting the growth of microbes [55]. The definition by Parker in 1974 is closer to the commonly accepted definition: organisms and substances which contribute to the intestinal microbial balance [56]. The accepted definition of probiotics is a slightly improved definition by Fuller in 1989 "live microbial feed supplement which beneficially affects the host by improving its intestinal microbial balance" [57]. Probiotics are added to food as adjunct cultures in concentrations of 10^7–10^8 CFU/g or mL, if not participating in the fermentative process, and if participating, they can reach a concentration between 10^8 and 10^9 CFU/g or mL after the fermentative process [58]. The most common probiotic bacteria are strains and species of Lactobacilli, Bifidobacteria, Enterococci and Lactococci. The species most thoroughly been studied are *Lactobacillus acidophilus*, *Lactobacillus casei*, *Lactobacillus lactis*, *Lactobacillus helveticus*, *Bifidobacterium longum*, *Bifidobacterium lactis*, *Bifidobacterium animalis* ssp. *lactis* and *Bifidobacterium bifidum*, *Bifidobacterium longum*, and *Bifidobacterium bifidus* [59].

Fermented dairy products, and especially yogurt, are ideal carriers for probiotic cultures to enter the human digestive system and ensure their survivability through the stomach [60]; therefore, the subject of probiotic yogurt is thoroughly studied from its medical and dairy science perspective. For the purpose of this work, only the effect of the probiotic cultures on the yogurt manufacture and characteristics is mentioned.

Most probiotic bacteria have no significant effect on the fermentation process or on the yogurt sensory properties [61]. Based on the studies of Allgeyer, *et al.* [61] and Atunes, *et al.* [62], the addition of *Bifidobacterium lactis* and *Lactobacillus acidophilus* resulted in no difference in the sensory evaluation of low and full fat probiotic and non-probiotic yogurt. The same results were noted by other researchers as well [63,64]. However, according to Akalin, *et al.* [65] the combination of probiotic culture (*Bifidobacterium animalis* ssp. *Lactis*) and fortification of WPC (whey protein concentrate) resulted in a stronger coagulum and increased firmness and adhesiveness values for the yogurt. This was attributed mostly to the effect of the fortification with WPC and not to the presence of the probiotic culture.

Overall, the incorporation of probiotic bacteria in yogurt manufacture is viable and this claim is shown by the amount of probiotic yogurt products available in the market; however, the actual challenge is to ensure that the probiotic culture reaches the intestines of the consumers alive and is able to establish itself among the native microflora [66].

3.6. Prebiotics and Dairy Products

According to Gibson and Roberfroid [67], prebiotics are classified as certain food ingredients that beneficially affect the host in a very specific way. Prebiotics are food components, non-digestible by humans, that selectively stimulate the growth and activity of certain bacterial species already existing in the human colon, and inductively improve the health of the host. Most prebiotics are oligosaccharides in general, fructooligosaccharides in particular. Most common oligosaccharides with prebiotic character are inulin, *trans*-galactooligosaccharide, lactulose, isomalt and oligofructose. They have been shown to stimulate the growth of endogenous Bifidobacteria, and make them the predominant species in human feces [68]. Based on the study of Cruza, *et al.* [69] who added oligofructose to plain yogurt, there was no influence on the pH, the proteolytic ability, or on the viability of *Streptococcus thermophilus* or *Lactobacillus bulgaricus*. However, the end product of this endeavor was characterized as a weak gel with thixotropic and pseudoplastic behavior. Finally, the oligofructose-fortified yogurts had a fairly high acceptance by consumers. Another study conducted by Pimentel [70] suggested that the addition of long chain inulin, another known prebiotic, in low fat yogurt can lead to interesting results in sensory properties and especially in texture characteristics. In particular, the replacement of native milk fats with long chain inulin created equally acceptable firmness and color as with yogurt containing native milk fats. Most studies on the addition of prebiotic oligosaccharides in yogurt agree that most of the characteristics of the final product and the process remain fairly close to the values of the originals. Several short chain prebiotics have a slightly negative effect on the firmness and creaminess of the yogurt whereas long chain prebiotics increase those values. Overall, the final choice remains with the consumer and their preference for texture.

4. CONCLUSIONS

Dairy processing and yogurt manufacture utilize several scientifically interesting procedures such as centrifugation, homogenization, heat treatment, and in the case of yogurt, manufacture and fermentation. Each procedure significantly affects the quality and sensory characteristics of the final product, whether it is milk or yogurt. Conventional heat treatment includes thermalization, low and high temperature pasteurization, ultra heat treatment and sterilization. The application of heat treatment in milk affects the flavor, the microbial content and the milk proteins. The more intense the heat treatment is, the more radical the changes that occur. Heat treatment also affects the texture of the produced yogurt, increasing the value of its texture characteristics (firmness, cohesiveness) and viscosity. Homogenization, typically used in dairy processing, is through application of pressure, reducing the milk fat globule size and preventing fat separation from the milk. Other treatments that cause the same homogenization effect with pressure in milk are ultra high pressure, ultrasound, microfluidization and pulsed electric fields. Each type of homogenization causes additional effects on milk and on the produced yogurt. Finally, conventional fermentation process, in yogurt manufacturing, included the utilization of the species *Streptococcus salivarius* subsp. *thermophilus* and *Lactobacillus delbrueckii* subsp. *bulgaricus*. Modern dairy science and nutrition have suggested the involvement of probiotic cultures and prebiotic ingredients in order to increase the nutritional value of dairy products, while minimizing detrimental effects on the sensory characteristics.

CONFLICTS OF INTEREST

The authors declare no conflict of interest.

REFERENCES

1. Tamime, A.Y.; Robisons, R.K. Chapter 1 Historical background. In Tamime and Robinson's Yogurt: Science and Technology, 3rd ed.; Woodhead Publishing LTD: Cambridge, UK, 2007; pp. 1–10.
2. Chandan, R.C. Chapter 1 History and consumption trends. In Manufacturing Yogurt and Fermented Milks; Chandan, R.C., Ed.; Blackwell Publishing: Ames, IA, USA, 2006; pp. 3–17.
3. Tamime, A.Y.; Robisons, R.K. Chapter 2 Backround to manufacturing practice. In Tamime and Robinson's Yogurt: Science and Technology, 3rd ed.; Woodhead Publishing LTD: Cambridge, UK, 2007; pp. 11–118.
4. Walstra, P.; Wouters, J.T.M.; Geurts, T.J. Chapter 7 Heat treatment. In Dairy Science and Technology; Taylor & Francis Group, LLC: Boca Raton, FL, USA, 2006; pp. 225–272.
5. Walstra, P.; Wouters, J.T.M.; Geurts, T.J. Chapter 22 Fermented milks. In Dairy Science and Technology; Taylor & Francis Group, LLC: Boca Raton, FL, USA, 2006; pp. 551–573.
6. Shaker, R.R.; Jumah, R.Y.; Abu-Jdayil, B. Rheological properties of plain yogurt during coagulation process: Impact of fat content and preheat treatment of milk. J. Food Eng. 2000, 44, 175–180.
7. Soukoulis, C.; Panagiotidis, P.; Koureli, R.; Tzia, C. Industrial yogurt manufacture: Monitoring of fermentation process and improvement of final product quality. J. Dairy Sci. 2007, 90, 2641–2654.
8. Kristo, E.; Biliaderis, C.G.; Tzanetakis, N. Modelling of the acidification process and rheological properties of milk fermented with a yogurt starter culture using response surface methodology. Food Chem. 2003, 83, 437–446.
9. Ion-Titapiccolo, G.; Alexander, M.; Corredig, M. Heating of milk before or after homogenization changes its coagulation behavior during acidification. Food Biophys. 2013, 8, 81–89.
10. Fox, P.P. Fat globules in milk. In Encyclopedia of Dairy Sciences, 2nd ed.; Elsevier Ltd., Academic Press: London, UK, 2011; pp. 1564–1548.
11. Walstra, P.; Wouters, J.T.M.; Geurts, T.J. Chapter 2 Milk components. In Dairy Science and Technology; Taylor & Francis Group, LLC: Boca Raton, FL, USA, 2006; pp. 17–108.
12. Walstra, P.; Wouters, J.T.M.; Geurts, T.J. Chapter 9 Homogenization. In Dairy Science and Technology; Taylor & Francis Group, LLC: Boca Raton, FL, USA, 2006; pp. 276–279.
13. Cano-Ruiz, M.E.; Richter, R.L. Effect of homogenization pressure on the milk fat globule membrane proteins. J. Dairy Sci. 1997, 11, 2732–2739.
14. Aguilera, J.M.; Kessler, H.G. Physicochemical and rheological properties of milk-fat globules with modified membranes. Milchwissenschaft 1988, 43, 411–415.

15. Cho, Y.H.; Lucey, J.A.; Singh, H. Rheological properties of acid milk gels as affected by the nature of the fat globule surface material and heat treatment of milk. Int. Dairy J. 1999, 9, 537–545.
16. Kilara, A. Chapter 5 Basic dairy processing principles. In Manufacturing Yogurt and Fermented Milks; Chandan, R.C., Ed.; Blackwell Publishing: Ames, IA, USA, 2006; pp. 73–89.
17. Lewis, M.J. Chapter 5 Improvements in the pasteurisation and sterilisation of milk. In Dairy Processing; Smith, G., Ed.; Woodhead Publishing LTD: Cambridge, UK, 2003; pp. 79–102.
18. Boelrijk, A.E.M.; de Jong, C.; Smit, G. Chapter 7 Flavour generation in dairy products. In Dairy Processing; Smith, G., Ed.; Woodhead Publishing LTD: Cambridge, UK, 2003; pp. 128–153.
19. Jaros, D.; Rohm, H. Chapter 8 Controlling the texture of fermented dairy products: The case of yoghurt. In Dairy Processing; Smith, G., Ed.; Woodhead Publishing LTD: Cambridge, UK, 2003.
20. Vedamuthu, E.R. Chapter 6 Starter cultures for yogurt and fermented milks. In Manufacturing Yogurt and Fermented Milks; Chandan, R.C., Ed.; Blackwell Publishing: Ames, IA, USA, 2006; pp. 89–117.
21. Horne, D.S. Formation and structure of acidified milk gels. Int. Dairy J. 1999, 9, 261–268.
22. Chandan, R.C.; O'Rell, K.R. Chapter 12 Principles of yogurt processing. In Manufacturing Yogurt and Fermented Milks; Chandan, R.C., Ed.; Blackwell Publishing: Ames, IA, USA, 2006; pp. 195–211.
23. De Brabandere, A.G.; de Baerdemaeker, J.G. Effects of process conditions on the pH development during yogurt fermentation. J. Food Eng. 1999, 41, 221–227.
24. Walstra, P.; Wouters, J.T.M.; Geurts, T.J. Chapter 11 Cooling and freezing. In Dairy Science and Technology; Taylor & Francis Group, LLC: Boca Raton, FL, USA, 2006; pp. 297–307.
25. Trujillo, A.J.; Capellas, M.; Saldo, J.; Gervilla, R.; Guamis, B. Applications of high-hydrostatic pressure on milk and dairy products: A review. Innov. Food Sci. Emerg. Technol. 2002, 4, 295–307.
26. Johnston, D.E.; Austin, B.A.; Murphy, R.J. Effects of high hydrostatic pressure on milk. Milchwissenschaft 1992, 47, 760–763.
27. Law, A.J.R.; Leaver, J.; Felipe, X.; Ferragut, V.; Pla, R.; Guamis, B. Comparison of the effects of high pressure and thermal treatments on the casein micelles in goat's milk. J. Agric. Food Chem. 1998, 46, 2523–2530.
28. Felipe, X.; Capellas, M.; Law, A.R. Comparison of the effects of high-pressure treatments and heat pasteurisation on the whey proteins in goat's milk. J. Agric. Food Chem. 1997, 45, 627–631.
29. Gervilla, R.; Ferragut, V.; Guamis, B. High hydrostatic pressure effects on colour and milk-fat globule of ewe's milk. J. Food Sci. 2001, 66, 880–885.
30. Ferragut, V.; Martiınez, V.M.; Trujillo, A.J.; Guamis, B. Properties of yoghurts made from whole ewe's milk treated by high hydrostatic pressure. Milchwissenschaft 2000, 55, 267–269.
31. Demirdöven, A.; Baysal, T. The use of ultrasound and combined technologies in food preservation. Food Rev. Int.2009, 25, 1–11.
32. Dolatowski, Z.J.; Stadnik, J.; Stasiak, D. Applications of ultrasound in food technology. Acta Sci. Pol. Technol. Aliment.2007, 6, 89–99.
33. Wu, H.; Hulbert, G.J.; Mount, J.R. Effects of ultrasound on milk homogenization and fermentation with yogurt starter. Innov. Food Sci. Emerg. Technol. 2009, 3, 211–218.
34. Nguyen, H.A.; Anema, S.G. Effect of ultrasonication on the properties of skim milk used in the formation of acid gels. Innov. Food Sci. Emerg. Technol. 2010, 11, 616–622.

35. Krešić, G.; Lelas, L.; Jambrak, A.R.; Herceg, Z.; Brnčić, S.R. Influence of novel food processing technologies on the rheological and thermophysical properties of whey proteins. J. Food Eng. 2008, 1, 64–73.
36. Chandrapala, J.; Zisu, B.; Palmer, M.; Kentish, S.; Ashokkumar, M. Effects of ultrasound on the thermal and structural characteristics of proteins in reconstituted whey protein concentrate. Ultrason. Sonochem. 2011, 18, 951–957.
37. Madadlou, A.; Mousavi, M.E.; Emam-Djomeh, Z. Comparison of pH-dependent sonodisruption of re-assembled casein micelles by 35 and 130 kHz ultrasounds. J. Food Eng. 2009, 95, 505–509.
38. Riener, J.; Noci, F.; Cronin, D.A. The effect of thermosonication of milk on selected physicochemical and microstructural properties of yoghurt gels during fermentation. Food Chem. 2009, 114, 905–911.
39. Bermúdez-Aguirre, D.; Corradini, M.G.; Mawson, R.; Barbosa-Canova, G.V. Modeling the inactivation of Listeria innocua in raw whole milk treated under thermo-sonication. Innov. Food Sci. Emerg. Technol. 2008, 10, 172–178.
40. Riener, J.; Noci, F.; Cronin, D.A.; Morgan, D.; Lyng, J.G. Characterisation of volatile compounds generated in milk by high intensity ultrasound. Int. Dairy J. 2009, 19, 269–272.
41. Vercet, A.; Oria, R.; Marquina, P.; Crelier, S.; López-Buesa, P. Rheological properties of yoghurt made with milk submitted to manothermosonication. J. Agric. Food Chem. 2002, 50, 6165–6171.
42. Morand, M.; Guyomarc'h, F.; Famelart, M.H. How to tailor heat-induced whey protein/κ-casein complexes as a means to investigate the acid gelation of milk—A review. Dairy Sci. Technol. 2011, 91, 97–126.
43. Ciron, C.I.E.; Gee, V.L.; Kelly, A.L.; Auty, M.A.E. Comparison of the effects of high-pressure microfluidization and conventional homogenization of milk on particle size, water retention and texture of non-fat and low-fat yoghurts. Int. Dairy J. 2010, 20, 314–320.
44. Kasaai, M.R.; Charlet, G.; Paquin, P.; Arul, J. Fragmentation of chitosan by microfluidization process. Innov. Food Sci. Emerg. Technol. 2003, 4, 403–413.
45. Skurtys, O.; Aguilera, J.M. Applications of microfluidic devices in food engineering. Food Biophys. 2008, 3, 1–15.
46. Ronkart, S.N.; Paquot, M.; Deroanne, C.; Fougnies, C.; Besbes, S.; Blecker, C.S. Development of gelling properties of inulin by microfluidization. Food Hydrocoll. 2010, 24, 318–324.
47. Da Cruz, A.G; de Assis Fonseca Faria, J.; Saad, S.M.I.; Bolini, H.M.A.; Sant'Ana, A.S.; Cristianini, M. High pressure processing and pulsed electric fields: Potential use in probiotic dairy foods processing. Trends Food Sci. Technol. 2010, 21, 483–493.
48. Ravishankar, S.; Zhang, H.; Kempkes, M.L. Pulsed electric fields. Food Sci. Technol. Int. 2008, 14, 429–432.
49. Wouters, P.C.; Bos, A.P.; Ueckert, J. Membrane permeabilization in relation to inactivation kinetics of Lactobacillus species due to pulsed electric fields. Appl. Environ. Microbiol. 2001, 67, 3092–3101.
50. Lin, S.; Clark, S.; Powers, J.R.; Luedecke, L.O.; Swanson, B.G. Thermal, ultra high pressure, and pulsed electric field attenuation of Lactobacillus: Part 2. Agro Food Ind. Hi-Tech 2002, 13, 6–11.
51. Calderon-Miranda, M.L.; Barbosa-Canovas, G.V.; Swanson, B.G. Transmission electron microscopy of Listeria innocua treated by pulsed electric fields and niasin in skimmed milk. Int. J. Food Microbiol. 1999, 51, 31–38.
52. Kelly, A.L.; Zeece, M. Applications of novel technologies in processing of functional foods. Aust. J. Dairy Technol.2009, 64, 12–16.
53. Metchnikoff, E. Essais Optimistes. Paris. The Prolongation of Life. Optimistic Studies; Mitchell, P.C., Ed.; Heinemann: London, UK, 1907.
54. Lilly, D.M.; Stillwell, R.H. Probiotics: Growth-promoting factors produced by microorganisms. Science 1965, 147, 747–748.
55. Sperti, G.S. Probiotics; Avi Publishing Co.: Westport, CT, USA, 1971.

56. Parker, R.B. Probiotics: The other half of the antibiotic story. Anim. Nutr. Health 1974, 29, 4–8.
57. Kneifel, W.; Mattila-Sandholm, T.; von Wright, A. Probiotic Bacteria detection and estimation in fermented and non-fermeneted dairy products. Encycl. Food Microbiol. 1999, 3, 1783–1789.
58. Vinderola, G.; Binetti, A.; Burns, P.; Reinheimer, J. Cell viability and functionality of probiotic bacteria in dairy products. Front. Microbiol. 2011, 2, 70.
59. Salminen, S.; Kenifel, W.; Ouwhand, A.C. Bacteria, beneficial. Probiotics, applications in dairy products. In Encyclopedia of Dairy Sciences, 2nd ed.; Elsevier Ltd., Academic Press: London, UK, 2011; pp. 412–419.
60. Elizaquível, P.; Sánchez, G.; Salvador, A.; Fiszman, S.; Dueñas, M.T.; López, P.; Fernández de Palencia, P.; Aznar, R. Evaluation of yogurt and various beverages as carriers of lactic acid bacteria producing 2-branched (1,3)-β-D-glucan. J. Dairy Sci. 2011, 94, 3271–3278.
61. Allgeyer, L.C.; Miller, M.J.; Lee, S.-Y. Sensory and microbiological quality of yogurt drinks with prebiotics and probiotics. J. Dairy Sci. 2010, 93, 4471–4479.
62. Atunes, A.E.; Cazetto, T.F.; Bolini, H.M. Viability of probiotic microorganisms during storage, post-acidification and sensory analysis of fat-free yogurts with added whey protein concentrate. Int. J. Dairy Technol. 2005, 58, 169–173.
63. Cruz, A.G.; Cadena, R.S.; Castro, W.F.; Esmerino, E.A.; Rodrigues, J.B.; Gaze, L.; Faria, J.A.F.; Freitas, M.Q.; Deliza, R.; Bolini, H.M.A. Consumer perception of probiotic yogurt: Performance of check all that apply (CATA), projective mapping, sorting and intensity scale. Food Res. Int. 2013, 54, 601–610.
64. Illupapalayam, V.V.; Smith, S.C.; Gamlath, S. Consumer acceptability and antioxidant potential of probiotic-yogurt with spices. LWT Food Sci. Technol. 2014, 55, 255–262.
65. Akalin, A.S.; Unal, G.; Dinkci, N.; Hayaloglu, A.A. Microstructural, textural, and sensory characteristics of probiotic yogurts fortified with sodium calcium caseinate or whey protein concentrate. J. Dairy Sci. 2012, 95, 3617–3628.
66. Hekmat, S.; Reid, G. Sensory properties of probiotic yogurt is comparable to standard yogurt. Nutr. Res. 2006, 26, 163–166.
67. Gibson, G.R.; Roberfroid, M.B. Dietary modulation of the human colonic microbiota: Introducing the concept of prebiotics. J. Nutr. 1995, 125, 1401–1412.
68. Roberfroid, M.B. Prebiotics: The concept revisited. J. Nutr. 2007, 137, 830–837.
69. Cruz, A.G.; Cavalcantia, R.N.; Guerreiroa, L.M.R; Sant'Anac, A.S.; Nogueirab, L.C.; Oliveirad, C.A.F.; Delizae, R.; Cunhaa, R.L.; Fariaa, J.A.F.; Bolinia, H.M.A. Developing a prebiotic yogurt: Rheological, physico-chemical and microbiological aspects and adequacy of survival analysis methodology. J. Food Eng. 2013, 114, 323–330.
70. Pimentel, T.C.; Cruz, A.G.; Prudencio, S.H. Short communication: Influence of long-chain inulin and Lactobacillus paracasei subspecies paracasei on the sensory profile and acceptance of a traditional yogurt. J. Dairy Sci. 2013, 96, 6233–6241.

CHAPTER 17

Dairy-Based Emulsions: Viscosity Affects Fat Difference Thresholds and Sweetness Perception

Susann Zahn, Karin Hoppert, Franziska Ullrich and Harald Rohm *

Institute of Food Technology and Bioprocess Engineering, Technische Universität Dresden, 01069 Dresden, Germany

ABSTRACT

In complex emulsions, viscosity or viscosity-associated sensory attributes such as creaminess are important for quality assessment and product differentiation. Two sets of emulsions with fat or locust bean gum content being varied at seven levels were developed; the two emulsions at each level had similar apparent viscosity. Additionally, sugar concentration was kept constant either with respect to total emulsion, or with respect to the aqueous phase. Series of two-alternative forced choice tests were performed with one constant stimulus, and just noticeable differences were calculated using probability regression. The results show that, when viscosity was not compensated, it was easy for the subjects to (a) distinguish emulsions with different fat content when the fat content was addressed in the question, and to (b) distinguish emulsions with different fat or locust bean gum content when creaminess was addressed. For the latter descriptor, it is of minor importance whether viscosity is altered by fat content or a thickener. Weber fractions that were calculated for viscosity were approximately 0.20. The quantitative effects of viscosity on sweetness, however, depend on how product rheology was modified.

Keywords: difference threshold; just noticeable difference (JND); fat; sugar; viscosity; model emulsion

1. INTRODUCTION

Food emulsions are complex systems which consist of at least three constituents: two immiscible liquids (oil and water), and an emulsifier which reduces interfacial tension, prevents phase separation and which is, among others, responsible for kinetic stability. The main factors that affect the rheology and texture of emulsions are the viscosity of the continuous phase, the volume fraction of the disperse phase, droplet size distribution, and type and chemical properties of the emulsifier [1,2].

The adjustment of the rheological properties of emulsions to specific needs is important because that directly determines the usage quality. In many cases, additional ingredients (for example, hydrocolloids) are used to generate desired characteristics such as a static yield value, or an extended stability against creaming or sedimentation [3,4]. Apart from the changes in the rheological properties of the product and the resulting impact on sensory texture, it is also necessary to consider that the sensory perception of, for example, viscosity interacts with flavor and taste perception [3,5,6]. As regards texture, mainly properties that can be expressed by sensory descriptors such as mouthfeel or creaminess are of special importance in product development [7,8]. Another important issue in product design is to know how much the magnitude of a particular property may be altered (for example, by modifying the formulation) before the resulting changes can be perceived by the human, regardless whether she/he is an expert or represents the average consumer. In many cases, such a difference threshold (or just noticeable difference, JND) is directly related to the magnitude of a reference stimulus, usually denoted as Weber fraction [9,10].

In a previous study [11], we have shown for a model dairy system that, when directly addressing the fat content, it is difficult to distinguish low differences as long as the viscosity of the product is kept at a constant level. We have also shown that sweetness discrimination is, to some extent, affected by the fat content, even at similar emulsion viscosity. The aim of the present study was to evaluate the response of trained subjects towards differences in viscosity of model emulsions. For that purpose, two different systems where viscosity was pairwise adjusted by (a) fat content or (b) a hydrocolloid were developed and subjected to rheological and sensory analysis.

2. MATERIALS AND METHODS

2.1. Materials

Raw materials for emulsion preparation were refined canola oil (Kaufland Warenhandels GmbH & Co KG, Neckarsulm, Germany), Viscogum GE locust bean gum (LBG; Cargill Texturizing Solutions, Hamburg, Germany), icing sugar (Pfeiffer & Langen GmbH & Co KG, Köln, Germany), a water-soluble butter-vanilla flavor (Dr. August Oetker KG, Bielefeld, Germany), β-carotene (Rudolf Wild GmbH & Co KG, Berlin, Germany), and skim milk powder (Käserei Champignon GmbH & Co KG, Heising, Germany).

2.2. Emulsion Preparation

Starting point was the formulation of Wendin & Hall [12], previously modified by Hoppert *et al.* [11] to ensure that emulsions with different fat content are perceived as iso-viscous and iso-sweet. In this study, the emulsions were prepared without adjusting viscosity. For that purpose, we mixed 16.5 g of an aqueous 3% (w/w) LBG dispersion with sugar, flavor, β-carotene, oil, 10% (w/w) reconstituted skim milk and water in the respective amounts (Table 1), and homogenized at 20,500 rpm for 15 min using a T15 ultra turrax (IKA Werke GmbH & Co KG, Staufen, Germany). Considering the lactose content of milk powder (51.8 g/100 g) and its relative sweetness of 0.33 [13], all emulsions had a sugar concentration in the aqueous phase, further denoted as sugar concentration index, of 0.05 (5%). The amount of added flavor was similarly adjusted to ensure a constant concentration in the aqueous phase. After production, the emulsions were stored at 5 °C and analyzed within 2 days.

Another set of seven emulsions was prepared at a constant fat content X_F = 1 g/100 g, and a constant skim milk content of 40 g/100 g; flavor and β-carotene were kept at 0.19 g/100g and 2.0 g/100 g, respectively. Compared to the formulations given in Table 1, sucrose and water addition had to be adjusted to achieve a total sugar concentration index of 0.047 (this corresponds to 0.050 when related to the aqueous phase;

see first line of Table 1). Target was to create emulsions that had apparent viscosities η_a at 50/s and 1000/s that were close to η_a of the corresponding emulsions given in Table 1. This was achieved by increasing the LBG content from X_{LBG} = 0.5 g/100 g (equivalent to 16.5 g/100 g of a 3 % solution) on six additional levels.

In a third set of emulsions, the content of LBG, sucrose, flavor, β-carotene and skim milk was kept constant (see last line of Table 1); when X_F was lower than 29 g/100 g, the amount of incorporated water was increased to compensate for the missing mass. This procedure gives, regardless of X_F, a total sugar concentration index of 0.033. Because of the different fat content, the sugar concentration related to the aqueous phase increased from 0.035 (1 g/100 g fat) to 0.050 (29 g/100 g fat).

Table 1. Formulation (g/100 g) of the emulsions with different fat content.

Oil	LBG [1]	Sucrose	Flavor	β-Carotene [2]	Skim milk [3]	Water	Sugar concentration index (−)	
							In entire system [4]	In aqueous phase [5]
1.0	16.5	4.00	0.19	2.44	48.83	27.04	0.047	0.050
5.0	16.5	3.82	0.18	2.40	47.79	24.31	0.045	0.050
10.0	16.5	3.60	0.17	2.34	46.40	20.99	0.042	0.050
15.0	16.5	3.38	0.16	2.28	44.90	17.78	0.040	0.050
20.0	16.5	3.16	0.15	2.21	43.28	14.70	0.038	0.050
25.0	16.5	2.94	0.14	2.13	31.53	11.77	0.035	0.050
29.0	16.5	2.67	0.13	2.00	40.00	9.70	0.033	0.050

[1] Aqueous solution of 3% (w/w) locust bean gum (LBG); [2] Aqueous solution of 1% (w/w) β-carotene. levels were adjusted to ensure similar color intensity; [3] Aqueous solution of 10% (w/w) skim milk powder; [4] Fraction of sugar (sucrose + 0.33 × lactose in skim milk); [5] Fraction of sugar (sucrose + 0.33 × lactose in skim milk) in the aqueous phase (sugar + water in LBG preparation + water in carotene + water in

2.3. Rheological Measurements

Steady and dynamic rheological experiments were carried out using an MCR300 rheometer (Anton Paar Germany GmbH, Ostfildern, Germany) equipped with a CC27 concentric cylinder geometry. All measurements were performed after equilibrium at 9 °C in duplicate. Flow curves were recorded by increasing shear rate from 0.1/s to 1.000/s (10 points per decade) within 600 s. Small deformation properties were analyzed by strain sweeps where strain γ was increased from 0.001 to 10 at a constant angular frequency of 1 Hz, and by sweeping frequency from 0.1 to 10 Hz at γ = 0.005.

2.4. Sensory Evaluation

2.4.1. General Conditions

Sensory analysis was performed in a laboratory with single booths at 22 ± 1 °C under red light. Ten milliliters samples were served on plastic spoons that came in petri dishes encoded with 3-digit random numbers. Tap water was provided *ad libitum* for mouth rinsing. The panel of fifteen that was recruited from students and PhD scientists of the institute was familiar with sensory investigations and had passed all of the standardized training procedures given in [14] with a score of at least 85%.

2.4.2. Product Ranking

Depending on emulsion formulation, the assessors were asked to rank sets of four samples either with respect to fat content, creaminess, or sweetness. A balanced incomplete block design was established with fourteen panel members so that each of the seven emulsions was assessed eight times. The serving order of the samples within each set was randomized. Differences in the sums of the ranks that were assigned were evaluated using Page tests [15] and, additionally, by Friedman's *F*-values. Multiple comparisons were performed by calculating the least significant difference.

2.4.3. Difference Testing and JND Calculation

Difference thresholds for the sensory descriptors "fat content", "creaminess" and "sweetness" were determined using sequences of two-alternative forced choice (AFC) tests. In each set, one sample was kept constant (reference), and the second sample was systematically varied. Per session, each of the fifteen panelists received four sample pairs in random configuration and was asked to pick the sample which was "higher"/"more intense" in the requested attribute. Each pair was evaluated twice by each panelist so that a total of 30 judgments were obtained. After plotting percentages of "higher"/"more intense" on probability ordinate against X_F or X_{LBG}, just noticeable differences were calculated from regression functions as half of the difference between stimulus magnitudes that correspond to 25% and 75% correct identification [9,16].

2.4.4. Estimation of Viscosity JNDs

Viscosity—concentration functions were used to calculate difference thresholds for apparent viscosity from JNDs that were obtained for fat content and creaminess (in the latter, the concentration parameter was

X_{LBG}). This transformation was based on the respective concentration dependent functions for η_a that was extracted from flow curves at 50/s. The viscosity JND was then calculated as geometric mean (Figure 1).

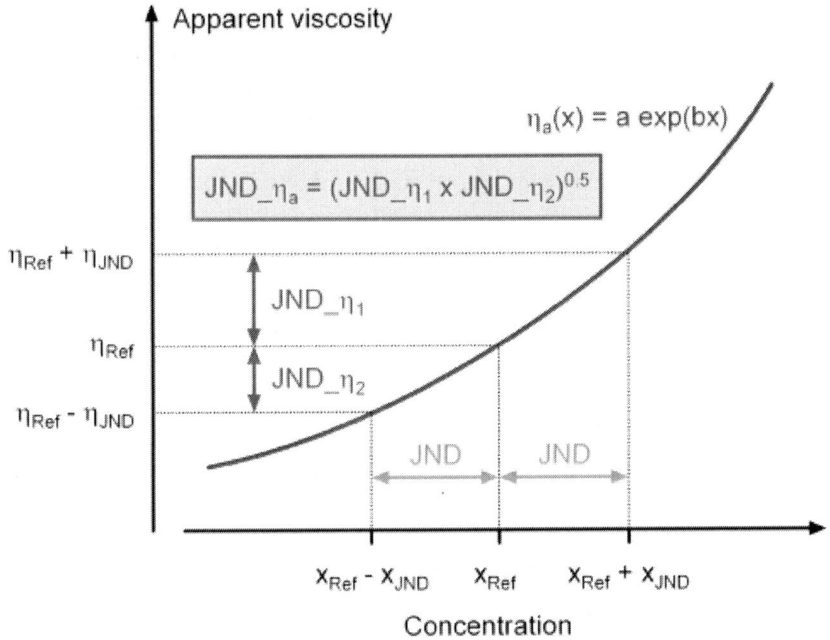

Figure 1. Schematic representation of the method for the calculation of just noticeable differences for viscosity (JND_η_a).

3. RESULTS AND DISCUSSION

3.1. Viscosity Adjustment of Emulsions Using Locust Bean Gum

For adjusting the viscosity of the seven emulsions with X_F = 1 g/100 g to that of the formulations given in Table 1, we first created systems where 0.02 g LBG was incorporated to mimic the viscosity effect of each g oil (*i.e.*, 0.1 g LBG to replace 5 g oil). Flow curves of these emulsions were recorded, and η_a taken at shear rates of 50/s and 1000/s showed an exponential relationship to X_{LBG} (Figure 2). The resulting viscosity equations ($R^2 > 0.98$) were then solved with respect to X_{LBG}. As two equations (for 50/s and 1000/s) were used, the final LBG concentration necessary to mimic a particular fat content was taken as the arithmetic mean of the two calculated values. The composition of the resulting emulsions is summarized in Table 2. All of these emulsions where viscosity was adjusted by LBG contained 1 g/100 g oil, 4.11 g/100 g sucrose, 0.19 g/100 g flavor, 2.0 g/100 g of 1% (w/w) aqueous β-carotene, and 40 g/100 g reconstituted skim milk (10% w/w), and the difference to a total of 100 g was made up with water. The sugar concentration index is constant at 0.05 in the entire system, and at 0.047 with respect to the aqueous phase.

With the given formulations it was possible to use seven pairs of emulsions (line 1 *vs.* line 1, line 2 *vs.* line 2, *etc.* from Table 1 and Table 2) that were characterized by almost identical flow curves. Figure 3 shows η_a at two shear rates, and the complex modulus G* and tan δ in the linear viscoelastic region as a function of fat content (emulsions from Table 1) and LBG content (emulsions from Table 2). Generally, η_a at 50/s and 1000/s increased exponentially ($R^2 > 0.98$) with increasing X_F or X_{LBG}. The accordance in the shear rate range which is relevant during oral processing [17] allows comparing the results of sensory experiments that are obtained by using these emulsions. Apparent viscosity at 50/s can be expressed as a function of X_F by $\eta_a = 0.21 \exp(0.05\ X_F)$ (see Figure 1). Small deformation stiffness for the two emulsion sets was also in good agreement. Whereas, especially at higher concentration, fat droplets interact with each other at low strain, it is the entanglement of LBG at the given concentrations (which are above coil overlap concentration [18,19]) that mimic that behavior. The phase shift in the linear viscoelastic region was, however, different, especially at higher polymer concentration. The frequency at the cross-over of storage and loss modulus—the reciprocal of which refers to an indirect measure of the relaxation time [20]—was, especially for systems with a higher viscosity, lower for the emulsions where the fat content was varied (data not shown).

3.2. Preliminary Sensory Tests

Three pairs of emulsions with corresponding viscosity (X_F = 5 g/100 g *vs.* X_{LBG} = 0.570 g/100 g; X_F = 15 g/100 g *vs.* X_{LBG} = 0.720 g/100 g; X_F = 25 g/100 g *vs.* X_{LBG} = 0.915 g/100 g; see lines 2, 4 and 6 in Table 1 and Table 2) were presented in two-AFCs to the panel in two replicate sessions. The results, expressed as the percentage of answers that were in accordance with the given sensory statement, and the respective significance levels [21] (Table 3) indicate that it was not possible to distinguish between the corresponding samples. It can also be concluded that the deviations in the rheological properties that were measured at low deformation (*i.e.*, in the linear viscoelastic region) are less relevant during the sensory assessment than shear viscosity.

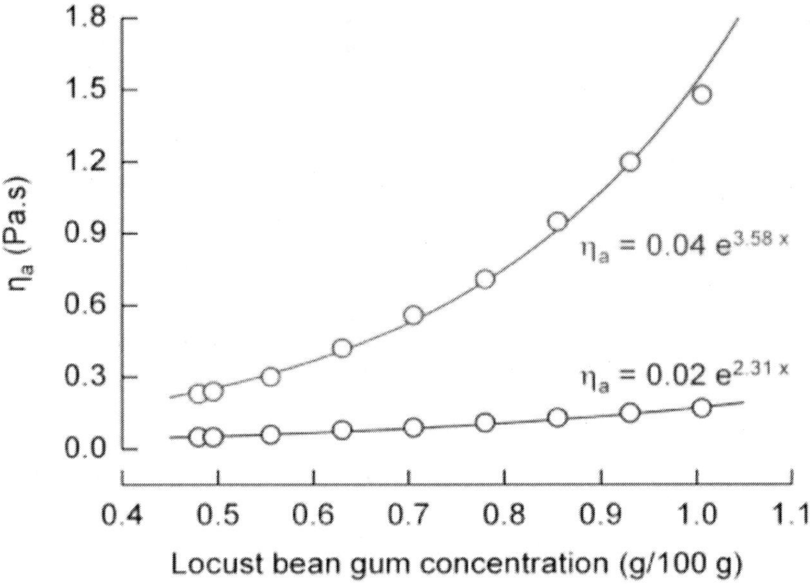

Figure 2. Apparent viscosity η_a as a function of locust bean gum concentration (data and respective fits) in emulsions with 1 g/100 g fat. Shear rate is 50/s (red) and 1000/s (blue).

Table 2. Formulation (g/100 g) of the emulsions with constant fat content.

Oil	LBG [1]	Sucrose	Flavor	β-Carotene [2]	Skim milk [3]	Water	Sugar concentration index (–)	
							In entire system [4]	In aqueous phase [5]
1.0	16.5	4.11	0.19	2.0	40.0	36.2	0.047	0.050
1.0	19.1	4.11	0.19	2.0	40.0	33.7	0.047	0.050
1.0	21.5	4.11	0.19	2.0	40.0	31.2	0.047	0.050
1.0	24.0	4.11	0.19	2.0	40.0	28.7	0.047	0.050
1.0	27.3	4.11	0.19	2.0	40.0	25.5	0.047	0.050
1.0	30.5	4.11	0.19	2.0	40.0	22.2	0.047	0.050
1.0	33.1	4.11	0.19	2.0	40.0	19.6	0.047	0.050

[1] Aqueous solution of 3% (w/w) locust bean gum (LBG); [2] Aqueous solution of 1% (w/w) β-carotene; [3] Aqueous solution of 10% (w/w) skim milk powder; [4] Fraction of sugar (sucrose + 0.33 × lactose in skim milk); [5] Fraction of sugar (sucrose + 0.33 × lactose in skim milk) in the aqueous phase (sugar + water in LBG gel + water in carotene + water in reconstituted skim milk considering skim milk moisture).

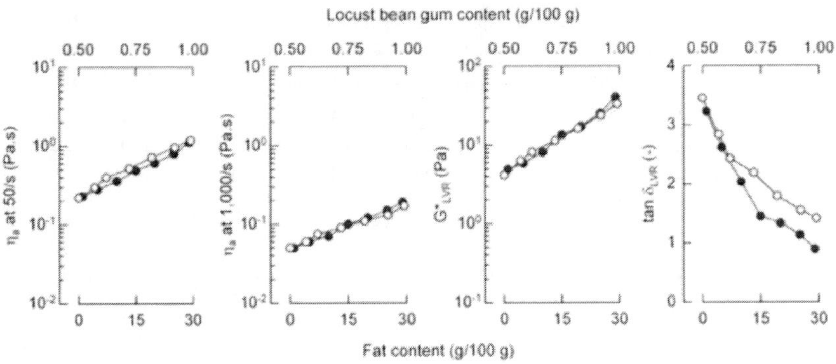

Figure 3. Apparent viscosity η_a at two shear rates, complex modulus G^*_{LVR} and tan δ_{LVR} in the linear viscoelastic region of emulsions with different fat content (lower *x*-axis, closed symbols) and different locust bean gum content (upper *x*-axis, open symbols).

Table 3. Percentages of coinciding statements in two-alternate force tests (n = 30) of three sample pairs with similar viscosity.

Sensory statement [1]	Sample pair [2]		
	F-5 *vs.* LBG-0.57	**F-15 *vs.* LBG-0.72**	**F-25 *vs.* LBG-0.91**
A has a higher fat content than B	50.0 [3]	63.3 (p = 0.20)	63.3 (p = 0.20)
A is more creamy than B	43.3 (p = 0.58)	63.3 (p = 0.20)	43.3 (p = 0.58)
A is sweeter than B	43.3 (p = 0.58)	50.0	33.3 (p = 0.10)

[1] A or B was assigned to the one or other sample randomly; [2] F-…, emulsions with given fat content (g/100 g) and 0.495 g/100 g locust bean gum (LBG); LBG-…, emulsions with given LBG content (g/100 g) and 1 g/100 g fat; [3] Number in brackets refer to probabilities (two-tailed). For (n = 30) and 50 % answers, no value is provided by Rössler *et al.* [21].

3.3. Sensory Evaluation of Emulsion Texture

Significant differences were obtained in the ranking of the emulsions from Table 1 with respect to the attribute "fat content". The panel placed all samples in correct order, and the mean ranks ranged from 1.13 (X_F = 1 g/100 g) to 4.0 (X_F = 29 g/100 g). For this set of emulsions, the subsequent procedure with two-AFCs for threshold determination was carried out using two constant stimuli: X_F = 5 g/100 g, and X_F = 15 g/100 g. The JNDs that were calculated using the regression lines of the *z*-coordinates of correct answers *vs.* X_F (Figure 4) were 4.06 and 4.32 g/100 g fat, and the ratios of JND to constant stimulus magnitude (Weber fractions) were 0.81 and 0.29, respectively. These JNDs are much smaller than the approximately 14 g/100 g fat that were analyzed for emulsions with a fat content similar to that of this study, but that had a viscosity which was kept constant by using varying amounts of LBG [11]. The differences in the JND therefore indicate that it is obviously emulsion viscosity which, despite the question towards fat content in the sensory assessments, has been used by the panelists to find their decision.

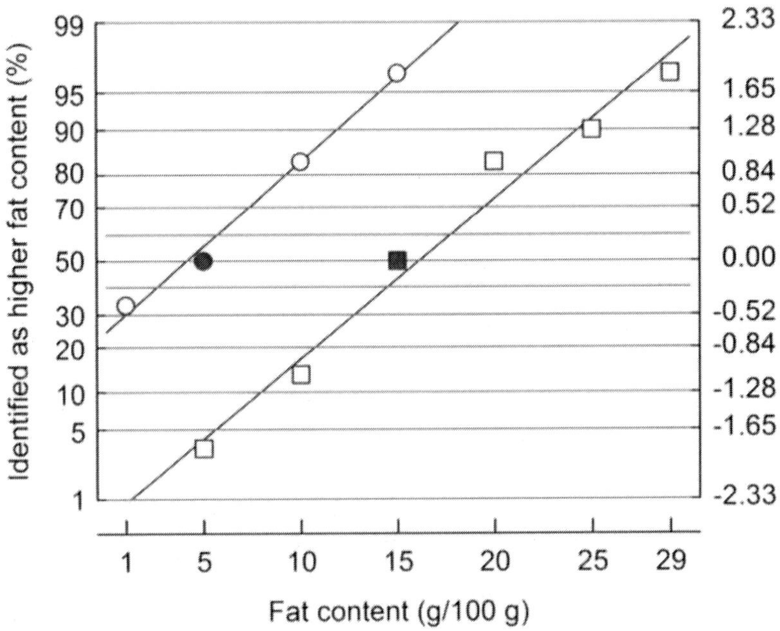

Figure 4. Plot of percentages of samples identified as being higher in fat content (left *y*-axis: percentage in normal probability grid, right *y*-axis: *z*-coordinates) as a function of emulsion fat content. Reference fat content (full symbols) was 5 g/100 g and 15 g/100 g.

Using the method illustrated in Figure 1, we subsequently calculated viscosity JNDs from those fat difference thresholds and, for η_a at a shear rate of 50/s, obtained 0.062 and 0.11 Pa.s at a reference X_F of 5 and 15 g/100 g. Because of the exponential relation between η_a and X_F, the resulting Weber fractions are similar (0.22 and 0.24, respectively). This can be taken as an additional argument for suggesting viscosity-driven decisions during the sensory assessments. It is, however, not possible to compare these Weber fractions with viscosity differences as low as 5% that have been correctly identified in a recent study using triangle tests [22] because absolute viscosities were not specified there. The Weber fraction obtained by non-oral tactile viscosity two-AFC comparisons reported in literature is 0.19 [23].

In another set of ranking experiments, the emulsions with different LBG content at constant X_F = 1 g/100 g (see Table 2) were presented to the panel in a similar setup. Here, the panel members were free to select a descriptor that, in their individual opinion, best described between-samples intensity differences. Possible choices were: (a) mouthfeel, (b) viscosity, (c) creaminess, and (d) persistence in the mouth. The panel ranked the emulsions correctly (*i.e.*, by increasing order of LBG content or viscosity; see also Figure 3), with mean rank sums ranging from 1.0 (X_{LBG} = 0.50 g/kg) to 3.88 (X_{LBG} = 0.99 g/kg); the Page coefficient was 5.98 and significantly ($p < 0.05$) higher than the critical value of 1.64. Five panelists reported that they had used (d) for their decision, four panelists each decided on the basis of (b) and (c), and two panelists decided by mouthfeel. Based on the number of panelists that had used creaminess to distinguish between the emulsions, by additionally considering the opinion of the panel members regarding the general meaning of the presented descriptors (some of them had problems to interpret "persistence in the mouth"), and because creaminess is related to both fat content and viscosity in complex foods [7,24], we decided to use creaminess as descriptor in the subsequent two-AFCs for threshold determination.

The results of these experiments are summarized in Table 4. The correlation of the *z*-coordinates of the percentages of correct answers with X_F or X_{LBG} of the respective systems was significant ($R^2 > 0.89$; $p < 0.05$). The JNDs obtained in the creaminess tests were, when expressed as fat content, similar for both references, but approximately 25% lower than the JNDs that were observed in the experiments where the decision was asked to be made with respect to fat content. This is presumably because creaminess, in contrast to fattiness, can be regarded as an attribute that is self-explaining and used more homogeneously, even by untrained panelists [25]. The JNDs that were calculated for LBG after analyzing emulsions with identical X_F and varying X_{LBG} were 0.07 g/100 g; this gives Weber fractions of 0.12 and 0.09, respectively.

Table 4. Just noticeable differences for creaminess expressed as fat (JND_F), locust bean gum (JND_{LBG}) or viscosity (JND_η), and corresponding Weber fractions K.

Reference (g/100 g)	Regression function [1]	R^2 (−)	JND_F, JND_{LBG} (g/100 g)	K_F or K_{LBG} (−) [2]	JND_η (Pa.s) [3]	K_η (−) [3]
Emulsions with different fat content (Table 1)						
$X_F = 5$	$Y = 0.228X_F - 0.89$	0.96	2.96	0.59	0.04	0.16
$X_F = 15$	$Y = 0.213X_F - 3.10$	0.91	3.16	0.21	0.08	0.17
Emulsions with different locust bean gum content (Section 3.1.)						
$X_{LBG} = 0.57$	$Y = 10.16X_{LBG} - 5.75$	0.89	0.07	0.12	0.07	0.22
$X_{LBG} = 0.72$	$Y = 10.10X_{LBG} - 7.25$	0.97	0.07	0.09	0.11	0.22

[1] Y is percentage in z-coordinates; [2] k = JND/X; [3] Calculated using the schematic presentation in Figure 1.

The JNDs that were finally calculated from the corresponding apparent viscosity at a shear rate of 50/s were related to the reference stimuli in a way so that the resulting Weber fractions were 0.16 (emulsions with constant X_{LBG} and varying X_F) and 0.22 (constant X_F, but varying X_{LBG}). Rheological properties at small deformation, especially the higher ratio of viscous to elastic contributions in emulsions with a higher X_{LBG} (see Figure 3), may serve as a possible explanation for Weber fraction differences: results of multivariate analysis showed that a higher sensory creaminess is apparently linked to a lower loss tangent at small deformations for comparable systems [26]. Emulsions with varying fat content exhibit a lower tan δ and, as indicated by the viscosity related Weber fraction, can be distinguished more easily.

3.4. Sweetness Difference Thresholds as Affected by Emulsion Viscosity

The formulations given in Table 1 (different fat content) and the systems where viscosity was adjusted with LBG (see Table 2) have a constant sugar concentration in the aqueous phase; such systems have been shown as being iso-sweet in emulsions with similar viscosity [11]. Initial ranking tests that were performed with respect to sweetness revealed, however, that the systems with higher X_F or X_{LBG} were perceived as significantly less sweet when viscosity was not adjusted. This is in line with cross-modal interactions reported in studies that refer to textured samples, e.g., [4,6,27], or even to low-viscosity foods [28]. At varying fat content, the mean ranks ranged between 3.38 (highest sweetness at X_F = 1 g/100 g, η_a (50/s) = 0.23 Pa.s) and 1.63 (lowest sweetness at X_F = 29 g/100 g, η_a (50/s) = 1.11 Pa.s). Mean sweetness ranks for samples with varying X_{LBG} ranged between 3.13 (low viscosity, low X_{LBG}) and 1.88 (high viscosity, high X_{LBG}). Figure 5 shows the results of the two-AFCs for the two emulsion types as a function of fat content or locust bean gum content (for apparent viscosity of these systems, see Figure 3), which confirm the results of the ranking experiments. Two possible mechanisms can be mentioned in context with these findings. It has been shown that the liberation of saliva during oral manipulation of systems with a higher viscosity [4,29] is higher so that concentration differences could be partly attenuated or compensated. A second explanation is related to the fact that, even at a constant X_{LBG} in the emulsions with different fat content, the amount of LBG in the aqueous phase varied from 0.54 to 0.74 g/100 g. That concentration is above coil overlap concentration [18,19]; at such concentrations it has been shown that polymeric carbohydrates results in a reduced intensity of taste and flavor perception [3,5]. The fact that the slope for sweetness reduction in LBG thickened systems is steeper might point on differences in the suppressing mechanisms; using different systems and methods, Hollowood *et al.* [30] showed comparable results. However, the statement that texture modification may compensate for a loss of sugar-induced sweet taste intensity because of the additive relationship that has been addressed for gel-like systems [31] is not supported.

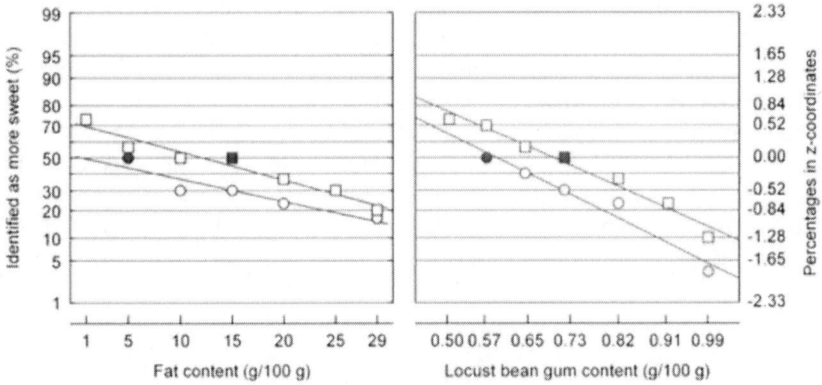

Figure 5. Judgment of sweetness of emulsions with different viscosity. Viscosity was adjusted by using different levels of fat or locust bean gum. The fraction of sugar in the aqueous phase was 0.05. Reference samples (full symbols) contained 5 or 15 g/100 g fat, or 0.57 or 0.72 g/100 g locust bean gum.

An effect of emulsion viscosity on perceived sweetness was also observed after subjecting samples to two-AFCs that had a total sugar concentration index of 0.033; in the samples of set 3, the sugar content related to the aqueous phase varied between 0.035 (for X_F = 1 g/100 g fat) and 0.05 (for X_F = 29 g/100 g fat). Despite similar total sugar, the samples with lower sugar content in the aqueous phase (*i.e.*, samples with higher viscosity) were judged as significantly less sweet. This indicates that, when considering different amounts of the disperse phase in a particular emulsion, when considering different amounts of thickeners, or when considering different system viscosity, it is absolutely necessary to ensure experimentally that no undesired side effects which result from improper formulation are present.

4. CONCLUSIONS

The difference thresholds for fat content obtained in this study were much lower than the thresholds that were analyzed by subjecting iso-viscous emulsions to sensory analysis. This indicates that, even when fat content is the given judgment attribute, subjects tend to judge such an attribute by a descriptor that is associated with fat content (here, apparent viscosity). For two sample sets with either varying fat content and constant thickener, or varying thickener content and constant fat, creaminess thresholds that were transferred into units of apparent viscosity resulted in Weber fractions of approximately 0.20, that correspond well with data from non-oral viscosity assessment. Sweetness judgments of emulsions with different viscosity indicate that interactions with viscosity have to be considered, but also that it is important by which means a viscosity modification is approached. The interactions can be provoked by changing fat content, but also by adjusting viscosity using a polymeric carbohydrate thickener; quantitative differences in the action on sweetness should, however, be considered.

ACKNOWLEDGMENTS

The study was performed within the project Innovative Strategies to Overcome Implicit Product-based and Personality-based Obstacles to Consume Healthy Food, funded by the Federal Ministry of Education and Research, Berlin (grant 0315670).

CONFLICTS OF INTEREST

The authors declare no conflict of interest.

REFERENCES

1. Flett, K.; Duizer, L.M.; Goff, H.D. Perceived creaminess and viscosity of aggregated particles of casein micelles and κ-carrageenan. J. Food Sci. 2010, 75, 5255–5262.
2. Kilcast, D.; Clegg, S. Sensory perception of creaminess and its relationship with food structure. Food Qual. Prefer.2002, 13, 609–623.
3. Cook, D.J.; Hollowood, T.A.; Linforth, R.S.T.; Taylor, A.J. Oral shear stress predicts flavour perception in viscous solutions. Chem. Senses 2003, 28, 11–23.
4. Tournier, C.; Sulmont-Rossé, C.; Sémon, E.; Vignin, A.; Issanchou, S.; Guichard, E. A study on texture-taste-aroma interactions: Physico-chemical and cognitive mechanisms. Int. Dairy J. 2009, 19, 450–458.
5. Hollowood, T.A.; Linforth, R.S.T.; Taylor, A.J. The effect of viscosity on the perception of flavour. Chem. Senses 2002, 27, 583–591.
6. Vingerhoeds, M.H.; de Wijk, R.A.; Zoet, F.D.; Nixdorf, R.R.; van Aken, G.A. How emulsion composition and structure affect sensory perception of low viscosity model emulsions. Food Hydrocoll. 2008, 22, 631–646.
7. De Wijk, R.A.; Prinz, J.F. Fatty versus creamy sensations for custard desserts, white sauces, and mayonnaises. Food Qual. Prefer. 2007, 18, 641–650.
8. Akhtar, M.; Murray, E.; Dickinson, E. Perception of creaminess of model oil-in-water dairy emulsions: Influence of the shear-thinning nature of a viscosity-controlling hydrocolloid. Food Hydrocoll. 2006, 20, 839–847.
9. Lawless, H.T.; Heymann, H. Sensory Evaluation of Food. Principles and Practices, 2nd ed.; Springer: New York, NY, USA, 2010; pp. 125–148.
10. Marks, L.E. Sensory Processes. The New Psychophysics; Academic Press: New York, NY, USA, 1974; pp. 32–58.
11. Hoppert, K.; Zahn, S.; Puschmann, A.; Ullmann, I.; Rohm, H. Quantification of sensory difference thresholds for fat in sweetness in dairy-based emulsions. Food Qual. Prefer. 2012, 26, 52–57.
12. Wendin, K.; Hall, G. Influence of fat, thickener and emulsifier contents on salad dressing: Static and dynamic sensory and rheological analysis. LWT Food Sci. Technol. 2001, 34, 222–233.
13. Parrish, F.W.; Talley, F.B.; Phillips, J.G. Sweetness of α-lactose, β-lactose, and equilibrium lactose relative to sucrose. J. Food Sci. 1981, 46, 933–935.
14. ISO 8586-1:2008. Sensory Analysis—General Guidance for the Selection, Training and Monitoring of Assessors—Part 1: Selected Assessors; International Organization for Standardization: Geneva, Switzerland, 2008.
15. ISO 8587:2006. Sensory Analysis—Methodology—Ranking; International Organization for Standardization: Geneva, Switzerland, 2006.
16. Rohm, H.; Raaber, S. Difference thresholds in texture evaluation of edible fats: Firmness and spreadability. J. Food Sci.1992, 57, 647–650.
17. Shama, F.; Sherman, P. Identification of stimuli controlling the sensory evaluation of viscosity. J. Texture Stud. 1973, 4, 111–118.
18. Launay, B.; Cuvelier, G.; Martinez-Reyes, S. Viscosity of locust been gum, guar and xanthan gum solutions in the Newtonian domain: A critical examination of the log (ηsp)0–log C[η]0 master curves. Carbohydr. Polym. 1997, 34, 385–395.
19. Richardson, P.H.; Wilmer, J.; Foster, T.J. Dilute solution properties of guar and locust bean gum in sucrose solutions. Food Hydrocoll. 1998, 12, 339–348.
20. Tadros, T. Application of rheology for assessment and prediction of the long-term physical stability of emulsions. In Adv. Colloid Interface Sci.; 2004; Volume 108–109, pp. 227–258.
21. Rössler, E.B.; Pangborn, R.M.; Sidel, J.L.; Stone, H. Expanded statistical tables for estimating significance in paired-comparison, paired-difference, duo-trio and triangle tests. J. Food Sci. 1973, 43, 940–943.

22. Kutter, A.; Hanesch, C.; Rauh, C.; Delgado, A. Impact of propriception and tactile sensations in the mouth on the perceived thickness of semi-solid foods. Food Qual. Prefer. 2011, 22, 193–197.
23. Jones, L.; Hunter, I.; Lafontaine, S. Viscosity discrimination: A comparison of an adaptive two-alternate forced-choice and an adjustment procedure. Perception 1997, 26, 1571–1578.
24. Kokini, J.L.; Cussler, E.L. Predicting the texture of liquid and melting semi-solid foods. J. Food Sci. 1983, 48, 1221–1225.
25. De Wijk, R.A.; Dijksterhuis, G.; Vereijken, P.; Prinz, J.F.; Weenen, H. PROP sensitivity reflects sensory discrimination between custard desserts. Food Qual. Prefer. 2007, 18, 597–604.
26. Ares, G.; Budelli, E.; Brozzone, F.; Gimenez, A.; Lema, P. Consumers' texture perception of milk desserts. I—Relationship with rheological measurements. J. Texture Stud. 2012, 43, 203–213.
27. Saint-Eve, A.; Paci-Kora, E.; Martin, N. Impact of the olfactory quality and chemical complexity of the flavouring agent on the texture of low fat stirred yoghurt assessed by three different sensory methodologies. Food Qual. Prefer.2004, 15, 655–668.
28. Pangborn, R.M.; Gibbs, Z.M.; Tassan, C. Effects of hydrocolloids on apparent viscosity and sensory propeties of selected beverages. J. Texture Stud. 1978, 9, 415–436.
29. Guinard, J.-X.; Zourmas-Morse, C.; Walchak, C. Relation between parotid saliva flow and composition and the perception of gustatory and trigeminal stimuli in foods. Physiol. Behav. 1998, 63, 109–118.
30. Hollowood, T.; Bayarri, S.; Marciani, L.; Busch, J.; Francis, S.; Spiller, R.; Taylor, A.; Hort, J. Modelling sweetness and texture perception in model emulsion systems. Eur. Food Res. Technol. 2008, 227, 537–545.
31. Knoop, J.-E.; Sala, G.; Smit, G.; Stieger, M. Combinatory effects of texture and aroma modification on taste perception of model gels. Chemosens. Percep. 2013, 6, 60–69.

Index